STUDENT'S SOLUTION MA

NOGGLE

PHYSICAL CHEMISTRY

THIRD EDITION

STUDENT'S SOLUTION MANUAL

NOGGLE

PHYSICAL CHEMISTRY

THIRD EDITION

Prepared by

Travis Adams
UNIVERSITY OF OREGON

Surita R. Bhatia
PRINCETON UNIVERSITY

Edward P. Hu
WASHINGTON UNIVERSITY SCHOOL OF MEDICINE

HarperCollins*CollegePublishers*

HarperCollins® and ■® are registered trademarks of HarperCollins Publishers Inc.

Student's Solution Manual to accompany Noggle, *Physical Chemistry,* Third Edition

ISBN 0-673-52343-8

96 97 98 99 00 9 8 7 6 5 4 3 2 1

CHAPTER 1 *Properties of Matter*

SECTION 1.1

1.1

To obtain molar volume, we divide the molecular weight by the density. This is easier to see if you just consider the units of density and molar volume.

$$\rho = 37.7 \cdot \frac{g}{liter}$$

Calculating molecular weight:

$$MW \quad 32.066 \cdot \frac{g}{mole} - 2 \cdot 15.9994 \cdot \frac{g}{mole} \qquad MW = 64.065 \cdot \frac{g}{mole}$$

Calculating molar volume:

$$V \quad \frac{MW}{\rho} \qquad V = 1.699 \cdot \frac{liter}{mole}$$

1.3

Given information:

$$R = 8.31451 \cdot \frac{joule}{mole \cdot K} \qquad T \quad 313.15 \cdot K \qquad V \quad 250 \cdot cm^3 \qquad V = 2.5 \cdot 10^{-4} \cdot m^3$$

For both RK equation and ideal gas, we must first calculate the molar volume:

$$n \quad \frac{58.0 \cdot g}{2 \cdot 12.011 \cdot \frac{g}{mole} \cdot 6 \cdot 1.00794 \cdot \frac{g}{mole}}$$

$$n = 1.929 \cdot mole$$

1

$$V_m \quad \frac{V}{n} \qquad V_m = 1.296 \cdot 10^{-4} \cdot \frac{m^3}{mole}$$

For the ideal gas law result:

$$P \quad \frac{R \cdot T}{V_m} \qquad P = 2.009 \cdot 10^7 \cdot Pa \qquad MPa \quad 10^6 \cdot Pa$$

$$P = 20.09 \cdot MPa$$

For the RK equation, we get the parameters from Table 1.1.

$$a = 9.882 \cdot \frac{Pa \cdot m^6}{mole^2 \cdot K^{0.5}} \qquad b = 56.84 \cdot 10^{-6} \cdot \frac{m^3}{mole}$$

$$P = \frac{R \cdot T}{V_m - b} - \frac{a}{\sqrt{T} \cdot V_m \cdot (V_m - b)} \qquad P = 1.267 \cdot 10^7 \cdot Pa$$

$$P = 12.67 \cdot MPa$$

The result from the RK equation is almost half that obtained using the ideal gas law.

1.5

We first convert the given density to molar volume, and then plug this volume into the RK equation to obtain a pressure. The constants a and b for neon can be obtained from Table 1.1.
Given information:

$$R = 8.31451 \cdot \frac{joule}{mole \cdot K} \qquad T = 298.15 \cdot K \qquad \rho = 10.0 \cdot \frac{g}{liter} \qquad MW = 20.179 \cdot \frac{g}{mole}$$

$$a = 0.1488 \cdot \frac{Pa \cdot m^6}{mole^2 \cdot K^{0.5}} \qquad b = 12.22 \cdot 10^{-6} \cdot \frac{m^3}{mole}$$

Calculating molar volume from density, converting liters to cubic meters:

$$V_m = \frac{MW}{\rho}$$

$$V_m = 0.002 \cdot \frac{m^3}{mole}$$

Calculation of pressure:

$$P := \frac{R \cdot T}{V_m - b} - \frac{a}{T \cdot V_m \cdot (V_m - b)}$$

$P = 1.234 \cdot 10^6 \cdot Pa \qquad MPa := 10^6 \cdot Pa$

$P = 1.23 \cdot MPa$

1.7

The easiest way to do this is to find the molar volume and convert this to a density. Given information:

$R := 8.31451 \cdot \dfrac{joule}{mole \cdot K} \qquad T := 400 \cdot K \qquad MW := 3 \cdot 12.011 \cdot \dfrac{g}{mole} - 6 \cdot 1.00794 \cdot \dfrac{g}{mole} \qquad MW = 42.081 \cdot \dfrac{g}{mole}$

$P := 10 \cdot atm \cdot \dfrac{101325 \cdot Pa}{1 \cdot atm} \qquad P = 1.013 \cdot 10^6 \cdot Pa \qquad a := 0.8391 \cdot \dfrac{Pa \cdot m^6}{mole} \qquad b := 82.01 \cdot 10^{-6} \cdot \dfrac{m^3}{mole}$

For the ideal gas case:

$V_m := \dfrac{R \cdot T}{P} \qquad V_m = 0.003282 \cdot \dfrac{m^3}{mole} \qquad \rho := \dfrac{MW}{V_m} \qquad \rho = 12.8 \cdot \dfrac{g}{liter}$

For the van der Waals equation, we rearragne it into a cubic equation in volume (eq. 1.6), and then solve for volume:

$V_m := root \left[P \cdot V_m^3 - (R \cdot T - b \cdot P) \cdot V_m^2 - a \cdot V_m - a \cdot b, V_m \right] \qquad V_m = 0.003105 \cdot \dfrac{m^3}{mole}$

$\rho := \dfrac{MW}{V_m} \qquad \rho = 13.6 \cdot \dfrac{g}{liter}$

1.9

Given information:

$$R = 8.31451 \cdot \frac{joule}{mole \cdot K} \qquad T = 300 \cdot K$$

$$P = 60 \cdot atm \cdot \frac{101325 \cdot Pa}{1 \cdot atm} \qquad P = 6.08 \cdot 10^6 \cdot Pa \qquad a = 0.1364 \cdot \frac{Pa \cdot m^6}{mole} \qquad b = 38.58 \cdot 10^{-6} \cdot \frac{m^3}{mole}$$

We solve the ideal gas case to use as an initial guess to solve the van der Waals equation:

$$V_m = \frac{R \cdot T}{P} \qquad V_m = 4.102892 \cdot 10^{-4} \cdot \frac{m^3}{mole}$$

We rearragne the van der Waals equation into a cubic equation in volume (eq. 1.6), and then solve for volume:

$$V_m = root \left[P \cdot V_m^3 - (R \cdot T - b \cdot P) \cdot V_m^2 - a \cdot V_m \quad a \cdot b, V_m \right] \qquad V_m = 398 \cdot \frac{cm^3}{mole}$$

SECTION 1.2

1.11
_____

The necessary equations come straight from Table 1.2. Given information:

$$R = 8.31451 \cdot \frac{joule}{mole \cdot K} \qquad T = 125 \cdot K$$

For the van der Waals equation:

$$a = 0.2283 \cdot \frac{Pa \cdot m^6}{mole} \qquad b = 42.69 \cdot 10^{-6} \cdot \frac{m^3}{mole}$$

$$B = b - \frac{a}{R \cdot T} \qquad B = -177 \cdot \frac{cm^3}{mole}$$

For the RK equation:

$$a = 3.194 \cdot \frac{Pa \cdot m^6}{mole \cdot K^{0.5}} \qquad b = 29.59 \cdot 10^{-6} \cdot \frac{m^3}{mole}$$

$$B = b - \frac{a}{R \cdot T^{1.5}} \qquad B = -245 \cdot \frac{cm^3}{mole}$$

For the Berthelot equation:

$$T_c = 190.6 \cdot K \qquad P_c = 4.641 \cdot 10^6 \cdot Pa$$

$$B = \frac{9 \cdot R \cdot T_c}{128 \cdot P_c} \cdot \left(1 - \frac{6 \cdot T_c^2}{T^2}\right) \qquad B = -311 \cdot \frac{cm^3}{mole}$$

As was noted in the text, the van der Waals equation gives the least accurate result, and the RK and Berthelot expressions are both fairly accurate.

1.13

At the Boyle temperature, the second virial coefficient equals zero. We use the equations in Table 1.2, set B equal to zero, and solve for temperature.

$$R = 8.31451 \cdot \frac{joule}{mole \cdot K}$$

For the van der Waals equation:

$$a = 0.1342 \cdot \frac{Pa \cdot m^6}{mole} \qquad\qquad b = 31.67 \cdot 10^{-6} \cdot \frac{m^3}{mole}$$

$$B = b - \frac{a}{R \cdot T_B} = 0 \qquad\qquad T_B = \frac{a}{b \cdot R} \qquad\qquad T_B = 510 \cdot K$$

For the Berthelot equation:

$$T_c = 151 \cdot K \qquad\qquad P_c = 4.955 \cdot 10^6 \cdot Pa$$

$$B = \frac{9 \cdot R \cdot T_c}{128 \cdot P_c} \cdot \left(1 - \frac{6 \cdot T_c^2}{T^2}\right) = 0 \qquad T_B = \sqrt{6} \cdot T_c \qquad\qquad T_B = 370 \cdot K$$

For the RK equation:

$$a = 1.671 \cdot \frac{Pa \cdot m^6}{mole \cdot K^{0.5}} \qquad\qquad b = 21.95 \cdot 10^{-6} \cdot \frac{m^3}{mole}$$

$$B = b - \frac{a}{R \cdot T_B^{1.5}} = 0 \qquad\qquad T_B = \left(\frac{a}{b \cdot R}\right)^{\frac{2}{3}} \qquad\qquad T_B = 438 \cdot K$$

1.15

At the Boyle temperature, the second virial coefficient equals zero. We set the expression for the Boyle temperature equal to zero and solve for temperature. As noted in Table 1.3, the Beattie-Bridgeman constants given are for volume in liters and pressure in atmospheres, so we must use a different value for R. From Table 1.2:

$$R = 0.08206 \qquad\qquad B(T) = B_0 - \frac{A_0}{R \cdot T_B} - \frac{c}{T_B^3} = 0$$

$$A_0 = 0.1975 \qquad\qquad B_0 = 0.02096 \qquad\qquad c = 0.050 \cdot 10^4$$

We can calculate B at several temperatures to get some idea of where the Boyle temperature is:

$$T = 400, 300 .. 100$$

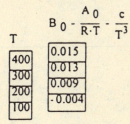

$$
\begin{array}{c|c}
T & B_0 - \dfrac{A_0}{R \cdot T} - \dfrac{c}{T^3} \\\hline
400 & 0.015 \\
300 & 0.013 \\
200 & 0.009 \\
100 & -0.004 \\
\end{array}
$$

The root lies somewhere between 100 K and 200 K.
We use 100 K as our first guess for the Boyle temperature.

$$T_B = 100 \cdot K$$

$$T_B = \text{root}\left(B_0 - \frac{A_0}{R \cdot T_B} - \frac{c}{T_B^3}, T_B \right)$$

$$T_B = 117 \cdot K$$

7

1.17

This is similar to example 1.6 in the text. The virial series can be written as:

$$z = 1 + Bc + Cc^2 + \ldots$$

where $c = 1/V$. From the data given, we must first calculate the concentration:

$$T := 295 \text{ K} \qquad R := 8.31451 \ \frac{\text{joule}}{\text{mole K}} \qquad V := \frac{RT}{P} \qquad c := \frac{1}{V}$$

Now, we can fit the data using a nonlinear regression program. Doing this, we obtain:

$B = -143 \text{ cm}^3/\text{mole}$
$C = 5.7 \ 10^3 \text{ cm}^6/\text{mole}^2$

1.19

$$i = 0..6 \qquad R = 8.31451 \cdot \frac{\text{joule}}{\text{mole} \cdot \text{K}} \qquad T = 700 \cdot \text{K} \qquad \text{MPa} = 1000000 \cdot \text{Pa}$$

The Virial Series typically takes the form $z = 1 + B(T) \cdot c^2 + C(T) \cdot c^3 + \ldots$, so therefore we need to relate concentration, c, to the compressibility factor, z. Recalling that $z = P \cdot V_m/(R \cdot T)$ and $c = 1/V_m$, $c = P/(z \cdot R \cdot T)$.

$P_i =$	$z_i =$	$\dfrac{P_i}{z_i \cdot R \cdot T}$ $\left(\dfrac{\text{mole}}{\text{cm}^3}\right)$
$0.1 \cdot \text{MPa}$	.9998	
$1.0 \cdot \text{MPa}$	.9986	
$2.0 \cdot \text{MPa}$	.9975	$1.719 \cdot 10^{-5}$
$4.0 \cdot \text{MPa}$	.9962	$1.721 \cdot 10^{-4}$
$6.0 \cdot \text{MPa}$	.9962	$3.445 \cdot 10^{-4}$
$8.0 \cdot \text{MPa}$	.9972	$6.899 \cdot 10^{-4}$
$10.0 \cdot \text{MPa}$	.9993	$1.035 \cdot 10^{-3}$
		$1.378 \cdot 10^{-3}$
		$1.719 \cdot 10^{-3}$

A least squares fit of the compressibility factor, z, versus concentration, c, to the form:

$z = 1 + B \cdot c + C \cdot c^2$ yields $B = -8.817$ cm^3/mol and $C = 4.9 \cdot 10^3$ cm^6/mol^2 .

$c = 0, .0001 ... .002$

$Z(c) = 1 - 8.817 \cdot c - 4.9 \cdot 10^3 \cdot c^2$

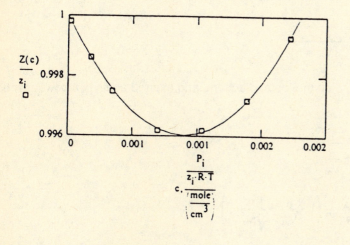

This plot shows compressibility factor versus concentration. The parabolic shape shows very clearly the importance of the third virial coefficient in this case.

1.21

The derivation is very similar to example 1.8 from the text. First, start with the RK equation. Multiply through by the molar volume and divide by RT to obtain the compressibility, z, on one side of the equation.

$$P = \frac{RT}{V_m - b} - \frac{a}{\sqrt{T} V_m (V_m + b)}$$

$$z = \frac{PV}{RT} = \frac{V_m}{V_m - b} - \frac{a}{RT^{3/2} (V_m + b)}$$

Divide the first term by molar volume on the top and bottom:

9

$$z = \frac{1}{1 - \dfrac{b}{V_m}} - \frac{a}{RT^{3/2}\left(V_m + b\right)}$$

Expand the first term using a power series ($x = b/V_m$):

$$\frac{1}{1 - x} = 1 + x + x^2 + x^3 + \ldots$$

$$z = 1 + \frac{b}{V_m} + \frac{b^2}{V_m^2} + \frac{b^3}{V_m^3} + \ldots - \frac{a}{RT^{3/2}\left(V_m + b\right)}$$

Divide the second term by $V_m RT^{3/2}$ on the top and bottom, and expand again in a power series ($x = -b/V_m$):

$$z = 1 + \frac{b}{V_m} + \frac{b^2}{V_m^2} + \frac{b^3}{V_m^3} + \ldots - \frac{a/V_m RT^{3/2}}{1 + \dfrac{b}{V_m}}$$

$$z = 1 + \frac{b}{V_m} + \frac{b^2}{V_m^2} + \frac{b^3}{V_m^3} + \ldots - \frac{a}{V_m RT^{3/2}}\left(1 - \frac{b}{V_m} + \frac{b^2}{V_m^2} - \ldots\right)$$

$$z = 1 + \frac{b}{V_m} + \frac{b^2}{V_m^2} + \frac{b^3}{V_m^3} + \ldots - \frac{a}{V_m RT^{3/2}} + \frac{ab}{V_m^2 RT^{3/2}} - \frac{ab^2}{V_m^3 RT^{3/2}} + \ldots$$

Collect like powers of Vm, and compare to the virial equation:

$$z = 1 + \frac{1}{V_m}\left(b - \frac{a}{RT^{3/2}}\right) + \frac{1}{V_m^2}\left(b^2 + \frac{ab}{RT^{3/2}}\right) + \frac{1}{V_m^3}\left(b^3 - \frac{ab^2}{RT^{3/2}}\right) - \ldots$$

$$z = 1 + \frac{B(T)}{V_m} + \frac{C(T)}{V_m^2} + \frac{D(T)}{V_m^3} + \ldots$$

Therefore:

$$B(T) = b - \frac{a}{RT^{3/2}} \qquad\qquad C(T) = b^2 + \frac{ab}{RT^{3/2}}$$

1.23

We use equation 1.11 and the expressions for the virial coefficients derived in problem 1.22:

$$\gamma = \frac{C - B^2}{RT} \qquad\qquad B(T) = b - \frac{a}{RT^2} \qquad\qquad C(T) = b^2$$

$$\gamma = \frac{C - B^2}{RT} = \frac{1}{RT}\left(b^2 - \left(b - \frac{a}{RT^2}\right)^2\right)$$

$$\gamma = \frac{1}{RT}\left(b^2 - \left(b^2 - \frac{2ab}{RT^2} + \frac{a^2}{R^2T^4}\right)\right)$$

$$\gamma = \frac{1}{RT}\left(b^2 - b^2 + \frac{2ab}{RT^2} - \frac{a^2}{R^2T^4}\right) = \frac{1}{RT}\left(\frac{2ab}{RT^2} - \frac{a^2}{R^2T^4}\right)$$

$$\gamma = \frac{2ab}{R^2T^3} - \frac{a^2}{R^3T^5}$$

1.25

Table 1.2 gives us B(T) and C(T) for the RK equation in terms of the constants a and b. Equation 1.1 from the text gives the relationship between the coefficients of the truncated virial series, β and γ, and B(T) and C(T). We plug the expressions from Table 1.2 into equations 1.11.

$$B(T) = b - \frac{a}{RT^{3/2}} \qquad\qquad C(T) = b^2 + \frac{ab}{RT^{3/2}}$$

$$\beta = B \qquad\qquad\qquad \beta = b - \frac{a}{RT^{3/2}}$$

$$\gamma = \frac{C - B^2}{RT} \qquad\qquad \gamma = \frac{1}{RT}\left(b^2 + \frac{ab}{RT^{3/2}} - \left(b - \frac{a}{RT^{3/2}}\right)^2\right)$$

$$\gamma = \frac{1}{RT}\left(b^2 + \frac{ab}{RT^{3/2}} - \left(b^2 - \frac{2ab}{RT^{3/2}} + \frac{a^2}{R^2T^3}\right)\right)$$

$$\gamma = \frac{1}{RT}\left(b^2 + \frac{ab}{RT^{3/2}} - b^2 + \frac{2ab}{RT^{3/2}} - \frac{a^2}{R^2T^3}\right)$$

$$\gamma = \frac{1}{RT}\left(\frac{3ab}{RT^{3/2}} - \frac{a^2}{R^2T^3}\right) = \frac{3ab}{R^2T^{5/2}} - \frac{a^2}{R^3T^4}$$

We plug the results for β and γ into the original equation to obtain:

$$V_m = \frac{RT}{P} + \left(b - \frac{a}{RT^{3/2}}\right) + \left(\frac{3ab}{R^2T^{5/2}} - \frac{a^2}{R^3T^4}\right)P$$

1.27

Table 1.2 gives us the expressions for the virial coefficients in terms of the Beattie-Bridgeman constants. We plug these in, along with R in the correct units, to obtain equations for the virial coefficients. The Beattie-Bridgeman constants for hydrogen can be found on Table 1.3.

$A_0 = 0.1975$ $\qquad$ $B_0 = 0.0296$ $\qquad$ $c = 500$

$a = -0.00506$ $\qquad$ $b = -0.04359$ $\qquad$ $R = 0.082057$

$$B = B_0 - \frac{A_0}{R \cdot T} - \frac{c}{T^3}$$ $\qquad$ $$C = \frac{A_0 \cdot a}{R \cdot T} - B_0 \cdot b - \frac{B_0 \cdot c}{T^3}$$

$$B = 2.96 \cdot 10^{-2} - \frac{2.407}{T} - \frac{500}{T^3}$$ $\qquad$ $$C = \frac{1.22 \cdot 10^{-2}}{T} + 1.29 \cdot 10^{-3} - \frac{14.8}{T^3}$$

Now, to obtain the equation for the molar volume as a function of pressure, we start with the virial series in pressure, equation 1.10. We divide this by pressure to obtain molar volume on one side of the equation by itself.

$$P \cdot V_m = R \cdot T + \beta \cdot P + \gamma \cdot P^2 + \dots$$

$$V_m = \frac{R \cdot T}{P} + \beta - \gamma \cdot P$$

We use equations 1.11 to convert the virial coefficients we have derived into the coefficients for this equation, and plug in 250 for the temperature:

$$T = 250 \qquad\qquad RT = (0.082057) \cdot (250)$$

$$RT = 20.51425$$

$$\beta = B \qquad\qquad \gamma = \frac{C - B^2}{R \cdot T}$$

$$\beta = 2.96 \cdot 10^{-2} - \frac{2.407}{T} - \frac{500}{T^3} \qquad \gamma = 12.1866 \cdot \frac{\left[\dfrac{1.22 \cdot 10^{-2}}{T} - 1.29 \cdot 10^{-3} - \dfrac{14.8}{T^3} - \left(2.96 \cdot 10^{-2} - \dfrac{2.407}{T} - \dfrac{500}{T^3} \right)^2 \right]}{T}$$

$$\beta = 1.994 \cdot 10^{-2} \qquad\qquad \gamma = 4.583 \cdot 10^{-5}$$

This gives us the expression we need to calculate the molar volume at 100 atm.

$$P = 100 \cdot atm$$

$$V_m = \frac{20.514 \cdot liter \cdot atm}{P} + 1.994 \cdot 10^{-2} \cdot liter + 4.583 \cdot 10^{-5} \cdot \frac{liter}{atm} \cdot P$$

$$V_m = 0.2297 \cdot \frac{liter}{mole}$$

SECTION 1.3

1.29

As was discussed in the text, we should avoid using the relationship $b = V_c/3$, since the critical volume is usually not known very accurately. The expressions for Pc and Tc, derived in the previous problem, are:

$$P_c = \sqrt{\frac{aR}{216b^3}} \qquad T_c = \sqrt{\frac{8a}{27bR}}$$

We solve these equations for a and b. There are many different ways to do this. Rearranging the expression for the critical temperature:

$$T_c^2 = \frac{8a}{27bR} \qquad a = \frac{27bRT_c^2}{8}$$

Plugging this into the equation for the critical pressure, we obtain:

$$P_c^2 = \frac{aR}{216b^3} = \left(\frac{27bRT_c^2}{8}\right)\left(\frac{R}{216b^3}\right)$$

$$P_c^2 = \frac{27bR^2T_c^2}{8 \times 216b^3} = \frac{R^2T_c^2}{64b^2}$$

$$b^2 = \frac{R^2T_c^2}{64P_c^2} \qquad b = \frac{RT_c}{8P_c}$$

We plug this result back into the expression for a derived from the critical temerparature, and simplify:

$$a = \frac{27bRT_c^2}{8} = \left(\frac{27RT_c^2}{8}\right)\left(\frac{RT_c}{8P_c}\right) = \frac{27R^2T_c^3}{64P_c}$$

$$a = \frac{27R^2T_c^3}{64P_c} \qquad b = \frac{RT_c}{8P_c}$$

14

1.31

The equations for a and b come from Table 1.4.

$$R = 8.31451 \cdot \frac{joule}{mole \cdot K} \qquad T_c = 318 \cdot K \qquad P_c = 5.88 \cdot 10^6 \cdot Pa$$

$$a = 0.42748 \cdot \frac{R^2 \cdot T_c^{2.5}}{P_c} \qquad a = 9.063 \cdot \frac{Pa \cdot m^6}{mole^2 \cdot K^{0.5}}$$

$$b = 0.086640 \cdot \frac{R \cdot T_c}{P_c} \qquad b = 3.9 \cdot 10^{-5} \cdot \frac{m^3}{mole}$$

1.33

Calculating the critical compressibility:

$$R = 8.31451 \cdot \frac{joule}{mole \cdot K} \qquad T_c = 536.4 \cdot K \qquad P_c = 5.5 \cdot 10^6 \cdot Pa \qquad V_c = 240 \cdot \frac{cm^3}{mole}$$

$$z_c = \frac{P_c \cdot V_c}{R \cdot T_c} \qquad z_c = 0.296$$

RK constants can be calculated from equations in Table 1.4:

$$a = 0.42748 \cdot \frac{R^2 \cdot T_c^{2.5}}{P_c} \qquad a = 35.8 \cdot \frac{Pa \cdot m^6}{mole^2 \cdot K^{0.5}}$$

$$b = 0.086640 \cdot \frac{R \cdot T_c}{P_c} \qquad b = 7.03 \cdot 10^{-5} \cdot \frac{m^3}{mole}$$

The value obtained for b can be used to recalculate the critical volume. Because a and b were calculated from Pc and Tc, calculating Vc from b and comparing it to the experimental value provides a measure for how accurate the RK equation is in the critical region. Again, from Table 1.4:

$$V_c = 3.847 \cdot b \qquad\qquad V_c = 270 \cdot \frac{cm^3}{mole}$$

SECTION 1.4

1.35

The critical constants are given in Table 1.1:

$$R = 8.31451 \cdot \frac{joule}{mole \cdot K} \qquad T_c = 190.6 \cdot K \qquad P_c = 4.955 \cdot 10^6 \cdot Pa$$

$$T = 229 \cdot K \qquad T_r = \frac{T}{T_c} \qquad T_r = 1.2$$

$$P = 14 \cdot 10^6 \cdot Pa \qquad P_r = \frac{P}{P_c} \qquad P_r = 2.83$$

From Figure 1.11, the critical compressibility is approximately 0.54. From this, we calculate the molar volume and density:

$$z = 0.54 \qquad V_m = \frac{z \cdot R \cdot T}{P} \qquad V_m = 7.344 \cdot 10^{-5} \cdot m^3 \qquad MW = 16.043 \cdot 10^{-3} \cdot \frac{kg}{mole}$$

$$\rho = \frac{MW}{V_m} \qquad \rho = 218 \cdot \frac{kg}{m^3}$$

SECTION 1.5

SECTION 1.6

1.37

We use equation 1.20. Given information:

$$R = 8.31451 \cdot \frac{joule}{mole \cdot K} \qquad T = 300 \cdot K \qquad M = 2 \cdot 14.0067 \cdot 10^{-3} \cdot \frac{kg}{mole} \qquad M = 0.02801 \cdot \frac{kg}{mole}$$

$$u = \sqrt{\frac{3 \cdot R \cdot T}{M}} \qquad u = 516.8 \cdot \frac{m}{sec}$$

1.39

We use equation 1.32 from the text. We are given:

$$R = 8.31451 \cdot \frac{joule}{mole \cdot K} \qquad T = 500 \cdot K \qquad M = 256 \cdot 10^{-3} \cdot \frac{kg}{mole} \qquad A = 1.235 \cdot 10^{-10} \cdot m^2$$

$$\Delta W = 1.234 \cdot 10^{-4} \cdot kg \qquad \Delta t = 6 \cdot min$$

$$\mu = \frac{\Delta W}{A \cdot \Delta t} \qquad \mu = 2.776 \cdot 10^3 \cdot \frac{kg}{m^2 \cdot sec}$$

$$P = \mu \cdot \left(\frac{2 \cdot \pi \cdot R \cdot T}{M}\right)^{0.5} \qquad P = 8.87 \cdot 10^5 \cdot Pa$$

$$P = 6.65 \cdot 10^3 \cdot torr$$

1.41

At this low pressure, argon can be treated as an ideal gas. We use equations 1.15b and 1.31 from the text. Given information:

$$R = 8.31451 \cdot \frac{joule}{mole \cdot K} \qquad L = 6.022137 \cdot 10^{23} \cdot \frac{1}{mole} \qquad T = 293.15 \cdot K \qquad M = 39.948 \cdot 10^{-3} \cdot \frac{kg}{mole}$$

$$P = 0.001 \cdot torr \qquad P = 0.133 \cdot Pa$$

$$nstar = \frac{P \cdot L}{R \cdot T} \qquad nstar = 3.294 \cdot 10^{19} \cdot m^{-3}$$

$$Z_{wall} = nstar \cdot \left(\frac{R \cdot T}{2 \cdot \pi \cdot M} \right)^{0.5} \qquad Z_{wall} = 3.246 \cdot 10^{21} \cdot \frac{1}{m^2 \cdot sec}$$

Zwall is the number of collisions per area per time. To get the number of collisions, we multiply by the area:

$$num = Z_{wall} \cdot 1 \cdot cm^2 \qquad num = 3.246 \cdot 10^{17} \cdot \frac{1}{sec}$$

1.43

We rearrange equation 1.32 from the text to solve for μ, and then solve for the ratio of $\Delta W/\Delta t$, here called "ratio." Assume an average molecular weight for air, similar to example 1.16 or problem 1.62.

$$R = 8.31451 \cdot \frac{joule}{mole \cdot K} \qquad T = 293.15 \cdot K \qquad M = 28.9321 \cdot 10^{-3} \cdot \frac{kg}{mole} \qquad A = 1.3 \cdot cm^2$$

$$P = 2.5 \cdot 10^{-6} \cdot torr \qquad L = 6.022137 \cdot 10^{23} \cdot \frac{1}{mole}$$

$$\mu = P \cdot \left(\frac{M}{2 \cdot \pi \cdot R \cdot T} \right)^{0.5} \qquad \mu = 4.581 \cdot 10^{-7} \cdot \frac{kg}{m^2 \cdot sec}$$

$$ratio = \mu \cdot A \qquad ratio = 5.956 \cdot 10^{-11} \cdot \frac{kg}{sec}$$

This gives us the flow rate in mass per time. We want a flow rate in molecules per time, so we divide by the molecular weight and multiply by Avogadro's number:

$$\text{flowrate} = \frac{\text{ratio} \cdot L}{M} \qquad \text{flowrate} = 1.24 \cdot 10^{15} \cdot \frac{1}{\text{sec}}$$

If this problem seems a little confusing, it helps to realize that there is a lot of information in the problem statement that you don't need. Even if you pick the right equation to use, it still maybe confusing as to what variable to solve for. If you think about the units of the answer that you want - a rate of molecules means molecules/sec - then is it a little easier to see that you need to solve for the ratio $\Delta W/\Delta t$, which has units of mass/sec.

1.45

We use equations 1.20, 1.35, and 1.34, respectively. Given information:

$$R = 8.31451 \cdot \frac{\text{joule}}{\text{mole} \cdot K} \qquad T = 423.15 \cdot K \qquad M = 4.002602 \cdot 10^{-3} \cdot \frac{\text{kg}}{\text{mole}}$$

For the rms speed:

$$u = \sqrt{\frac{3 \cdot R \cdot T}{M}} \qquad u = 1624 \cdot \frac{m}{\text{sec}}$$

We show four significant figures in the answer due to the broadness of the distribution. This is discussed in detail in example 1.13 in the text.

For the average speed:

$$v = \sqrt{\frac{8 \cdot R \cdot T}{\pi \cdot M}} \qquad v = 1496 \cdot \frac{m}{\text{sec}}$$

For the most probable speed:

$$v_p = \sqrt{\frac{2 \cdot R \cdot T}{M}} \qquad v_p = 1326 \cdot \frac{m}{\text{sec}}$$

Here we have made the substitution that $k/m = R/M$.

1.47

This is similar to example 1.21 from the text. To determine the percent of molecules traveling with a certain range of speeds, we integrate the distribution function. It is easiest to do this intergral in terms of dimensionless variables. This integral is given in equation 1.37b, and must be solved numerically. We assume an average molecular weight for air, as in example 1.16.

$$R = 8.31451 \cdot \frac{joule}{mole \cdot K} \qquad M = 28.9321 \cdot 10^{-3} \cdot \frac{kg}{mole} \qquad T = 293.15 \cdot K \qquad v_1 = 0 \cdot \frac{mi}{hr} \qquad v_2 = 55 \cdot \frac{mi}{hr}$$

$$i = 1 .. 2 \qquad w_i = \frac{v_i}{\sqrt{\frac{2 \cdot R \cdot T}{M}}} \qquad w_1 = 0$$

$$w_2 = 0.0598992$$

Recall that k/m = R/M. Also note that it is usually a good idea to kep lots of significant figures in intermediate results when you are doing an intergral numerically to avoid round–off errors. When we get the final result, we keep that to the correct amount of significant figures.

$$P = \frac{4}{\sqrt{\pi}} \cdot \int_{w_1}^{w_2} e^{-w^2} \cdot w^2 \, dw \qquad P = 0.00016$$

$$P = 0.016 \cdot \%$$

1.49

This is similar to example 1.21 from the text. To determine the percent of molecules traveling with a certain range of speeds, we integrate the distribution function. However, this problem asks for molecules with speeds greater than 1500 m/s - essentially this means in the range of 1500 m/s to infinity. It is possible to integrate numerically out to a large value, since the distribution function drops off. However, it is much easier to remember that you are integrating to find a percent of molecules. If we find the percent of molecules travelling with speeds between 0 and 1500 m/s and subtract this percentage from 100%, we will find the percent of molecules with speeds greater than 1500. We do the integral in terms of dimensionless variables, as shown in equation 1.37b.

$$R = 8.31451 \cdot \frac{joule}{mole \cdot K} \qquad T = 750 \cdot K \qquad v_1 = 0 \cdot \frac{m}{sec} \qquad v_2 = 1500 \cdot \frac{m}{sec}$$

$$M = 1 \cdot 12.011 \cdot 10^{-3} \cdot \frac{kg}{mole} + 4 \cdot 1.00794 \cdot 10^{-3} \cdot \frac{kg}{mole} \qquad M = 1.60428 \cdot 10^{-2} \cdot \frac{kg}{mole}$$

$$i = 1 .. 2 \qquad w_i = \frac{v_i}{\sqrt{\frac{2 \cdot R \cdot T}{M}}} \qquad w_1 = 0$$

$$w_2 = 1.7012449$$

Recall that k/m = R/M. Also note that it is usually a good idea to kep lots of significant figures in intermediate results when you are doing an intergral numerically to avoid round-off errors. When we get the final result, we keep that to the correct amount of significant figures.

$$P_1 = \frac{4}{\sqrt{\pi}} \cdot \int_{w_1}^{w_2} e^{-w^2} \cdot w^2 \, dw \qquad P_1 = 0.87763$$

$$P_1 = 87.8 \cdot \%$$

This is the percent of molecules travelling with speeds less than 1500 m/s. The rest of the molecules must be travelling with speeds greater than 1500 m/s, so to get the answer we want, we subtract this from 100%.

$$P_2 = 100 \cdot \% - P_1 \qquad P_2 = 12.2 \cdot \%$$

1.51

This problem is very similar to example 1.19 from the text, except that a different distribution function is used. The average speed is calculated using equation 1.30:

$$\underline{v} = \int_0^\infty v\, f(v)\, dv$$

Our distribution function is:

$$f(v)\, dv = \frac{m}{kT} \exp\left(\frac{mv^2}{2kT}\right) v\, dv$$

Thus, we have:

$$\underline{v} = \int_0^\infty v^2 \frac{m}{kT} \exp\left(\frac{mv^2}{2kT}\right) dv$$

From an integral table (such as the one found on the inside front cover of your P-Chem book):

$$\int_0^\infty x^2\, e^{-ax^2}\, dx = \frac{1}{4a}\sqrt{\frac{\pi}{a}}$$

Using a = m/2kT, we obtain the average speed:

$$\underline{v} = \frac{m}{kT}\left(\frac{2kT}{4m}\right)\sqrt{\frac{2kT}{m}\pi} = \frac{1}{2}\sqrt{\frac{2\pi kT}{m}} = \sqrt{\frac{\pi kT}{m}} = \sqrt{\frac{\pi RT}{M}}$$

Recall that k/m = R/M.

SECTION 1.7

1.53

To find the distance at which $U(r)$ is a minimum, we take the first derivative of $U(r)$, set it equal to zero, and solve for r:

$$U(r) = 4\varepsilon \left[\left(\frac{\sigma}{r}\right)^{12} - \left(\frac{\sigma}{r}\right)^{6} \right] \qquad \frac{dU}{dr} = 4\varepsilon \left[\frac{-12\sigma^{12}}{r^{13}} + \frac{6\sigma^{6}}{r^{7}} \right] = 0$$

$$\frac{-12\sigma^{12}}{r^{13}} + \frac{6\sigma^{6}}{r^{7}} = 0 \qquad \frac{6\sigma^{6}}{r^{7}} = \frac{12\sigma^{12}}{r^{13}} \qquad 1 = \frac{2\sigma^{6}}{r^{6}}$$

$$r^{6} = 2\sigma^{6} \qquad\qquad r = 2^{1/6}\sigma$$

To find the value of U at this minimum, we just plug this value of r back into the Lennard-Jones potential:

$$U_{min} = 4\varepsilon \left[\frac{\sigma^{12}}{\left(2^{1/6}\sigma\right)^{12}} - \frac{\sigma^{6}}{\left(2^{1/6}\sigma\right)^{6}} \right] = 4\varepsilon \left[\frac{\sigma^{12}}{4\sigma^{12}} - \frac{\sigma^{6}}{2\sigma^{6}} \right] = 4\varepsilon \left[\frac{1}{4} - \frac{1}{2} \right] = 4\varepsilon \left[\frac{1}{4} \right] = -\varepsilon$$

$$U_{min} = -\varepsilon$$

1.55

Equation 1.38 allows us to compute the second virial coefficient, given an expression for the potential. The Sutherland potential, with $n = 6$, is given by:

$$U = \infty \text{ for } 0 < r < \sigma$$

$$U = -\varepsilon \left(\frac{\sigma}{r}\right)^6 \text{ for } \sigma < r < \infty$$

We plug this into equation 1.38. Since the Sutherland potential is discontinuous at $r = \sigma$, we split the integral into two integrals:

$$B(T) = 2\pi L \int_0^\infty \left(1 - e^{-U/kT}\right) r^2 \, dr \qquad B(T) = 2\pi L \int_0^\sigma \left(1 - e^{-\infty/kT}\right) r^2 \, dr + 2\pi L \int_\sigma^\infty \left(1 - e^{\varepsilon(\sigma/r)^6/kT}\right) r^2 \, dr$$

The first of these integrals is equal to zero, since e raised to negative infinity tends toward zero. For the second integral, we make the substitution that $\rho = r/\sigma$, and plug in the definition of b_0:

$$B(T) = 2\pi L \sigma^3 \int_1^\infty \left(1 - e^{\frac{\varepsilon}{kT}\rho^{-6}}\right) \rho^2 \, d\rho \qquad b_0 = \frac{2\pi L \sigma^3}{3}$$

$$B(T) = 3b_0 \int_1^\infty \left(1 - e^{\frac{\varepsilon}{kT}\rho^{-6}}\right) \rho^2 \, d\rho$$

Now, expand the exponential in a power series.

$$B(T) = 3b_0 \int_1^\infty \left(1 - \left(\begin{array}{c} 1 + \frac{\varepsilon}{kT}\frac{1}{\rho^6} + \left(\frac{\varepsilon}{kT}\right)^2 \frac{1}{2\rho^{12}} + \left(\frac{\varepsilon}{kT}\right)^3 \frac{1}{6\rho^{18}} + \left(\frac{\varepsilon}{kT}\right)^4 \frac{1}{24\rho^{24}} + \\ \left(\frac{\varepsilon}{kT}\right)^5 \frac{1}{120\rho^{30}} + \left(\frac{\varepsilon}{kT}\right)^6 \frac{1}{720\rho^{36}} + \ldots \end{array}\right)\right) \rho^2 \, d\rho$$

Let $x = \varepsilon/kT$. Cancel the leading ones, multiply through by ρ^2, and integrate:

$$B(T) = 3b_0 \int_1^\infty \left(1 - \left(1 + \frac{x}{\rho^6} + \frac{x^2}{2\rho^{12}} + \frac{x^3}{6\rho^{18}} + \frac{x^4}{24\rho^{24}} + \frac{x^5}{120\rho^{30}} + \frac{x^6}{720\rho^{36}} + \ldots\right)\right)\rho^2 \, d\rho$$

$$B(T) = 3b_0 \int_1^\infty \left(-\frac{x}{\rho^4} - \frac{x^2}{2\rho^{10}} - \frac{x^3}{6\rho^{16}} - \frac{x^4}{24\rho^{22}} - \frac{x^5}{120\rho^{28}} - \frac{x^6}{720\rho^{34}} - \ldots\right) d\rho$$

$$B(T) = 3b_0 \left[-\frac{x}{3\rho^3} - \frac{x^2}{18\rho^9} - \frac{x^3}{90\rho^{15}} - \frac{x^4}{504\rho^{21}} - \frac{x^5}{3240\rho^{27}} - \frac{x^6}{23760\rho^{33}} - \ldots\right]_1^\infty$$

The terms at infinity disappear nicely to give us:

$$B(T) = b_0 \left[-x - \frac{x^2}{6} - \frac{x^3}{30} - \frac{x^4}{168} - \frac{x^5}{1080} - \frac{x^6}{7920} - \ldots\right]$$

1.57

This is discussed in section 1.7 of the text, and particularly in the caption for Figure 1.20. The Lennard-Jones potential works reasonably well for closed-shell, neutral, nonpolar, fairly symmetric molecules. Thus, it would not work well for open shell molecules, such as the methyl radical, any ionic species, such as a hydroxide ion, and polar species, such as HCl, and any long molecule, such as octane.

1.59

The Lennard-Jones and square-well potential constants are given in Table 1.7. For the Lennard-Jones potential, we will use Figure 1.21 to determine B^* and C^*, and then calculate B and C. (B^* is defined in terms of an integral, and in theory we could just perform the integration to get B^*. However, this integral is difficult to do numerically, and if you try this you may have problems getting it to converge. Just use Figure 1.21).

$$\varepsilon\_k := 243 \cdot K \qquad b_0 := 77.97 \cdot \frac{cm^3}{mole} \qquad T := 295 \cdot K \qquad Tstar := \frac{T}{\varepsilon\_k} \qquad Tstar = 1.214$$

For $T^* = 1.214$, B^* can be read off of Figure 1.21(a) as approximately -1.58.

$$Bstar := -1.58 \qquad B := b_0 \cdot Bstar \qquad B = -123 \cdot \frac{cm^3}{mole}$$

26

C° for this reduced temperature is approximately 0.61.

$$Cstar := 0.61 \qquad C := b_0^2 \cdot Cstar \qquad C = 3.7 \cdot 10^3 \cdot \left(\frac{cm^3}{mole}\right)^2$$

For the square-well potential, the integration can be carried out to yield an expression for B, which is given by equation 1.44.

$$\varepsilon\_k := 224 \cdot K \qquad b_0 := 55.72 \cdot \frac{cm^3}{mole} \qquad R := 1.652 \qquad B := b_0 \cdot \left[1 - (R^3 - 1) \cdot \left(e^{\frac{\varepsilon\_k}{T}} - 1\right)\right]$$

$$B = -167 \cdot \frac{cm^3}{mole}$$

SECTION 1.8

1.61

We use the definition of partial pressure, equation 1.45:

$MPa = 10^6 \cdot Pa \qquad P = 8.45 \cdot MPa \qquad x_i = 0.2095 \qquad P_i = x_i \cdot P$

$P_i = 1.77 \cdot MPa$

1.63

We solve the RK equation just as we did before, but now we use equation 1.47 to calculate a and b for the mixture of gases. Given information:

$R = 8.31451 \cdot \dfrac{joule}{mole \cdot K} \qquad T = 500 \cdot K \qquad ntot = 3 \cdot mole - 2 \cdot mole - 1 \cdot mole \quad ntot = 6 \cdot mole$

$V = 2.4 \cdot liter \qquad V_m = \dfrac{V}{ntot} \qquad V_m = 4 \cdot 10^{-4} \cdot \dfrac{m^3}{mole}$

We will let ammonia be species 1, nitrogen be species 2, and hydrogen be species 3. The RK constants for the individual species are obtained from Table 1.1.

$i = 1..3 \qquad n_1 = 3 \cdot mole \qquad\qquad n_2 = 2 \cdot mole \qquad\qquad n_3 = 1 \cdot mole$

$$a_1 - 8.650 \cdot \dfrac{Pa \cdot m^6}{mole^2 \cdot K^{0.5}} \qquad a_2 = 1.551 \cdot \dfrac{Pa \cdot m^6}{mole^2 \cdot K^{0.5}} \qquad a_3 = 0.1447 \cdot \dfrac{Pa \cdot m^6}{mole^2 \cdot K^{0.5}}$$

$$b_1 = 25.85 \cdot 10^{-6} \cdot \dfrac{m^3}{mole} \qquad b_2 = 26.74 \cdot 10^{-6} \cdot \dfrac{m^3}{mole} \qquad b_3 = 18.44 \cdot 10^{-6} \cdot \dfrac{m^3}{mole}$$

Calculating mole fractions and RK constants:

$$x = \frac{n}{ntot} \qquad amix = \left(\sum_i x_i \cdot \sqrt{a_i} \right)^2 \qquad bmix = \sum_i x_i \cdot b_i$$

$$amix = 3.799 \cdot \frac{Pa \cdot m^6}{mole^2 \cdot K^{0.5}} \qquad bmix = 2.491 \cdot 10^{-5} \cdot \frac{m^3}{mole}$$

$$P = \frac{R \cdot T}{V_m - bmix} - \frac{amix}{\sqrt{T} \cdot V_m \cdot (V_m + bmix)} \qquad P = 1.008 \cdot 10^7 \cdot Pa \qquad MPa = 10^6 \cdot Pa$$

$$P = 10.08 \cdot MPa$$

SECTION 1.9

1.65
_____

Equation 1.49 defines the coefficient of thermal expansion; however, this definition is in terms of the total volume, while the equation we are given is for the specific volume. We use the total mass, m, to convert from total volume to specific volume:

$$V = mv \qquad \left(\frac{\partial V}{\partial T}\right)_P = m\left(\frac{\partial v}{\partial T}\right)_P \qquad \alpha = \frac{1}{V}\left(\frac{\partial V}{\partial T}\right)_P = \frac{1}{mv}\left[m\left(\frac{\partial v}{\partial T}\right)_P\right] = \frac{1}{v}\left(\frac{\partial v}{\partial T}\right)_P$$

In this case, the factor of mass cancels out, and so it does not affect how we do the derivation. However, it is always good to check. Taking the derivative of specific volume:

$$v = \exp\left(-6.0781 + 1.01257 \ln T + \frac{280.663}{T}\right)$$

$$\left(\frac{\partial v}{\partial T}\right)_P = \left(\frac{1.01257}{T} - \frac{280.663}{T^2}\right) \times \left[\exp\left(-6.0781 + 1.01257 \ln T + \frac{280.663}{T}\right)\right]$$

$$\alpha = \frac{1}{v}\left(\frac{\partial v}{\partial T}\right)_P$$

$$\alpha = \frac{\left(\frac{1.01257}{T} - \frac{280.663}{T^2}\right) \times \left[\exp\left(-6.0781 + 1.01257 \ln T + \frac{280.663}{T}\right)\right]}{\exp\left(-6.0781 + 1.01257 \ln T + \frac{280.663}{T}\right)}$$

$$\alpha = \frac{1.01257}{T} - \frac{280.663}{T^2}$$

1.67

We use equation 1.50, the definition of the isothermal compressibility:

$$\kappa_T = -\frac{1}{V}\left(\frac{\partial V}{\partial P}\right)_T$$

$$V = \frac{RT}{P} - B(T) \qquad\qquad \left(\frac{\partial V}{\partial P}\right)_T = -\frac{RT}{P^2}$$

$$\kappa_T = -\frac{1}{V}\left(\frac{\partial V}{\partial P}\right)_T = -\frac{1}{\frac{RT}{P}-B}\left(-\frac{RT}{P^2}\right) = \frac{RT}{P^2\left(\frac{RT}{P}-B\right)} = \frac{RT}{PRT - P^2B}$$

$$\kappa_T = \frac{RT}{PRT - P^2B}$$

1.69

We start with equation 1.50, the definition of the isothermal compressibility.

$$\kappa_T = -\frac{1}{V}\left(\frac{\partial V}{\partial P}\right)_T$$

However, since the van der Waals equation is explicit in pressure, not in volume, we use the following property of partial derivatives, which are reviewed in Appendix II of the text:

$$\left(\frac{\partial V}{\partial P}\right)_T = \frac{1}{\left(\frac{\partial P}{\partial V}\right)_T}$$

$$P = \frac{RT}{V_m - b} - \frac{a}{V_m^2}$$

$$\left(\frac{\partial P}{\partial V_m}\right)_T = \frac{-RT}{(V_m - b)^2} + \frac{2a}{V_m^3} = \frac{-RTV_m^3 + 2a(V_m - b)^2}{V_m^3 (V_m - b)^2}$$

$$\left(\frac{\partial V_m}{\partial P}\right)_T = \frac{V_m^3 (V_m - b)^2}{-RTV_m^3 + 2a(V_m - b)^2} = \frac{V_m^3 (V_m - b)^2}{-RTV_m^3 + 2a(V_m^2 - 2bV_m + b^2)} = \frac{V_m^3 (V_m - b)^2}{-RTV_m^3 + 2aV_m^2 - 4abV_m + 2ab^2}$$

$$\kappa_T = -\frac{1}{V_m}\left(\frac{\partial V_m}{\partial P}\right)_T = \frac{V_m^2 (V_m - b)^2}{RTV_m^3 - 2aV_m^2 + 4abV_m - 2ab^2}$$

$$\kappa_T = \frac{V_m^2 (V_m - b)^2}{RTV_m^3 - 2aV_m^2 + 4abV_m - 2ab^2}$$

1.71

This problem is similar to example 1.27 from the text. We use equation 1.51. The isothermal compressibility of benzene and its molar volume at one atm are given in Table 1.7.

$$V_0 = 89 \cdot cm^3 \qquad \kappa_T = 63.5 \cdot 10^{-6} \cdot \frac{1}{atm} \qquad P = 1000 \cdot atm \qquad P_0 = 1 \cdot atm$$

$$V = V_0 \cdot [1 - \kappa_T \cdot (P - P_0)]$$

$$V = 83 \cdot cm^3$$

1.73

This problem is similar to example 1.28 from the text. We use equation 1.53, rearranged and written in difference form instead of as a derivative. The coefficients of thermal expansion and compressibility for CCl4 are given in Table 1.7.

$$\alpha = 1.236 \cdot 10^{-3} \cdot \frac{1}{K} \qquad \kappa_T = 91.0 \cdot 10^{-6} \cdot \frac{1}{atm} \qquad \Delta T = 1 \cdot K$$

$$\Delta P = \frac{\alpha}{\kappa_T} \cdot \Delta T \qquad \Delta P = 13.6 \cdot atm$$

If you were a little confused about which equation to use, it helps to remember just what the meaning of a partial derivative is. (dP/dT)v indicates a change in pressure with resepct to temperature at a constant volume, which is exactly what this problem asks for.

CHAPTER 2 *The First Law of Thermodynamics*

SECTION 2.3

2.1

He is a monoatomic gas, so we use Cvm = 3/2 R:

$$R = 8.31451 \cdot \frac{joule}{mole \cdot K} \qquad C_{vm} = \frac{3}{2} \cdot R \qquad C_{vm} = 12.47 \cdot \frac{joule}{mole \cdot K} \qquad C_{vm}(obs) = 12.55 \cdot \frac{joule}{mole \cdot K}$$

For this simple monatomic, equipartition theory gives a good result. Oxygen is a diatomic, so we must take into account rotation and vibration.

$$C_{vm} = \frac{7}{2} \cdot R \qquad C_{vm} = 29.1 \cdot \frac{joule}{mole \cdot K} \qquad C_{vm}(obs) = 20.81 \cdot \frac{joule}{mole \cdot K}$$

The calculated value is incorrect, but is not that far off. The remaining molecules are polyatomics, so we use equation 2.7. CO_2 is linear, so we have:

$$N = 3 \qquad C_{vm} = \frac{5}{2} \cdot R + (3 \cdot N - 5) \cdot R \qquad C_{vm} = 54.04 \cdot \frac{joule}{mole \cdot K} \qquad C_{vm}(obs) = 28.09 \cdot \frac{joule}{mole \cdot K}$$

Ammonia is nonlinear, so we have:

$$N = 4 \qquad C_{vm} = 3 \cdot R + (3 \cdot N - 6) \cdot R \qquad C_{vm} = 74.83 \cdot \frac{joule}{mole \cdot K} \qquad C_{vm}(obs) = 28.47 \cdot \frac{joule}{mole \cdot K}$$

For ethylene we again use the linear formula:

$$N = 4 \qquad C_{vm} = \frac{5}{2} \cdot R + (3 \cdot N - 5) \cdot R \qquad C_{vm} = 78.99 \cdot \frac{joule}{mole \cdot K} \qquad C_{vm}(obs) = 31.16 \cdot \frac{joule}{mole \cdot K}$$

Ethane is nonlinear:

$$N = 8 \qquad C_{vm} = 3 \cdot R + (3 \cdot N - 6) \cdot R \qquad C_{vm} = 174.6 \cdot \frac{joule}{mole \cdot K} \qquad C_{vm}(obs) = 39.46 \cdot \frac{joule}{mole \cdot K}$$

None of the heat capacities for the polyatomics came close to that predicted by equipartition theory, and this is not surprising. As was mentioned in the text, equipartion theory usually does not give good results for polyatomics.

SECTION 2.4

2.3
_____

We use equation 2.15, and assume methane to be an ideal gas at these temperatures. With this assumption, the second term in equation 2.15 is equal to zero, and simplifies to equation 2.16. We are told to assume a constant heat capacity, which means that we can pull the heat capacity out of the integral for internal energy.

$$n = 1 \cdot mole \qquad C_{vm} = 30.86 \cdot \frac{joule}{mole \cdot K} \qquad C_v = n \cdot C_{vm} \qquad T_1 = 300 \cdot K \qquad T_2 = 400 \cdot K$$

$$R = 8.31451 \cdot \frac{joule}{mole \cdot K} \qquad kJ = 10^3 \cdot joule$$

(a) At constant pressure:

$$\Delta U = \int_{T_1}^{T_2} C_v \, dT \qquad \Delta U = 3.086 \cdot 10^3 \cdot joule \qquad \Delta U = 3.086 \cdot kJ$$

$$w = -\int_{V_1}^{V_2} P \, dV = -P \cdot (V_2 - V_1) = -n \cdot R \cdot (T_2 - T_1)$$

Since the pressure is constant, we can pull it out of the integral for work. We then use the ideal gas law to convert the pressure and volume terms into a change in temerpature, which we know.

$$w = -n \cdot R \cdot (T_2 - T_1) \qquad w = -0.831 \cdot kJ$$

$$q = \Delta U - w \qquad q = 3.917 \cdot kJ$$

(b) At constant volume, the change in internal energy will still be the same. However, since dV equals zero, and the only type of work that can be done is PV work, the work done will be zero. Thus, q will be equal to the change in internal energy.

$$\Delta U = 3.086 \cdot kJ$$

$$w = 0 \cdot kJ \qquad q = \Delta U - w$$

$$q = 3.086 \cdot kJ$$

2.5

We use equation 2.13:

$$\left(\frac{\partial U}{\partial V}\right)_T = T\left(\frac{\partial P}{\partial T}\right)_V - P$$

The virial series is easily rearranged to obtain an equation explicit in pressure:

$$z = \frac{PV_m}{RT} = 1 + \frac{B(T)}{V_m} \qquad PV_m = RT + \frac{RT\,B(T)}{V_m} \qquad P = \frac{RT}{V_m} + \frac{RT\,B(T)}{V_m^2}$$

Keep in mind that B is a function of temerpature when you take the derivative:

$$\left(\frac{\partial P}{\partial T}\right)_V = \frac{R}{V_m} + \frac{BR}{V_m^2} + \frac{RT}{V_m^2}\left(\frac{dB}{dT}\right)$$

$$\left(\frac{\partial U}{\partial V}\right)_T = T\left(\frac{\partial P}{\partial T}\right)_V - P = T\left[\frac{R}{V_m} + \frac{BR}{V_m^2} + \frac{RT}{V_m^2}\left(\frac{dB}{dT}\right)\right] - P = \frac{RT}{V_m} + \frac{BRT}{V_m^2} + \frac{RT^2}{V_m^2}\left(\frac{dB}{dT}\right) - \frac{RT}{V_m} - \frac{BRT}{V_m^2}$$

Terms cancel to give us:

$$\left(\frac{\partial U}{\partial V}\right)_T = \frac{RT^2}{V_m^2}\left(\frac{dB}{dT}\right)$$

2.7

This problem is similar to example 2.4. Using equation 2.14:

$$P = 1 \cdot atm \qquad T = 273.15 \cdot K \qquad \alpha = -0.0547 \cdot 10^{-3} \cdot \frac{1}{K} \qquad \kappa_T = 47 \cdot 10^{-6} \cdot \frac{1}{atm}$$

$$intpress = \frac{T \cdot \alpha}{\kappa_T} - P \qquad\qquad intpress = -319 \cdot atm$$

2.9

Assuming methane obeys the RK equation, we use the result derived in problem 2.8 and the RK constants given in Table 1.1:

$$R = 8.31451 \cdot \frac{joule}{mole \cdot K} \qquad T = 100 \cdot K \qquad V_m = 0.0351 \cdot liter \qquad MPa = 10^6 \cdot Pa$$

$$a = 3.194 \cdot \frac{Pa \cdot m^6}{mole^2 \cdot K^{0.5}} \qquad b = 29.59 \cdot 10^{-6} \cdot \frac{m^3}{mole}$$

$$intpress = \frac{3 \cdot a}{2 \cdot \sqrt{T} \cdot V_m \cdot (V_m + b)} \qquad intpress = 211 \cdot MPa$$

SECTION 2.5

2.11

Our starting point is equation 2.22. Since the process takes places at constant pressure, the dP term is zero, and the equation simplifies to equation 2.23. Table 2.2 gives us an expression for the dependence of Cpm on temperature, which we can integrate.

$$T_1 = 300 \cdot K \qquad T_2 = 700 \cdot K \qquad mass = 1 \cdot kg \qquad MW = 2 \cdot (12.011 + 1.00794) \cdot 10^{-3} \cdot \frac{kg}{mole}$$

$$MW = 0.026 \cdot \frac{kg}{mole} \qquad n = \frac{mass}{MW} \qquad n = 38.406 \cdot mole \qquad kJ = 10^3 \cdot joule$$

$$c_1 = 47.18 \cdot \frac{joule}{mole \cdot K} \qquad c_2 = 25.91 \cdot 10^{-3} \cdot \frac{joule}{mole \cdot K^2} \qquad c_3 = -4.23 \cdot 10^{-6} \cdot \frac{joule}{mole \cdot K^3} \qquad c_4 = -9.37 \cdot 10^5 \cdot \frac{joule \cdot K}{mole}$$

$$\Delta H = \int_{T_1}^{T_2} n \cdot \left(c_1 + c_2 \cdot T + c_3 \cdot T^2 + \frac{c_4}{T^2} \right) dT \qquad \Delta H = 838.151 \cdot kJ$$

2.13

Our starting point is equation 2.22. Since the process takes places at constant pressure, the dP term is zero, and the equation simplifies to equation 2.23. Table 2.2 gives us an expression for the dependence of Cpm on temperature, which we can integrate. Don't forget to convert temperatures to Kelvins.

$$T_1 = 298.15 \cdot K \qquad T_2 = 600.652 \cdot K \qquad mass = 1 \cdot kg \qquad MW = 207.2 \cdot 10^{-3} \cdot \frac{kg}{mole}$$

$$n = \frac{mass}{MW} \qquad n = 4.826 \cdot mole \qquad kJ = 10^3 \cdot joule$$

$$c_1 = 22.13 \cdot \frac{joule}{mole \cdot K} \qquad c_2 = 11.72 \cdot 10^{-3} \cdot \frac{joule}{mole \cdot K^2} \qquad c_4 = 0.96 \cdot 10^5 \cdot \frac{joule \cdot K}{mole}$$

$$\Delta H = \int_{T_1}^{T_2} n \cdot \left(c_1 + c_2 \cdot T + \frac{c_4}{T^2} \right) dT \qquad \Delta H = 40.78 \cdot kJ$$

2.15

Our starting point is equation 2.22. Since the process takes places at constant pressure, the dP term is zero, and the equation simplifies to equation 2.23. Table 2.2 gives us an expression for the dependence of Cpm on temperature, which we can integrate. Don't forget to convert temperatures to Kelvins.

$$T_1 = 298.15 \cdot K \qquad T_2 = 933.52 \cdot K \qquad mass = 1 \cdot kg \qquad MW = 26.98154 \cdot 10^{-3} \cdot \frac{kg}{mole}$$

$$n = \frac{mass}{MW} \qquad n = 37.062 \cdot mole \qquad kJ = 10^3 \cdot joule$$

$$c_1 = 20.67 \cdot \frac{joule}{mole \cdot K} \qquad c_2 = 12.38 \cdot 10^{-3} \cdot \frac{joule}{mole \cdot K^2} \qquad c_4 = 0 \cdot \frac{joule \cdot K}{mole}$$

$$\Delta H = \int_{T_1}^{T_2} n \cdot \left(c_1 + c_2 \cdot T + \frac{c_4}{T^2} \right) dT \qquad \Delta H = 666.3 \cdot kJ$$

2.17

Our starting point is equation 2.22. Since the process takes places at constant pressure, the dP term is zero, and the equation simplifies to equation 2.23. Table 2.3 gives us an expression for the dependence of Cpm on temperature, which we can integrate. Don't forget to convert temperatures to Kelvins.

$$T_1 = 330.15 \cdot K \qquad T_2 = 673.15 \cdot K \qquad n = 3.52 \cdot mole \qquad kJ = 10^3 \cdot joule$$

$$a = 8.468 \cdot \frac{joule}{mole \cdot K} \qquad b = 269.45 \cdot 10^{-3} \cdot \frac{joule}{mole \cdot K^2} \qquad c = -143.45 \cdot 10^{-6} \cdot \frac{joule}{mole \cdot K^3} \qquad d = 29.63 \cdot 10^{-9} \cdot \frac{joule}{mole \cdot K^4}$$

$$\Delta H = \int_{T_1}^{T_2} n \cdot \left(a + b \cdot T + c \cdot T^2 + d \cdot T^3 \right) dT \qquad \Delta H = 133.18 \cdot kJ$$

2.19

We use equation 2.21:

$$\left(\frac{\partial H}{\partial P}\right)_T = V - T\left(\frac{\partial V}{\partial T}\right)_P$$

$$V_m = \frac{RT}{P} + B(T) \qquad \left(\frac{\partial V}{\partial T}\right)_P = \frac{R}{P} + \frac{dB}{dT}$$

Don't forget that B is a function of T.

$$\left(\frac{\partial H}{\partial P}\right)_T = V - T\left(\frac{\partial V}{\partial T}\right)_P = \frac{RT}{P} + B - T\left(\frac{R}{P} + \frac{dB}{dT}\right) = \frac{RT}{P} + B - \frac{RT}{P} - T\frac{dB}{dT}$$

Terms cancel to give us:

$$\left(\frac{\partial H}{\partial P}\right)_T = B - T\frac{dB}{dT}$$

2.21

The only thing we can do with the data we are given is approximate the partial derivative as a ratio of differences.

$$kJ = 10^3 \cdot joule \qquad MPa = 10^6 \cdot Pa$$

$$P_1 = 0.1 \cdot MPa \qquad P_2 = 1.0 \cdot MPa \qquad H_1 = 1391.7 \cdot \frac{kJ}{kg} \qquad H_2 = 1375.7 \cdot \frac{kJ}{kg}$$

$$\Delta P = P_1 - P_2 \qquad \Delta H = H_1 - H_2$$

$$derivative = \frac{\Delta H}{\Delta P} \qquad derivative = -17.78 \cdot \frac{cm^3}{gm}$$

To convert this to a molar basis:

$$MW = 30.0696400 \cdot \frac{gm}{mole} \qquad derivmol = MW \cdot derivative \qquad derivmol = -535 \cdot \frac{joule}{mole \cdot MPa}$$

2.23

To determine the final temerpature, we must do an energy balance around the system, the system being the two gases we are mixing. Since no heat can escape, what this amounts to is saying that the enthaply of the gases before they are mixed must equal the enthaply of the gases after they are mixed. However, we cannot calculate absolute enthalphies, only differences in enthalpies. What we will do is pick a reference state of 0 K, and calculate the enthalpies of the initial and final states relative to these reference states. Let species 1 be methane and species 2 be ethylene.

$$m_1 = 1 \cdot gm \qquad m_2 = 1 \cdot gm \qquad MW_1 = 16.04276 \cdot \frac{gm}{mole} \qquad MW_2 = 28.05376 \cdot \frac{gm}{mole}$$

$$n_1 = \frac{m_1}{MW_1} \qquad n_2 = \frac{m_2}{MW_2}$$

To calculate the enthalphy change, we start with equation 2.22. For a change at constant pressure, this reduces to equation 2.23. We are told to assume constant heat capacities. For methane, we use the heat capacity at 15 C, since this is the temperature closest to 300 K that Table 2.1 lists. The enthalpy change before mixing is given by:

$$T_{ref} = 0 \cdot K \qquad T_{1initial} = 300 \cdot K \qquad T_{2initial} = 600 \cdot K$$

$$C_{p1} = 35.46 \cdot \frac{joule}{mole \cdot K} \qquad C_{p2} = 42.17 \cdot \frac{joule}{mole \cdot K}$$

$$\Delta H_{initial} = n_1 \cdot \int_{T_{ref}}^{T_{1initial}} C_{p1}\, dT + n_2 \cdot \int_{T_{ref}}^{T_{2initial}} C_{p2}\, dT \qquad \Delta H_{initial} = 1.565 \cdot 10^3 \cdot joule$$

To find the final temperature, we integrate symbollically the expression for the change in enthaply after mixing (no great feat with constant heat capacities), set it equal to the change in enthaply before mixing, and solve for the final temperature. Since the system will have reached equilibrium, the two gases will have the same final temperature.

$$\Delta H_{final} = n_1 \cdot \int_{T_{ref}}^{T_{final}} C_{p1}\, dT + n_2 \cdot \int_{T_{ref}}^{T_{final}} C_{p2}\, dT$$

$$\Delta H_{final} = n_1 \cdot \left(C_{p1} \cdot T_{final} - C_{p1} \cdot T_{ref}\right) + n_2 \cdot \left(C_{p2} \cdot T_{final} - C_{p2} \cdot T_{ref}\right)$$

$$\Delta H_{final} = \left(n_1 \cdot C_{p1} \cdot T_{final} - n_1 \cdot C_{p1} \cdot T_{ref} + n_2 \cdot C_{p2} \cdot T_{final}\right) - n_2 \cdot C_{p2} \cdot T_{ref}$$

$$\Delta H_{final} = \Delta H_{initial}$$

$$T_{final} = \frac{\left(\Delta H_{initial} - n_1 \cdot C_{p1} \cdot T_{ref} - n_2 \cdot C_{p2} \cdot T_{ref}\right)}{\left(-n_1 \cdot C_{p1} - C_{p2} \cdot n_2\right)} \qquad\qquad T_{final} = 421 \cdot K$$

SECTION 2.6

2.25

We start with equation 2.25:

$$\left(\frac{\partial C_p}{\partial P}\right)_T = -T\left(\frac{\partial^2 V}{\partial T^2}\right)_P$$

Taking the derivative of volume with respect to temperature:

$$V = \frac{RT}{P} + B(T) \qquad \left(\frac{\partial V}{\partial T}\right)_P = \frac{R}{P} + \frac{dB}{dT} \qquad \frac{\partial}{\partial T}\left(\frac{\partial V}{\partial T}\right)_P = 0 + \frac{d^2B}{dT^2}$$

$$\left(\frac{\partial C_p}{\partial P}\right)_T = -T\left(\frac{\partial^2 V}{\partial T^2}\right)_P = -T\frac{d^2B}{dT^2}$$

2.27

Starting with equation 2.25:

$$\left(\frac{\partial C_p}{\partial P}\right)_T = -T\left(\frac{\partial^2 V}{\partial T^2}\right)_P$$

Taking the derivative of volume with respect to temperature:

$$V = \frac{RT}{P} + B(T) \qquad \left(\frac{\partial V}{\partial T}\right)_P = \frac{R}{P} + \frac{dB}{dT} \qquad \frac{\partial}{\partial T}\left(\frac{\partial V}{\partial T}\right)_P = 0 + \frac{d^2B}{dT^2}$$

$$\left(\frac{\partial C_p}{\partial P}\right)_T = -T\left(\frac{\partial^2 V}{\partial T^2}\right)_P = -T\frac{d^2B}{dT^2}$$

We use the RK form of the second virial coefficient from Table 1.2 and take the derivative with respect to temperature:

$$B = b - \frac{a}{RT^{3/2}} \qquad \frac{dB}{dT} = -\frac{3}{2}\left(\frac{a}{RT^{5/2}}\right) = \frac{3a}{2RT^{5/2}} \qquad \frac{d^2B}{dT^2} = -\frac{5}{2}\left(\frac{3a}{2RT^{7/2}}\right) = -\frac{15a}{4RT^{7/2}}$$

$$\left(\frac{\partial C_p}{\partial P}\right)_T = -T\left(\frac{15a}{4RT^{7/2}}\right) = \frac{15a}{4RT^{5/2}}$$

To calculate the difference between the zero-pressure heat capacity and the heat capacity at 1 bar, we write this derivative as a finite difference, and plug in values for a and b from Table 1.1:

$$\frac{\Delta C_p}{\Delta P} = \frac{15a}{4RT^{5/2}} = \frac{15\left(1.551\frac{Pa\ m^6}{mole^2 K^{1/2}}\right)}{4\left(8.31451\frac{joule}{mole\ K}\right)(298.15\ K)^{5/2}} = 4.557 \times 10^{-7}\frac{joule}{mole\ Pa\ K}$$

$$\Delta C_p = \left(4.557 \times 10^{-7}\frac{joule}{mole\ Pa\ K}\right)\Delta P = \left(4.557 \times 10^{-7}\frac{joule}{mole\ Pa\ K}\right)\left(\frac{10^5\ Pa}{1\ bar}\right)(1\ bar)$$

$$\Delta C_p = 0.046\frac{joule}{mole\ K}$$

2.29

The result from the previous problem is:

$$\left(\frac{\partial C_v}{\partial V}\right)_T = T\left(\frac{\partial^2 P}{\partial T^2}\right)_V$$

Since the van der Waals equation is explicit in pressure, taking the derivative is straightforward:

$$P = \frac{RT}{V_m - b} - \frac{a}{V_m^2} \qquad \left(\frac{\partial P}{\partial T}\right)_V = \frac{R}{V_m - b} \qquad \left(\frac{\partial^2 P}{\partial T^2}\right)_V = 0$$

$$\left(\frac{\partial C_v}{\partial V}\right)_T = T\left(\frac{\partial^2 P}{\partial T^2}\right)_V = T \times 0 = 0$$

The derivative of Cv with respect to volume is zero, so Cv of a van der Waals gas is not a function of volume.

SECTION 2.7

2.31
_____

Equation 2.27 gives us an expression for Cpm - Cvm:

$$C_{pm} - C_{vm} = T\left(\frac{\partial V_m}{\partial T}\right)_P\left(\frac{\partial P}{\partial T}\right)_{V_m}$$

The virial series is explicit in volume, and one of the derivatives we have to take requires an expression which is explicit in pressure. The virial series can be solved for pressure, but we can also use some properties of partial derivatives to transform the derivative into one which is a little easier to compute. The second approach is shown here, since it is a useful trick to know in case an equation cannot be solved explicitly for both pressure and volume, which is often the case. Properties of partial derivatives are reviewed in Appendix II of the text.

$$\left(\frac{\partial P}{\partial T}\right)_{V_m} = -\left(\frac{\partial P}{\partial V_m}\right)_T\left(\frac{\partial V_m}{\partial T}\right)_P = -\frac{\left(\frac{\partial V_m}{\partial T}\right)_P}{\left(\frac{\partial V_m}{\partial P}\right)_T}$$

$$V_m = \frac{RT}{P} + B \qquad \left(\frac{\partial V_m}{\partial T}\right)_P = \frac{R}{P} + \frac{dB}{dT} = \frac{R + B'}{P} \qquad \left(\frac{\partial V_m}{\partial P}\right)_T = -\frac{RT}{P^2}$$

$$\left(\frac{\partial P}{\partial T}\right)_{V_m} = -\frac{\left(\frac{\partial V_m}{\partial T}\right)_P}{\left(\frac{\partial V_m}{\partial P}\right)_T} = -\left(\frac{R + B'P}{P}\right)\left(-\frac{P^2}{RT}\right) = P\left(\frac{R + B'P}{RT}\right)$$

$$C_{pm} - C_{vm} = T\left(\frac{R + B'P}{P}\right) \times P\left(\frac{R + B'P}{RT}\right) = PT\left(\frac{(R + B'P)^2}{PRT}\right) = \frac{(R + B'P)^2}{R}$$

$$C_{pm} - C_{vm} = \frac{(R + B'P)^2}{R}$$

2.33

We use equation 2.28 and calculate Cvm at 20 C, since this is the temperature at which the other data are given:

$$T = 293 \cdot K \qquad V_m = 14.8 \cdot cm^3 \qquad \alpha = 0.18 \cdot 10^{-3} \cdot \frac{1}{K} \qquad \kappa_T = 3.9 \cdot 10^{-6} \cdot \frac{1}{atm} \qquad C_{pm} = 27.8 \cdot \frac{joule}{mole \cdot K}$$

$$C_{vm} = C_{pm} - \frac{T \cdot V_m \cdot \alpha^2}{\kappa_T} \qquad C_{vm} = 24.1 \cdot \frac{joule}{mole \cdot K}$$

The observed value is 23.4 joule/mole K.

2.35

Equation 2.27 gives us an expression for Cpm - Cvm:

$$C_{pm} - C_{vm} = T \left(\frac{\partial V_m}{\partial T}\right)_P \left(\frac{\partial P}{\partial T}\right)_{V_m}$$

The van der Waals equation is explicit in pressure, so we must use the cyclic rule for partial derivatives to evaluate one of the derivatives. Properties of partial derivatives are reviewed in Appendix II of the text.

$$\left(\frac{\partial V_m}{\partial T}\right)_P = -\left(\frac{\partial V_m}{\partial T}\right)_P \left(\frac{\partial P}{\partial T}\right)_{V_m} = -\frac{\left(\frac{\partial P}{\partial T}\right)_{V_m}}{\left(\frac{\partial P}{\partial V_m}\right)_T}$$

$$P = \frac{RT}{V_m - b} - \frac{a}{V_m^2} \qquad \left(\frac{\partial P}{\partial T}\right)_{V_m} = \frac{R}{V_m - b} \qquad \left(\frac{\partial P}{\partial V_m}\right)_T = -\frac{RT}{(V_m - b)^2} + \frac{2a}{V^3} = \frac{-RTV^3 + 2a(V_m - b)^2}{V^3(V_m - b)^2}$$

$$\left(\frac{\partial V_m}{\partial T}\right)_P = -\frac{\left(\frac{\partial P}{\partial T}\right)_{V_m}}{\left(\frac{\partial P}{\partial V_m}\right)_T} = -\left(\frac{R}{V_m - b}\right)\left(\frac{V^3(V_m - b)^2}{-RTV^3 + 2a(V_m - b)^2}\right) = \frac{RV^3(V_m - b)}{RTV^3 - 2a(V_m^2 + 2bV_m + b^2)}$$

$$\left(\frac{\partial V_m}{\partial T}\right)_P = \frac{RV^3(V_m - b)}{RTV^3 - 2aV_m^2 + 4abV_m - 2ab^2}$$

$$C_{pm} - C_{vm} = T\left(\frac{\partial V_m}{\partial T}\right)_P\left(\frac{\partial P}{\partial V_m}\right)_T = T\left(\frac{RV^3(V_m - b)}{RTV^3 - 2aV_m^2 + 4abV_m - 2ab^2}\right)\left(\frac{R}{V_m - b}\right)$$

$$C_{pm} - C_{vm} = \frac{R^2TV^3}{RTV^3 - 2aV_m^2 + 4abV_m - 2ab^2}$$

SECTION 2.8

2.37

The enthalpy imperfection is most easily calculated from eq. 2.34:

$H_i = U_i + PV_m - RT$

The internal energy imperfection for a Redlich-Kwong gas was derived in problem 2.36 as:

$$U_{im} = \frac{3a}{2b\sqrt{T}} \ln\left[\frac{V_m}{V_m + b}\right]$$

The PV_m term in eq. 2.34 is easily evaluated for the Redlich-Kwong equation of state:

$$PV_m = \frac{RTV_m}{V_m - b} - \frac{a}{\sqrt{T}(V_m + b)}$$

Therefore, the enthalpy imperfection is:

$$H_i = \frac{3a}{2b\sqrt{T}} \ln\left[\frac{V_m}{V_m + b}\right] + \frac{RTV_m}{V_m - b} - \frac{a}{\sqrt{T}(V_m + b)} - \frac{RT(V_m - b)}{(V_m - b)}$$

$$H_i = \frac{bRT}{V_m - b} - \frac{a}{\sqrt{T}(V_m + b)} + \frac{3a}{2b\sqrt{T}} \ln\left[\frac{V_m}{V_m + b}\right]$$

2.39

The energy imperfection is given by equation 2.31:

$$U_i = \int_{\infty}^{V} \left[T\left(\frac{\partial P}{\partial T}\right)_V - P \right] dV$$

$$P = \frac{RT}{V_m - b} - \frac{a}{TV_m^2} \qquad\qquad \left(\frac{\partial P}{\partial T}\right)_{V_m} = \frac{R}{V_m - b} + \frac{a}{T^2 V_m^2}$$

$$T\left(\frac{\partial P}{\partial T}\right)_{V_m} - P = \frac{RT}{V_m - b} + \frac{aT}{T^2 V_m^2} - \frac{RT}{V_m - b} + \frac{a}{TV_m^2} = \frac{2a}{TV_m^2}$$

$$U_i = \int_{\infty}^{V_m}\left[T\left(\frac{\partial P}{\partial T}\right)_{V_m} - P\right]dV_m = \int_{\infty}^{V_m}\frac{2a}{TV_m^2}\,dV_m = \left[-\frac{2a}{TV_m}\right]_{\infty}^{V_m} = -\frac{2a}{TV_m}$$

$$U_i = -\frac{2a}{TV_m}$$

2.41

Use equation 2.35 to calculate the enthalpy change of a real gas. Since the compression is isothermal, dT is equal to zero, so the only contribution to the enthalpy change comes from the enthalpy imperfection. The enthalpy imperfection for a van der Waals gas is derived in example 2.15 in the text.

$$R = 8.31451 \cdot \frac{joule}{mole \cdot K} \qquad a = 0.5581 \cdot \frac{Pa \cdot m^6}{mole^2} \qquad b = 65.14 \cdot 10^{-6} \cdot \frac{m^3}{mole} \qquad kJ = 10^3 \cdot joule$$

$$V_{m1} = 25.6228 \cdot \frac{liter}{mole} \qquad V_{m2} = 1.11902 \cdot \frac{liter}{mole} \qquad T = 310 \cdot K$$

$$H_{i1} = \frac{b \cdot R \cdot T}{V_{m1} - b} - \frac{2 \cdot a}{V_{m1}} \qquad H_{i2} = \frac{b \cdot R \cdot T}{V_{m2} - b} - \frac{2 \cdot a}{V_{m2}}$$

$$\Delta H = H_{i2} - H_{i1} \qquad\qquad \Delta H = -0.801 \cdot \frac{kJ}{mole}$$

2.43

We use the RK equation to calculate the pressure for each volume given, and from this, the compressibility factor. The equations for the energy and enthalpy imperfections are given in problems 2.36 and 2.37.

$$R = 8.31451 \cdot \frac{joule}{mole \cdot K} \qquad a = 1.551 \cdot \frac{Pa \cdot m^6}{mole^2 \cdot \sqrt{K}} \qquad b = 26.74 \cdot 10^{-6} \cdot \frac{m^3}{mole} \qquad T = 300 \cdot K$$

$j = 0 .. 3$ $V_{m_0} = 1 \cdot liter$ $V_{m_1} = 10 \cdot liter$ $V_{m_2} = 100 \cdot liter$ $V_{m_3} = 1000 \cdot liter$ $bar = 10^5 \cdot Pa$

$$P_j = \frac{R \cdot T}{V_{m_j} - b} - \frac{a}{\sqrt{T} \cdot V_{m_j} \cdot \left(V_{m_j} + b\right)} \qquad z_j = \frac{P_j \cdot V_{m_j}}{R \cdot T}$$

$$U_{i_j} = \frac{-3 \cdot a}{2 \cdot b \cdot \sqrt{T}} \cdot \ln\left(\frac{V_{m_j} + b}{V_{m_j}}\right) \qquad H_{i_j} = \frac{b \cdot R \cdot T}{V_{m_j} - b} - \frac{a}{\sqrt{T} \cdot \left(V_{m_j} + b\right)} + \frac{3 \cdot a}{2 \cdot b \cdot \sqrt{T}} \cdot \ln\left(\frac{V_{m_j}}{V_{m_j} + b}\right)$$

$$V_m = \begin{bmatrix} 1 \\ 10 \\ 100 \\ 1000 \end{bmatrix} \cdot liter \qquad P = \begin{bmatrix} 24.757 \\ 2.492 \\ 0.249 \\ 0.025 \end{bmatrix} \cdot bar \qquad z = \begin{bmatrix} 0.99251 \\ 0.999101 \\ 0.999909 \\ 0.999991 \end{bmatrix}$$

$$V_m = \begin{bmatrix} 1 \\ 10 \\ 100 \\ 1000 \end{bmatrix} \cdot liter \qquad H_i = \begin{bmatrix} -151.239 \\ -15.657 \\ -1.571 \\ -0.157 \end{bmatrix} \cdot joule \qquad U_i = \begin{bmatrix} -132.556 \\ -13.414 \\ -1.343 \\ -0.134 \end{bmatrix} \cdot joule$$

SECTION 2.9

2.45
_____

Isothermal, reversible expansions are discussed in the first part of section 2.9 of the text. If we assume the methane is an ideal gas at 400 K, then we can calculate the work of a reversible isothermal change using equation 2.36:

$$R = 8.31451 \cdot \frac{joule}{mole \cdot K} \qquad T = 400 \cdot K \qquad m = 1 \cdot kg \qquad MW = 16.04276 \cdot \frac{gm}{mole} \qquad n := \frac{m}{MW}$$

$$P_1 = 1 \cdot atm \qquad V_1 = \frac{R \cdot T}{P_1} \qquad\qquad kJ = 10^3 \cdot joule$$

$$P_2 = 10 \cdot atm \qquad V_2 = \frac{R \cdot T}{P_2}$$

$$w = -n \cdot R \cdot T \cdot \ln\left(\frac{V_2}{V_1}\right) \qquad w = 477 \cdot kJ$$

2.47
_____

Isothermal, reversible expansions are discussed in the first part of section 2.9 of the text. Since we are concerned with a reversible expansion, we can substitute the pressure given by the van der Waals equation for Pex.

$$w = -\int_{V_1}^{V_2} P_{ex} \, dV \qquad P_{ex} = P \qquad P = \frac{n \cdot R \cdot T}{V - n \cdot b} - \frac{n^2 \cdot a}{V^2}$$

$$w = -\int_{V_1}^{V_2} \frac{n \cdot R \cdot T}{V - n \cdot b} - \frac{n^2 \cdot a}{V^2} \, dV$$

$$w = -n \cdot \frac{\left(\ln\left(V_2 - n \cdot b\right) \cdot R \cdot T \cdot V_2 + n \cdot a\right)}{V_2} - n \cdot \frac{\left(\ln\left(V_1 - n \cdot b\right) \cdot R \cdot T \cdot V_1 + n \cdot a\right)}{V_1}$$

$$w = -n \cdot \ln\left(V_2 - n \cdot b\right) \cdot R \cdot T - \frac{n^2}{V_2} \cdot a + n \cdot \ln\left(V_1 - n \cdot b\right) \cdot R \cdot T - \frac{n^2}{V_1} \cdot a$$

$$w = -n \cdot R \cdot T \cdot \ln\left(\frac{V_2 - n \cdot b}{V_1 - n \cdot b}\right) - n^2 \cdot a \cdot \left(\frac{1}{V_2} - \frac{1}{V_1}\right)$$

Evaluating this for the reversible, isothermal expansion stated in the problem:

$$R = 8.31451 \cdot \frac{joule}{mole \cdot K} \qquad T = 303.15 \cdot K \qquad n = 6.00 \cdot mole \qquad a = 0.6849 \cdot \frac{Pa \cdot m^6}{mole^2} \qquad b = 56.76 \cdot 10^{-6} \cdot \frac{m^3}{mole}$$

$$V_1 = 10 \cdot liter \qquad V_2 = 150 \cdot liter \qquad kJ = 10^3 \cdot joule$$

$$w = -n \cdot R \cdot T \cdot \ln\left(\frac{V_2 - n \cdot b}{V_1 - n \cdot b}\right) - n^2 \cdot a \cdot \left(\frac{1}{V_2} - \frac{1}{V_1}\right) \qquad w = -39.14 \cdot kJ$$

For the ideal gase case, we use equation 2.36:

$$w = -n \cdot R \cdot T \cdot \ln\left(\frac{V_2}{V_1}\right) \qquad w = -41 \cdot kJ$$

2.49

We plug the conditions given into the result of problem 2.48:

$$R = 8.31451 \cdot \frac{joule}{mole \cdot K} \qquad T = 300 \cdot K \qquad n = 1 \cdot mole \qquad a = 7.895 \cdot \frac{Pa \cdot m^6}{mole^2 \cdot \sqrt{K}} \qquad b = 35.43 \cdot 10^{-6} \cdot \frac{m^3}{mole}$$

$$V_1 = 24 \cdot liter \qquad V_2 = 1.22 \cdot liter$$

$$w = -n \cdot R \cdot T \cdot \ln\left(\frac{V_2 - n \cdot b}{V_1 - n \cdot b}\right) - \frac{n^2 \cdot a}{b \cdot \sqrt{T}} \cdot \ln\left[\frac{\left(V_2 + n \cdot b\right) \cdot V_1}{\left(V_1 + n \cdot b\right) \cdot V_2}\right] \qquad w = 7152 \cdot \frac{joule}{mole}$$

For the ideal gase case, we use equation 2.36:

$$w = -n \cdot R \cdot T \cdot \ln\left(\frac{V_2}{V_1}\right) \qquad w = 7431 \cdot \frac{joule}{mole}$$

2.51

This problem is similar to example 2.17 from the text. We use equation 2.38 and Cvm = 3/2 R, since we have an ideal monatomic gas::

$$R = 8.31451 \cdot \frac{joule}{mole \cdot K} \qquad T_1 = 1200 \cdot K \qquad V_1 = 1 \cdot liter \qquad V_2 = 22 \cdot liter \qquad C_{vm} = \frac{3}{2} \cdot R$$

$$C_{vm} \cdot \ln\left(\frac{T_2}{T_1}\right) = -R \cdot \ln\left(\frac{V_2}{V_1}\right) \qquad T_2 = \exp\left(-R \cdot \frac{\ln\left(\frac{V_2}{V_1}\right)}{C_{vm}}\right) \cdot T_1 \qquad T_2 = 153 \cdot K$$

2.53

Remember that assuming ideal gas does NOT mean that Cvm = 3/2 R; this is only true for ideal, monatomic gases, and ammonia is polyatomic. The ideal gas approximation allows us to use equation 2.39a to evaluate the final temperature. As a first approximation to the final temperature, we assume a constant heat capacity and integrate equation 2.39a to obtain equation 2.39b. The heat capacity of ammonia can be found on Table 2.1.

$$R = 8.31451 \cdot \frac{joule}{mole \cdot K} \qquad T_1 = 293.15 \cdot K \qquad MPa = 10^6 \cdot Pa \qquad P_1 = 0.10 \cdot MPa \qquad P_2 = 2.0 \cdot MPa$$

$$C = K$$

$$C_{pm} = 37.29 \cdot \frac{joule}{mole \cdot K}$$

$$C_{pm} \cdot \ln\left(\frac{T_2}{T_1}\right) = R \cdot \ln\left(\frac{P_2}{P_1}\right) \qquad T_2 = \exp\left[R \cdot \frac{\ln\left(\frac{P_2}{P_1}\right)}{(C_{pm})}\right] \cdot T_1 \qquad T_2 = 572 \cdot K$$

However, since ammonia is a polyatomic gas, it would be more accurate to use the temperature-dependent heat capacities given on Table 2.1. We go back to equation 2.39a and integrate it with a temperature-dependent heat capacity. This will give us an equation for the final temperature.

$$c_1 = 29.75 \cdot \frac{joule}{mole \cdot K} \qquad c_2 = 25.10 \cdot 10^{-3} \cdot \frac{joule}{mole \cdot K^2} \qquad c_4 = -1.55 \cdot 10^5 \cdot \frac{joule \cdot K}{mole}$$

$$\int_{T_1}^{T_2} \frac{c_1 + c_2 \cdot T - \dfrac{c_4}{T^2}}{T} \, dT = R \cdot \ln\left(\frac{P_2}{P_1}\right)$$

$$\frac{1}{2} \cdot \frac{\left(-c_4 + 2 \cdot c_1 \cdot \ln(T_2) \cdot T_2^2 + 2 \cdot c_2 \cdot T_2^3\right)}{T_2^2} - \frac{1}{2} \cdot \frac{\left(-c_4 + 2 \cdot c_1 \cdot \ln(T_1) \cdot T_1^2 + 2 \cdot c_2 \cdot T_1^3\right)}{T_1^2} = R \cdot \ln\left(\frac{P_2}{P_1}\right)$$

The final temperature must be solved for numerically. We use the result obtained assuming a constnat heat capacity as our first guess.

$$T_2 = root\left[\frac{1}{2} \cdot \frac{\left(-c_4 + 2 \cdot c_1 \cdot \ln(T_2) \cdot T_2^2 + 2 \cdot c_2 \cdot T_2^3\right)}{T_2^2} - \frac{1}{2} \cdot \frac{\left(-c_4 + 2 \cdot c_1 \cdot \ln(T_1) \cdot T_1^2 + 2 \cdot c_2 \cdot T_1^3\right)}{T_1^2} \dots, T_2 \right.$$
$$\left. + -R \cdot \ln\left(\frac{P_2}{P_1}\right) \right]$$

$$T_2 = 555 \cdot K \qquad T_2 = T_2 - 273.15 \cdot K \qquad T_2 = 282 \cdot C$$

2.55
___

We will assume ideal gas, which allows us to use equation 2.39a to evaluate the final temperature. We also assume that air is an ideal mixture, so that the heat capacity of the mixture is just a weighted average of the heat capacity of oxygen and nitrogen. Assume the air is 21% oxygen and 79% nitrogen. As a first approximation to the final temperature, we assume a constant heat capacity and integrate equation 2.39a to obtain equation 2.39b. Heat capacities can be found on Table 2.1.

$$R = 8.31451 \cdot \frac{joule}{mole \cdot K} \qquad T_1 = 295.15 \cdot K \qquad P_1 = 14.7 \cdot psi \qquad P_2 = 65 \cdot psi + 14.7 \cdot psi \qquad C = K$$

$$C_{pm} = 0.79 \cdot 29.04 \cdot \frac{joule}{mole \cdot K} + 0.21 \cdot 29.16 \cdot \frac{joule}{mole \cdot K}$$

$$C_{pm} \cdot \ln\left(\frac{T_2}{T_1}\right) = R \cdot \ln\left(\frac{P_2}{P_1}\right) \qquad T_2 = \exp\left[R \cdot \frac{\ln\left(\frac{P_2}{P_1}\right)}{(C_{pm})}\right] \cdot T_1 \qquad T_2 = 479 \cdot K$$

However, it would be more accurate to use the temperature-dependent heat capacities given on Table 2.1. We go back to equation 2.39a and integrate it with a temperature-dependent heat capacity. This will give us an equation for the final temperature.

$$c_1 = 25.79 \cdot \frac{joule}{mole \cdot K} \qquad c_2 = 8.09 \cdot 10^{-3} \cdot \frac{joule}{mole \cdot K^2} \qquad c_3 = -1.46 \cdot 10^{-6} \cdot \frac{joule}{mole \cdot K^3} \qquad c_4 = 0.88 \cdot 10^5 \cdot \frac{joule \cdot K}{mole}$$

$$c_5 = 29.30 \cdot \frac{joule}{mole \cdot K} \qquad c_6 = 6.14 \cdot 10^{-3} \cdot \frac{joule}{mole \cdot K^2} \qquad c_7 = 0.88 \cdot 10^{-6} \cdot \frac{joule}{mole \cdot K^3} \qquad c_8 = -1.59 \cdot 10^5 \cdot \frac{joule \cdot K}{mole}$$

$$\int_{T_1}^{T_2} \frac{0.79 \cdot \left(c_1 + c_2 \cdot T - c_3 \cdot T^2 + \frac{c_4}{T^2}\right) + 0.21 \cdot \left(c_5 + c_6 \cdot T + c_7 \cdot T^2 + \frac{c_8}{T^2}\right)}{T} \, dT = R \cdot \ln\left(\frac{P_2}{P_1}\right)$$

$$.395 \cdot T_2^2 \cdot c_3 - \frac{.105}{T_2^2} \cdot c_8 + .21 \cdot T_2 \cdot c_6 - \frac{.395}{T_2^2} \cdot c_4 - .79 \cdot c_1 \cdot \ln(T_2) + .21 \cdot \ln(T_2) \cdot c_5 \ldots \qquad = R \cdot \ln\left(\frac{P_2}{P_1}\right)$$

$$+ .79 \cdot T_2 \cdot c_2 - .105 \cdot T_2^2 \cdot c_7 + \frac{.395}{T_1^2} \cdot c_4 - .79 \cdot c_1 \cdot \ln(T_1) - .395 \cdot T_1^2 \cdot c_3 - .21 \cdot \ln(T_1) \cdot c_5 \ldots$$

$$+ .79 \cdot T_1 \cdot c_2 - .105 \cdot T_1^2 \cdot c_7 - .21 \cdot T_1 \cdot c_6 + \frac{.105}{T_1^2} \cdot c_8$$

$$T_2 = \text{root}\left[.395 \cdot T_2^2 \cdot c_3 - \frac{.105}{T_2^2} \cdot c_8 + .21 \cdot T_2 \cdot c_6 - \frac{.395}{T_2^2} \cdot c_4 - .79 \cdot c_1 \cdot \ln(T_2) + .21 \cdot \ln(T_2) \cdot c_5 \ldots \qquad , T_2 \right.$$

$$+ .79 \cdot T_2 \cdot c_2 - .105 \cdot T_2^2 \cdot c_7 + \frac{.395}{T_1^2} \cdot c_4 - .79 \cdot c_1 \cdot \ln(T_1) - .395 \cdot T_1^2 \cdot c_3 - .21 \cdot \ln(T_1) \cdot c_5 \ldots$$

$$\left. + .79 \cdot T_1 \cdot c_2 - .105 \cdot T_1^2 \cdot c_7 - .21 \cdot T_1 \cdot c_6 + \frac{.105}{T_1^2} \cdot c_8 - R \cdot \ln\left(\frac{P_2}{P_1}\right) \right]$$

$$T_2 = 475 \cdot K \qquad T_2 = T_2 - 273.15 \cdot K \qquad T_2 = 202 \cdot C$$

2.57

We use equation 2.43:

$$T_1 = 295.15 \cdot K \qquad V_1 = 9 \qquad V_2 = 1 \qquad \gamma = 1.4 \qquad C = K$$

$$T_2 = T_1 \cdot \left(\frac{V_1}{V_2}\right)^{\gamma - 1} \qquad T_2 = 711 \cdot K$$

$$T_2 = T_2 - 273.15 \cdot K$$

$$T_2 = 438 \cdot C$$

2.59

This problem is similar to example 2.20 from the text.
(a) Ideal gas: If an ideal gas expands into a vacuum, it does no work. Therefore, its internal energy and hence temperature do not change. The final temperature is 15 C.

$$T_2 = 288.15 \cdot K$$

(b) van der Waals gas: We write the expression for the change of internal energy, and find the final temperature at which this change is zero. We use the result for the energy imperfection derived in example 2.14 of the text:

$$R = 8.31451 \cdot \frac{joule}{mole \cdot K} \qquad T_1 = 288.15 \cdot K \qquad V_{m1} = 4.0 \cdot liter \qquad V_{m2} = 20.0 \cdot liter$$

$$a = 0.1364 \cdot \frac{Pa \cdot m^6}{mole^2} \qquad C_{vm} = 29.04 \cdot \frac{joule}{mole \cdot K}$$

$$C_{vm} \cdot T_2 - C_{vm} \cdot T_1 + \frac{-a}{V_{m2}} - \frac{-a}{V_{m1}} = 0$$

$$T_2 = \frac{-\left(-C_{vm} \cdot T_1 - \frac{a}{V_{m2}} + \frac{a}{V_{m1}}\right)}{C_{vm}} \qquad T_2 = 287 \cdot K$$

(c) RK gas: This is similar to part (b), except we use the expression derived in problem 2.36 for the energy imperfection of an RK gas. To find the final temperature, we must solve the resulting equation numerically.

$$a = 1.551 \cdot \frac{Pa \cdot m^6}{mole^2 \cdot \sqrt{K}} \qquad b = 26.74 \cdot 10^6 \cdot \frac{m^3}{mole}$$

$$T_2 = root\left(C_{vm} \cdot T_2 - C_{vm} \cdot T_1 + \frac{-3 \cdot a}{2 \cdot b \cdot \sqrt{T_2}} \cdot \ln\left(\frac{V_{m2} + b}{V_{m2}}\right) - \frac{-3 \cdot a}{2 \cdot b \cdot \sqrt{T_1}} \cdot \ln\left(\frac{V_{m1} - b}{V_{m1}}\right), T_2\right)$$

$$T_2 = 288 \cdot K$$

2.61

To do this problem, we must carefully define the system we are working with. A gas is being compressed reversibly and giving off heat to its surroundings, the water bath. The compression of the gas is not adiabatic; however, the water bath and the gas together are adiabatic. To calculate the amount of heat given off to the surroundings by the gas, we define the system as the gas and treat it as an isothermal compression:

$$R = 8.31451 \cdot \frac{\text{joule}}{\text{mole} \cdot \text{K}} \qquad T = 300 \cdot K \qquad P_1 = 1 \cdot atm \qquad P_2 = 10 \cdot atm \qquad n = 5 \cdot mole$$

$$w = -n \cdot R \cdot T \cdot \ln\left(\frac{V_2}{V_1}\right) = -n \cdot R \cdot T \cdot \ln\left[\frac{n \cdot R \cdot \frac{T}{P_2}}{n \cdot R \cdot \frac{T}{P_1}}\right] = -n \cdot R \cdot T \cdot \ln\left(\frac{P_1}{P_2}\right) \qquad w = -n \cdot R \cdot T \cdot \ln\left(\frac{P_1}{P_2}\right)$$

$$q = w \qquad\qquad q = 2.872 \cdot 10^4 \cdot \text{joule}$$

Now, we consider the water bath and the gas together to be the system. Since the system is adiabatic, all of the energy released into the system goes into heating up its contents. The temperature change is simply the heat relased divided by the heat capacity of the system.

$$C_p = 9000 \cdot \frac{\text{joule}}{K} \qquad \Delta T = \frac{q}{C_p} \qquad \Delta T = 3.2 \cdot K$$

$$T_2 = T + \Delta T \qquad\qquad T_2 = 303 \cdot K$$

2.63

This problem is similar to example 2.22 from the text. We start with the result of example 2.21, which gives us the relationship between the second virial coefficient and the Joule-Thomson coefficient.

$$\mu = \frac{T \cdot Bprime - B}{C_{pm}}$$

We plug in the expressions for B and dB/dT. The expression for B for a Berthelot gas can be found on Table 1.2

$$B = \frac{9 \cdot R \cdot T_c}{128 \cdot P_c} \cdot \left[1 - \left(\frac{6 \cdot T_c^2}{T^2}\right)\right] \qquad\qquad Bprime = \frac{27}{32} \cdot R \cdot \frac{T_c^3}{\left(P_c \cdot T^3\right)}$$

$$\mu = \frac{\left[\dfrac{27}{(32 \cdot T^2)} \cdot R \cdot \dfrac{T_c^{\,3}}{P_c} - \dfrac{9}{128} \cdot R \cdot \dfrac{T_c}{P_c} \cdot \left(1 - 6 \cdot \dfrac{T_c^{\,2}}{T^2} \right) \right]}{C_{pm}}$$

At the Joule-Thomsom inversion temperature, the JT coefficient is equal to zero. We set the expression equal to zero and solve for temperature.

$$0 = \frac{\left[\dfrac{27}{(32 \cdot T_i^{\,2})} \cdot R \cdot \dfrac{T_c^{\,3}}{P_c} - \dfrac{9}{128} \cdot R \cdot \dfrac{T_c}{P_c} \cdot \left(1 - 6 \cdot \dfrac{T_c^{\,2}}{T_i^{\,2}} \right) \right]}{C_{pm}}$$

$$T_c = 154.3 \cdot K \qquad T_i = 3 \cdot \sqrt{2} \cdot T_c \qquad T_i = 655 \cdot K$$

2.65

This problem is similar to example 2.22 from the text. We start with the result of example 2.21, which gives us the relationship between the second virial coefficient and the Joule-Thomson coefficient.

$$\mu = \frac{T \cdot Bprime - B}{C_{pm}}$$

We plug in the expressions for B and dB/dT. The expression for B for a Berthelot gas can be found on Table 1.2

$$B = b - \frac{a}{R \cdot T^{1.5}} \qquad\qquad Bprime = 1.5 \cdot \frac{a}{(R \cdot T^{2.5})}$$

$$\mu = \frac{\left[2.5 \cdot \dfrac{a}{(R \cdot T^{1.5})} - b \right]}{C_{pm}}$$

At the Joule-Thomsom inversion temperature, the JT coefficient is equal to zero. We set the expression equal to zero and solve for temperature.

$$0 = \frac{\left[2.5 \cdot \dfrac{a}{\left(R \cdot T_i^{1.5} \right)} - b \right]}{C_{pm}}$$

$$R = 8.31451 \cdot \frac{joule}{mole \cdot K} \qquad a = 9.882 \cdot \frac{Pa \cdot m^6}{mole^2 \cdot \sqrt{K}} \qquad b = 45.15 \cdot 10^{-6} \cdot \frac{m^3}{mole}$$

$$T_i = 1.842015749320193303 \cdot \frac{a^{\left(\frac{2}{3} \right)}}{\left[b^{\left(\frac{2}{3} \right)} \cdot R^{\left(\frac{2}{3} \right)} \right]} \qquad\qquad T_i = 1630 \cdot K$$

CHAPTER 3   *The Second Law of Thermodynamics*

SECTION 3.1

3.1

To calculate the work obtainable by a steam engine, we rearrange equation 3.2. The temperature of the hot reservoir for each case is given, and the temperature of the cold reservoir is the condensing temperature given. For steam at 1 atm:

$$T_1 = 300.15 \cdot K \qquad T_2 = 373.15 \cdot K \qquad q_2 = 10^6 \cdot joule \qquad kJ = 10^3 \cdot joule$$

$$w = -q_2 \cdot \frac{T_2 - T_1}{T_2}$$

$$w = -196 \cdot kJ$$

For steam at 15.3 atm:

$$T_2 = 473.15 \cdot K$$

$$w = -q_2 \cdot \frac{T_2 - T_1}{T_2}$$

$$w = -366 \cdot kJ$$

3.3

Equation (3.3) gives us an expression for the work:

$$w = q_1 \cdot \left(\frac{T_2 - T_1}{T_1} \right)$$

Student's Solutions Manual

Since power = work/time, we have:

$$P = \frac{q_1}{t} \cdot \left(\frac{T_2 - T_1}{T_1} \right)$$

Using q1/t = (500 J/sec K * delta T) gives us:

$$P = 500 \cdot \frac{\text{joule}}{\text{sec} \cdot K} \cdot \left[\frac{(T_2 - T_1)^2}{T_1} \right]$$

(a)　　$T_2 = 303 \cdot K$　　　$T_1 = 298 \cdot K$　　　$P = 500 \cdot \frac{\text{joule}}{\text{sec} \cdot K} \cdot \left[\frac{(T_2 - T_1)^2}{T_1} \right]$　　　$P = 41.946 \cdot \text{watt}$

(b)　　$T_2 = 303 \cdot K$　　　$T_1 = 293 \cdot K$　　　$P = 500 \cdot \frac{\text{joule}}{\text{sec} \cdot K} \cdot \left[\frac{(T_2 - T_1)^2}{T_1} \right]$　　　$P = 170.648 \cdot \text{watt}$

3.5

For step 1 of the Carnot cycle,

$$q_2 = R \cdot T_2 \cdot \ln\left(\frac{V_2}{V_1} \right) \qquad \Delta S_1 = \frac{q_2}{T_2} = R \cdot \ln\left(\frac{V_2}{V_1} \right)$$

For step 2 of the cycle,

$$w = \Delta U \quad \text{so} \quad dq = 0 \quad \Delta S_2 = 0$$

Similarly, for steps 3 and 4 of the cycle,

$$\Delta S_3 = R \cdot \ln\left(\frac{V_3}{V_4} \right) \qquad \Delta S_4 = 0$$

The plot is shown below. For a PV plot, the area enclosed is the work. For a TS plot, the area enclosed is the net heat, since dq = TdS.

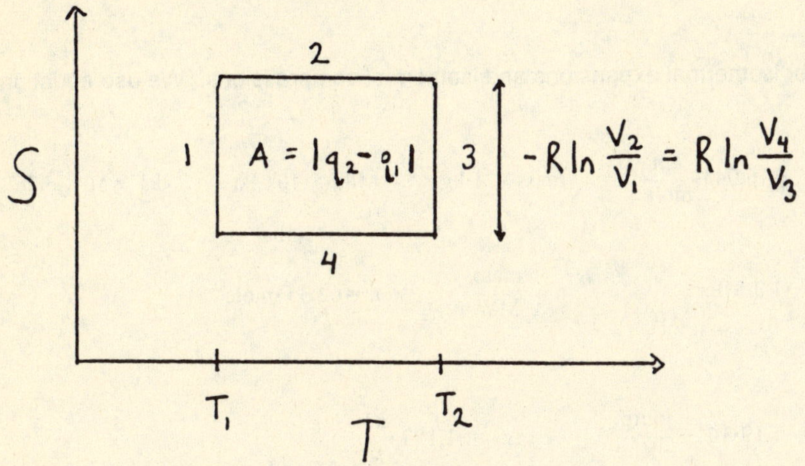

SECTION 3.3

3.7

The same equations apply for isothermal expansions and isothermal compressions. We use equation 3.14b:

$$R = 8.31451 \cdot \frac{joule}{mole \cdot K} \qquad MW = 16.043 \cdot \frac{gm}{mole} \qquad mass = 1 \cdot kg \qquad MPa = 10^6 \cdot Pa \qquad kJ = 10^3 \cdot joule$$

$$P_1 = 0.1 \cdot MPa \qquad P_2 = 1.0 \cdot MPa \qquad n = \frac{mass}{MW} \qquad n = 62.33 \cdot mole$$

$$\Delta S = -n \cdot R \cdot \ln\left(\frac{P_2}{P_1}\right) \qquad \Delta S = -1.19 \cdot 10^3 \cdot \frac{joule}{K} \qquad \Delta S = -1.193 \cdot \frac{kJ}{K}$$

3.9

We assume that the heat capacity is constant, so that equations 3.13 can be used. From Table 2.1:

$$C_{pm} = 20.93 \cdot \frac{joule}{mole \cdot K} \qquad C_{vm} = 12.59 \cdot \frac{joule}{mole \cdot K} \qquad T_1 = 298.15 \cdot K \qquad T_2 = 373.15 \cdot K$$

For changes at constant volume:

$$\Delta S = C_{vm} \cdot \ln\left(\frac{T_2}{T_1}\right) \qquad \Delta S = 2.825 \cdot \frac{joule}{mole \cdot K}$$

This is only the entropy change for one mole (note the units). We must multiply by the total number of moles:

$$\text{mass} = 1 \cdot \text{kg} \qquad\qquad MW = 40.33 \cdot \frac{\text{gm}}{\text{mole}} \qquad\qquad n = \frac{\text{mass}}{MW}$$

$$\Delta S_{total} = n \cdot \Delta S \qquad\qquad \Delta S_{total} = 70.05 \cdot \frac{\text{joule}}{K}$$

For the entropy change at constant pressure:

$$\Delta S = C_{pm} \cdot \ln\left(\frac{T_2}{T_1}\right) \qquad\qquad \Delta S = 4.696 \cdot \frac{\text{joule}}{\text{mole} \cdot K}$$

$$\Delta S_{total} = n \cdot \Delta S \qquad\qquad \Delta S_{total} = 116.45 \cdot \frac{\text{joule}}{K}$$

3.11

We use eq. (3.9) and the heat capacity from Table 2.2. Assume Cv = Cp - R.

$$T_1 = 310 \cdot K \qquad\qquad T_2 = 867 \cdot K \qquad\qquad R = 8.31451 \cdot \frac{\text{joule}}{\text{mole} \cdot K}$$

$$C_{vm}(T) = 29.75 \cdot \frac{\text{joule}}{\text{mole} \cdot K} + 25.10 \cdot 10^{-3} \cdot \frac{\text{joule}}{\text{mole} \cdot K^2} \cdot T - 1.55 \cdot 10^5 \cdot \frac{\text{joule} \cdot K}{\text{mole}} \cdot \frac{1}{T^2} - R$$

$$\Delta S = \int_{T_1}^{T_2} \frac{C_{vm}(T)}{T} \, dT \qquad\qquad \Delta S = 35.323 \cdot \frac{\text{joule}}{\text{mole} \cdot K}$$

3.13

We integrate eq. (3.11a) numerically:

$i = 0 .. 4$ $C_{pm_i} =$ $T_i =$

$n = 0 .. 3$

C_{pm_i}
$8.53 \cdot \dfrac{joule}{mole \cdot K}$
$14.63 \cdot \dfrac{joule}{mole \cdot K}$
$21.54 \cdot \dfrac{joule}{mole \cdot K}$
$23.84 \cdot \dfrac{joule}{mole \cdot K}$
$24.54 \cdot \dfrac{joule}{mole \cdot K}$

T_i
$298 \cdot K$
$500 \cdot K$
$1000 \cdot K$
$1500 \cdot K$
$2000 \cdot K$

$$\Delta S = \sum_n \frac{1}{2} \cdot \left(\frac{C_{pm_n}}{T_n} - \frac{C_{pm_{n+1}}}{T_{n+1}} \right) \cdot \left(T_{n+1} - T_n \right) \qquad \Delta S = 34.945 \cdot \frac{joule}{mole \cdot K}$$

3.15

(a) Since the gas is ideal and monatomic, we know that Cvm = 3/2R and Cpm = 5/2R. We use equations 3.14b and 2.36 to calculate the entropy change and the work. Because enthalpy and internal energy are only functions of temperature for an ideal gas, they are zero for an isothermal process:

$$R = 8.31451 \cdot \frac{joule}{mole \cdot K} \qquad C_{vm} = \frac{3}{2} \cdot R \qquad bar = 10^5 \cdot Pa \qquad P_1 = 1 \cdot bar \qquad P_2 = 10 \cdot bar$$

$$n = 1 \cdot mole \qquad T = 400 \cdot K \qquad V_1 = \frac{n \cdot R \cdot T}{P_1} \qquad V_2 = \frac{n \cdot R \cdot T}{P_2}$$

$$\Delta S = -n \cdot R \cdot \ln\left(\frac{P_2}{P_1}\right) \qquad \Delta S = -19.14 \cdot \frac{joule}{K}$$

$$\Delta U = 0 \cdot joule$$

$$\Delta H = 0 \cdot joule$$

$$w = -\int_{V_1}^{V_2} \frac{n \cdot R \cdot T}{V} \, dV \qquad w = 7658 \cdot joule$$

$$q = \Delta U - w \qquad q = -7658 \cdot joule$$

(b) Before calculating the change in internal energy or enthalphy, we must find the final temperature. This is most easily done using equation 2.39b:

$$T_1 = 400 \cdot K \qquad C_{pm} = \frac{5}{2} \cdot R$$

$$T_2 = T_1 \cdot e^{\frac{R}{C_{pm}} \cdot \ln\left(\frac{P_2}{P_1}\right)} \qquad T_2 = 1005 \cdot K$$

Equations 2.16 and 2.23 can now be used to find changes in internal energy and enthalpy:

$$\Delta U = n \cdot \int_{T_1}^{T_2} C_{vm} \, dT \qquad \Delta U = 7542 \cdot joule$$

$$\Delta H = n \cdot \int_{T_1}^{T_2} C_{pm} \, dT \qquad \Delta H = 12571 \cdot joule$$

Since the process is adiabatic, q = 0, and we can thus find w:

$$q = 0 \cdot joule$$

$$w = \Delta U - q \qquad w = 7542 \cdot joule$$

Finally, as was discussed in Example 3.7, for an adiabatic, reversible expansion, the change in entropy is zero. This may become more apparent if the reader reviews Section 2.7, which deals with adiabatic expansions.

$$\Delta S = 0 \cdot \frac{joule}{K}$$

(c) The heat capacities are constant, so we use equation 3.13 to calculate the entropy. Again, equation 2.16 can be used to calculate the change in internal energy.

$$T_2 = 273 \cdot K$$

$$\Delta S = n \cdot C_{vm} \cdot \ln\left(\frac{T_2}{T_1}\right) \qquad \Delta S = -4.76 \cdot \frac{joule}{K}$$

$$\Delta U = \int_{T_1}^{T_2} n \cdot C_{vm} \, dT \qquad \Delta U = -1584 \cdot joule$$

Since the process occurs at constant volume, the work done is zero (see the definition of work, eq. 2.9). The change in internal energy is equal to the heat plus the work, so once we know work, we can find q:

$$w = 0 \cdot joule$$

$$q = \Delta U - w \qquad\qquad q = -1584 \cdot joule$$

Finally, the change in enthalpy is determined using the definition of enthaply and the ideal gas law:

$$\Delta PV = n \cdot R \cdot (T_2 - T_1)$$

$$\Delta H = \Delta U + \Delta PV \qquad\qquad \Delta H = -2640 \cdot joule$$

(b) We use equations 3.13, 2.16, and 2.9 (we could also use 2.23 to calculate the enthalpy change, the result will be the same):

$$C_{pm} = \frac{5}{2} \cdot R$$

$$\Delta S = n \cdot C_{pm} \cdot \ln\left(\frac{T_2}{T_1}\right) \qquad\qquad \Delta S = -7.94 \cdot \frac{joule}{K}$$

$$\Delta U = \int_{T_1}^{T_2} n \cdot C_{vm}\, dT \qquad\qquad \Delta U = -1584 \cdot joule$$

for ideal gases, this reduces to $w = -n \cdot R \cdot (T_2 - T_1)$

$$w = -\int_{V_1}^{V_2} P_{ex}\, dV \qquad\qquad w = 1056 \cdot joule$$

$$q = \Delta U - w \qquad\qquad q = -2640 \cdot joule$$

$$\Delta H = \Delta U + \Delta PV \qquad\qquad \Delta H = -2640 \cdot joule$$

3.17

This is similar to example 3.2. A heat balance on the system gives us:

$$\Delta H_1 + \Delta H_2 = 0$$

$$Cp_1 \cdot \left(T_{final} - T_1\right) = Cp_2 \cdot \left(T_2 - T_{final}\right)$$

$$T_1 = 273 \cdot K \qquad T_2 = 373 \cdot K \qquad Cp_1 = 34.5 \cdot \frac{joule}{K} \qquad Cp_2 = 17.5 \cdot \frac{joule}{K}$$

$$T_{final} = \frac{\cdot \left(Cp_1 \cdot T_1 - Cp_2 \cdot T_2\right)}{\left(Cp_1 - Cp_2\right)} \qquad T_{final} = 306.7 \cdot K$$

The entropy change for the system is:

$$\Delta S_{system} = \Delta S_1 - \Delta S_2$$

$$\Delta S_{system} = Cp_1 \cdot \ln\left(-\frac{T_{final}}{T_1}\right) + Cp_2 \cdot \ln\left(-\frac{T_{final}}{T_2}\right)$$

$$\Delta S_{system} = 0.583 \cdot \frac{joule}{K}$$

3.19

Since U, H, and S are all state variables, we can break up the change into two steps: a constant pressure heating and an isothermal change in pressure. For an ideal gas, the internal energy and enthalpy depend only on temperature. For the entropy change, we combine eq. (3.11) and (3.14b).

$$R = 8.31451 \cdot \frac{joule}{mole \cdot K} \qquad T_1 = 500 \cdot K \qquad T_2 = 300 \cdot K \qquad bar = 10^5 \cdot Pa$$

$$C_{vm} = \frac{3}{2} \cdot R \qquad C_{pm} = C_{vm} - R \qquad P_1 = 1 \cdot bar \qquad P_2 = 3 \cdot bar$$

$$\Delta U \cdot C_{vm} \cdot (T_2 \quad T_1) \qquad\qquad \Delta U = -2494 \cdot \frac{joule}{mole}$$

$$\Delta H \cdot C_{pm} \cdot (T_2 - T_1) \qquad\qquad \Delta H = -4157 \cdot \frac{joule}{mole}$$

$$\Delta S = C_{pm} \cdot \ln\left(\frac{T_2}{T_1}\right) - R \cdot \ln\left(\frac{P_2}{P_1}\right) \qquad\qquad \Delta S = -19.75 \cdot \frac{joule}{mole \cdot K}$$

3.21

Step (a): Adiabatic expansion. Since this process is reversible and adiabatic, q = 0 and the entropy change is 0. We can find the final temperature by rearranging eq. 2.38.

$$T_1 = 400 \cdot K \qquad bar = 10^5 \cdot Pa \qquad P_1 = 1 \cdot bar \qquad R = 8.31451 \cdot \frac{joule}{mole \cdot K} \qquad C_{pm} = \frac{5}{2} \cdot R$$

$$V_1 = \frac{R \cdot T_1}{P_1} \qquad V_1 = 33.258 \cdot liter \qquad V_2 = 2 \cdot V_1$$

$$C_{vm} = \frac{3}{2} \cdot R$$

$$C_{vm} \cdot \ln\left(\frac{T_2}{T_1}\right) = -R \cdot \ln\left(\frac{V_2}{V_1}\right) \qquad T_2 = \exp\left(-R \cdot \frac{\ln\left(\frac{V_2}{V_1}\right)}{C_{vm}}\right) \cdot T_1 \qquad T_2 = 251.98 \cdot K$$

$$\Delta U_a = C_{vm} \cdot (T_2 - T_1) \qquad \Delta U_a = -1846 \cdot \frac{joule}{mole}$$

$$\Delta H_a = C_{pm} \cdot (T_2 - T_1) \qquad \Delta H_a = -3077 \cdot \frac{joule}{mole}$$

$$\Delta S_a = 0 \cdot \frac{joule}{mole \cdot K} \qquad q_a = 0 \cdot \frac{joule}{mole} \qquad w_a = \Delta U_a - q_a \qquad w_a = -1846 \cdot \frac{joule}{mole}$$

Step (b). Constant volume heating. Since this is a constant volume process, no PV work is done.

$T_1 = 251.98 \cdot K$ $T_2 = 400 \cdot K$ $P_2 = \dfrac{R \cdot T_2}{V_2}$ $P_2 = 0.5 \cdot bar$

$\Delta U_b = C_{vm} \cdot (T_2 - T_1)$ $\Delta U_b = 1846 \cdot \dfrac{joule}{mole}$

$\Delta H_b = C_{pm} \cdot (T_2 - T_1)$ $\Delta H_b = 3077 \cdot \dfrac{joule}{mole}$

$\Delta S_b = C_{vm} \cdot \ln\left(\dfrac{T_2}{T_1}\right)$ $\Delta S_b = 5.763 \cdot \dfrac{joule}{mole}$

$w_b = 0 \cdot \dfrac{joule}{mole}$ $q_b = \Delta U_b - w_b$ $q_b = 1846 \cdot \dfrac{joule}{mole}$

Step (c): Isothermal expansion. Since the process is isothermal, and U and H for an ideal gas only depend on temperature, the change in internal energy and enthalpy is zero.

$P_1 = 0.5 \cdot bar$ $P_2 = 1 \cdot bar$

$\Delta U_c = 0 \cdot \dfrac{joule}{mole}$ $\Delta H_c = 0 \cdot \dfrac{joule}{mole}$ $V_1 = \dfrac{R \cdot T_2}{P_1}$ $V_2 = \dfrac{R \cdot T_2}{P_2}$

$\Delta S_c = -R \cdot \ln\left(\dfrac{P_2}{P_1}\right)$ $\Delta S_c = -5.763 \cdot \dfrac{joule}{mole \cdot K}$

$w_c = -R \cdot T_2 \cdot \ln\left(\dfrac{V_2}{V_1}\right)$ $w_c = 2305 \cdot \dfrac{joule}{mole}$ $q_c = \Delta U_c - w_c$ $q_c = -2305 \cdot \dfrac{joule}{mole}$

For the overall process, we simply add the steps together. Note that for the state variables, the overall change is zero, since the overall final and initial states are the same.

$\Delta U = \Delta U_a + \Delta U_b + \Delta U_c$ $\Delta U = 0 \cdot \dfrac{joule}{mole}$

$\Delta H = \Delta H_a + \Delta H_b + \Delta H_c$ $\Delta H = 0 \cdot \dfrac{joule}{mole}$ $q = q_a + q_b + q_c$ $q = -459 \cdot \dfrac{joule}{mole}$

$\Delta S = \Delta S_a + \Delta S_b + \Delta S_c$ $\Delta S = 0 \cdot \dfrac{joule}{mole}$ $w = w_a + w_b + w_c$ $w = 459 \cdot \dfrac{joule}{mole}$

3.23

We calculate the number of moles of each species, and then use eq. 3.15:

$$R = 8.31451 \cdot \frac{joule}{mole \cdot K} \qquad T = 356 \cdot K \qquad P = 1 \cdot atm \qquad V_1 = 3.5 \cdot liter \qquad V_2 = 5.6 \cdot liter$$

$$n_1 = \frac{P \cdot V_1}{R \cdot T} \qquad n_2 = \frac{P \cdot V_2}{R \cdot T} \qquad n_1 = 0.12 \cdot mole \qquad n_2 = 0.192 \cdot mole$$

$$X_1 = \frac{n_1}{n_1 + n_2} \qquad X_2 = \frac{n_2}{n_1 + n_2} \qquad X_1 = 0.385 \qquad X_2 = 0.615$$

$$\Delta S = -R \cdot (n_1 \cdot \ln(X_1) + n_2 \cdot \ln(X_2)) \qquad \Delta S = 1.73 \cdot \frac{joule}{mole \cdot K}$$

3.25

(a) This is straightforward:

$$\Delta H_{fus} = 7070 \cdot \frac{joule}{mole} \qquad T_{fus} = 505.05 \cdot K \qquad \Delta S = \frac{\Delta H_{fus}}{T_{fus}} \qquad \Delta S = 14 \cdot \frac{joule}{mole \cdot K}$$

(b) This is similar to exampe 3.9 in the text.

$$Cp_{solid} = 28.1 \cdot \frac{joule}{mole \cdot K} \qquad Cp_{liq} = 30.2 \cdot \frac{joule}{mole \cdot K}$$

$$T = T_{fus} - 55 \cdot K \qquad T = 450.05 \cdot K$$

$$\Delta S = Cp_{liq} \cdot \ln\left(\frac{T_{fus}}{T}\right) - \frac{\Delta H_{fus}}{T_{fus}} + Cp_{solid} \cdot \ln\left(\frac{T}{T_{fus}}\right) \qquad \Delta S = -13.8 \cdot \frac{joule}{mole \cdot K}$$

SECTION 3.5

3.27

There is a pattern to deriving Maxwell equations which is a little clearer after looking at a few derivations. This derivation is similar to example 3.11 in the text. We start with the equation of state for U, since T and P multiply the partial derivatives in this equation, and these are the variables which appear in the top of the partial derivatives in the final solution:

$$dU = TdS - PdV$$

Next, we write the total derivative for U in terms of S and V, since these are the variables which are in the bottom of the partial derivatives in the final solution:

$$dU = \left(\frac{\partial U}{\partial S}\right)_V dS + \left(\frac{\partial U}{\partial V}\right)_S dV$$

Comparing these two equations, we can see that:

$$\left(\frac{\partial U}{\partial S}\right)_V = T \qquad \left(\frac{\partial U}{\partial V}\right)_S = -P$$

Taking partial derivatives, we obtain:

$$\left(\frac{\partial}{\partial V}\right)_S \left(\frac{\partial U}{\partial S}\right)_V = \left(\frac{\partial T}{\partial V}\right)_S \qquad \left(\frac{\partial}{\partial S}\right)_V \left(\frac{\partial U}{\partial V}\right)_S = -\left(\frac{\partial P}{\partial S}\right)_V$$

By the properties of partial derivatives, the left hand sides of these equations are equal, so we have the Maxwell relation:

$$\left(\frac{\partial T}{\partial V}\right)_S = -\left(\frac{\partial P}{\partial S}\right)_V$$

3.29

"Readily measured quantities" just means obtain this partial derivative in terms of T, P, and V, which are quantities you can measure. We can begin by writing the total derivative for H in terms of S and P:

$$dH = \left(\frac{\partial H}{\partial S}\right)_P dS + \left(\frac{\partial H}{\partial P}\right)_S dP$$

We compare this with the equation of state for enthalpy:

$$dH = TdS + VdP$$

Based on this comparison, we can see that:

$$dG = \left(\frac{\partial G}{\partial T}\right)_P dT + \left(\frac{\partial G}{\partial P}\right)_T dP$$

Comparing these two equations, we can see that:

$$\left(\frac{\partial H}{\partial S}\right)_P = T$$

3.31

This is similar to example 3.12. We begin with the total derivative for H, written in terms of S and P:

$$dH = \left(\frac{\partial H}{\partial S}\right)_P dS + \left(\frac{\partial H}{\partial P}\right)_S dP$$

We can compare this with the equation of state for enthalpy:

$$dH = TdS + VdP$$

Comparing these two equations, we can see that:

$$T = \left(\frac{\partial H}{\partial S}\right)_P$$

We use the chain rule on the partial derivative to obtain:

$$T = \left(\frac{\partial H}{\partial S}\right)_P = \frac{\left(\frac{\partial H}{\partial T}\right)_P}{\left(\frac{\partial S}{\partial T}\right)_P}$$

By definition, the partial derivative in the numerator is the constant pressure heat capacity, so we have:

$$T = \left(\frac{\partial H}{\partial S}\right)_P = \frac{C_P}{\left(\frac{\partial S}{\partial T}\right)_P}$$

$$T\left(\frac{\partial S}{\partial T}\right)_P = C_P$$

3.33

First, we write the partial derivative in terms of P, V, and T by using the second Maxwell relation in Table 3.1:

$$\left(\frac{\partial S}{\partial V}\right)_T = \left(\frac{\partial P}{\partial T}\right)_V$$

We can simply take the derivative of the van der Waals equation with respect to temperature, treating V as a constant:

$$P = \frac{RT}{V_m - b} - \frac{a}{V_m^2}$$

$$\left(\frac{\partial P}{\partial T}\right)_V = \frac{R}{V_m - b}$$

3.35

First we write μ and Cp in terms of their definitions, found on Table 3.1:

$$-\mu C_p = -\left(\frac{\partial T}{\partial P}\right)_H \left(\frac{\partial H}{\partial T}\right)_P$$

We can rewrite the partial derivatives using the triple product rule to obtain the solution:

$$-\left(\frac{\partial T}{\partial P}\right)_H \left(\frac{\partial H}{\partial T}\right)_P = \left(\frac{\partial H}{\partial P}\right)_T$$

$$-\mu C_p = \left(\frac{\partial H}{\partial P}\right)_T$$

3.37

We begin with the definitions of alpha and Cp found on Table 3.1, and get the expression in terms of state variables. In the derivation below, the derivatives are all partial derivatives:

$$\frac{\alpha V T}{C_p} = \frac{\frac{1}{V} \cdot \left(\frac{dV}{dT}\right)_P \cdot V}{\left(\frac{C_p}{T}\right)} = \frac{\left(\frac{dV}{dT}\right)_P}{\left(\frac{dS}{dT}\right)_P}$$

We can flip the derivative in the denominator:

$$\frac{\left(\frac{dV}{dT}\right)_P}{\left(\frac{dS}{dT}\right)_P} = \left(\frac{dV}{dT}\right)_P \cdot \left(\frac{dT}{dS}\right)_P$$

Next, we use the chain rule:

$$\left(\frac{dV}{dT}\right)_P \cdot \left(\frac{dT}{dS}\right)_P = \left(\frac{dV}{dS}\right)_P$$

And a Maxwell relation:

$$\left(\frac{dV}{dS}\right)_P = \left(\frac{dT}{dP}\right)_S$$

Thus:

$$\frac{\alpha V T}{C_p} = \left(\frac{dT}{dP}\right)_S$$

To calculate this quantity for CCl4, we use the data given on Table 1.8:

$$\alpha = 1.236 \cdot 10^{-3} \cdot \frac{1}{K} \qquad V = 97 \cdot \frac{cm^3}{mole} \qquad T = 293.15 \cdot K \qquad C_p = 132 \cdot \frac{joule}{mole \cdot K}$$

$$answer = \frac{\alpha \cdot V \cdot T}{C_p} \qquad answer = 0.027 \cdot \frac{K}{atm}$$

3.39

This can be a frustrating derivation, because it can lead to a lot of dead-ends. We begin with the derived relationships for Cp and Cv, found in Table 3.1:

$$\frac{C_p}{C_v} = \frac{T\left(\frac{\partial S}{\partial T}\right)_P}{T\left(\frac{\partial S}{\partial T}\right)_v} = \left(\frac{\partial S}{\partial T}\right)_P\left(\frac{\partial T}{\partial S}\right)_v$$

We use the triple product rule on both derivatives to obtain:

$$\frac{C_p}{C_v} = \left(\frac{\partial S}{\partial P}\right)_T\left(\frac{\partial P}{\partial T}\right)_S\left(\frac{\partial T}{\partial V}\right)_S\left(\frac{\partial V}{\partial S}\right)_T$$

We replace each of the partial derivatives with the Maxwell relations:

$$\frac{C_p}{C_v} = \left(\frac{\partial V}{\partial T}\right)_P\left(\frac{\partial S}{\partial V}\right)_P\left(\frac{\partial P}{\partial S}\right)_v\left(\frac{\partial T}{\partial P}\right)_v$$

Now, use the triple product rule twice, on the first two partial derivatives, and rearrange:

$$\frac{C_p}{C_v} = \left(\frac{\partial V}{\partial P}\right)_T \left(\frac{\partial P}{\partial T}\right)_v \left(\frac{\partial S}{\partial P}\right)_v \left(\frac{\partial P}{\partial V}\right)_S \left(\frac{\partial P}{\partial S}\right)_v \left(\frac{\partial T}{\partial P}\right)_v$$

$$\frac{C_p}{C_v} = \frac{\left(\frac{\partial V}{\partial P}\right)_T}{\left(\frac{\partial P}{\partial V}\right)_S} \left(\frac{\partial P}{\partial T}\right)_v \left(\frac{\partial S}{\partial P}\right)_v \left(\frac{\partial P}{\partial S}\right)_v \left(\frac{\partial T}{\partial P}\right)_v = \frac{\left(\frac{\partial V}{\partial P}\right)_T}{\left(\frac{\partial P}{\partial V}\right)_S} \left(\frac{\partial P}{\partial T}\right)_v \left(\frac{\partial S}{\partial P}\right)_v \left(\frac{\partial P}{\partial S}\right)_v \left(\frac{\partial T}{\partial P}\right)_v$$

Cancel out some of the partial derivatives, and multiply top and bottom by -1/V:

$$\frac{C_p}{C_v} = \frac{-\frac{1}{V}\left(\frac{\partial V}{\partial P}\right)_T}{-\frac{1}{V}\left(\frac{\partial P}{\partial V}\right)_S}$$

Finally, we use the definitions of κ_S and κ_T to obtain the solution:

$$\frac{C_p}{C_v} = \frac{\kappa_T}{\kappa_T}$$

3.41

We use eq. (3.25), and rearrange to solve for the adiabatic compressibility.

$$V_m = 89.8 \cdot \frac{cm^3}{mole} \qquad C_{pm} = 134 \cdot \frac{joule}{mole \cdot K} \qquad \alpha = 1.24 \cdot 10^{-3} \cdot \frac{1}{K} \qquad c = 1295 \cdot \frac{m}{sec}$$

$$M = 78.1136 \cdot \frac{gm}{mole} \qquad T = 298.15 \cdot K \qquad \rho = \frac{M}{V_m} \qquad \rho = 0.87 \cdot \frac{gm}{cm^3}$$

$$c = \frac{1}{\sqrt{\rho \cdot \kappa_S}} \qquad \kappa_S = \frac{1}{\rho \cdot c^2} \qquad \kappa_S = 6.855 \cdot 10^{-10} \cdot \frac{1}{Pa}$$

Now, we use eq. (3.27) ro find the isothermal compressibility:

$$\kappa_T = \kappa_S + \frac{T \cdot V_m \cdot \alpha^2}{C_{pm}} \qquad\qquad \kappa_T = 9.927 \cdot 10^{-10} \cdot \frac{1}{Pa}$$

Finally, we use eq. (3.26) to compute Cv:

$$\gamma = \frac{\kappa_T}{\kappa_S} \qquad C_{vm} = \frac{C_{pm}}{\gamma} \qquad C_{vm} = 92.5 \cdot \frac{joule}{mole \cdot K}$$

SECTION 3.6

3.43

The Debye T^3 relation is given by eq. (3.29):

$$C_{pm} = aT^3$$

From Table 3.1, one of the basic equations for entropy is:

$$dS = \frac{C_P}{T} dT + \left(\frac{\partial V}{\partial T}\right)_P dP$$

The second term on the right-hand side is zero, since we are dealing with a solid. Thus, we have:

$$\Delta S = \int_0^T \frac{C_P}{T} dT$$

$$\Delta S = \int_0^T aT^2 dT$$

$$\Delta S = \frac{aT^3}{3} = \frac{C_{pm}}{3}$$

3.45

This is similar to example 3.21. We will integrate eq. (3.28) numerically. We can calculate the entropy at 13 K using the Debye law, eq. (3.30).

$i = 0..22 \qquad n = 0..21$

$T_i =$	$C_{p_i} =$
13·K	2.866
14·K	3.474
15·K	4.167
20·K	8.368
25·K	13.159
30·K	17.991
35·K	22.531
40·K	26.527
50·K	32.991
60·K	37.928
70·K	41.735
80·K	44.978
90·K	47.823
100·K	50.417
120·K	55.689
140·K	61.505
160·K	67.906
180·K	75.396
200·K	83.722
220·K	93.387
240·K	104.098
260·K	116.148
278.69·K	128.7

$$C_{pm_i} = C_{p_i} \cdot \frac{\text{joule}}{\text{mole} \cdot \text{K}}$$

$$S_o = \frac{C_{pm_0}}{3}$$

$$\Delta S = S_o \cdot \sum_n \frac{C_{pm_n} + C_{pm_{n-1}}}{2} \cdot \ln\left(\frac{T_{n+1}}{T_n}\right)$$

$$\Delta S = 129.085 \cdot \frac{\text{joule}}{\text{mole} \cdot \text{K}}$$

SECTION 3.7

3.47

In example 3.23, the formulas for the entropy imperfection of a gas obeying the virial series is derived:

$$S_i = -\left(B'P - \frac{\gamma}{2}P^2\right)$$

We ignore the γ term. The RK form of the second virial coefficient is given on Table 1.2:

$$B(T) = b - \frac{a}{RT^{3/2}}$$

The derivative with respect to temperature is:

$$B' = \frac{3a}{2RT^{5/2}}$$

Thus, the energy imperfection is given by:

$$S_i = -\frac{3aP}{2RT^{5/2}}$$

The constant a for acetylene in the RK equation can be found on Table 1.1. The standard entropy is given on Table 3.2.

$$\text{bar} := 10^5 \text{ Pa} \qquad P := 1 \text{ bar} \qquad T := 298.15 \text{ K}$$

$$R := 8.31451 \frac{\text{joule}}{\text{mole K}} \qquad a := 7.895 \frac{\text{Pa m}^6 \text{ K}^{1/2}}{\text{mole}^2} \qquad S^\theta = 200.94 \frac{\text{joule}}{\text{mole K}}$$

$$S := S^\theta - \frac{3aP}{2RT^{5/2}} \qquad S = 200.85 \frac{\text{joule}}{\text{mole K}}$$

3.49

The standard entropy of Ar at 298.15 K is given on Table 3.2. We will assume ideal gas and use eq. (3.31).

$$R = 8.31451 \cdot \frac{joule}{mole \cdot K} \qquad T_1 = 298.15 \cdot K \qquad T_2 = 775 \cdot K \qquad S_{298} = 154.853 \cdot \frac{joule}{mole \cdot K}$$

$$C_{pm} = 2.5 \cdot R$$

$$S = S_{298} + C_{pm} \cdot \ln\left(\frac{T_2}{T_1}\right) \qquad S = 174.709 \cdot \frac{joule}{mole \cdot K}$$

3.51

Table 3.2 gives the standard entropy of ethane at 298.15 K. We use eq. (3.31) and the heat capacity given on Table 2.2:

$$T_1 = 298.15 \cdot K \qquad T_2 = 2500 \cdot K \qquad S_{298} = 5.740 \cdot \frac{joule}{mole \cdot K}$$

$$C_p(T) = 14.22 \cdot \frac{joule}{mole \cdot K} + 0.00922 \cdot \frac{joule}{mole \cdot K^2} \cdot T - 1.87 \cdot 10^{-6} \cdot \frac{joule}{mole \cdot K^3} \cdot T^2 - 7.51 \cdot 10^5 \cdot \frac{joule \cdot K}{mole} \cdot \frac{1}{T^2}$$

$$S = S_{298} + \int_{T_1}^{T_2} \frac{C_p(T)}{T} \, dT \qquad S = 46.354 \cdot \frac{joule}{mole \cdot K}$$

3.53

Since the liquid is incompressible, we use eq. (3.39): The properties of benzene can be found on Table 1.8.

$$MPa = 10^6 \cdot Pa \qquad P_{ref} = 0.1 \cdot MPa \qquad P = 15 \cdot MPa \qquad \alpha = 1.237 \cdot 10^{-3} \cdot \frac{1}{K} \qquad V = 89 \cdot cm^3$$

$$\Delta S = -\alpha \cdot V \cdot (P - P_{ref}) \qquad \Delta S = -1.64 \cdot \frac{joule}{mole \cdot K}$$

3.55

We use eq. (3.39):

$$MPa = 10^6 \cdot Pa \qquad P_{ref} = 0.1 \cdot MPa \qquad P = 1 \cdot MPa \qquad \alpha = 1.236 \cdot 10^{-3} \cdot \frac{1}{K} \qquad V = 97 \cdot cm^3$$

$$S_{ref} = 216.40 \cdot \frac{joule}{mole \cdot K}$$

$$S = S_{ref} - \alpha \cdot V \cdot (P - P_{ref}) \qquad S = 216.292 \cdot \frac{joule}{mole \cdot K}$$

3.57

A reversible, adiabatic process is isentropic. We can set the change in entropy, eq. (3.37), equal to zero and solve for T. As a first guess, we set T2 = 500 K, since we know the gas will probably cool upon expansion.

$$R = 8.31451 \cdot \frac{joule}{mole \cdot K} \qquad V_1 = 0.75 \cdot \frac{liter}{mole} \qquad V_2 = 50.6 \cdot \frac{liter}{mole} \qquad T_1 = 1492 \cdot K$$

$$C_{pm}(T) = 25.79 \cdot \frac{joule}{mole \cdot K} - 0.00809 \cdot \frac{joule}{mole \cdot K^2} \cdot T - 1.49 \cdot 10^{-6} \cdot \frac{joule}{mole \cdot K^3} \cdot T^2 + 0.88 \cdot 10^5 \cdot \frac{joule \cdot K}{mole} \cdot \frac{1}{T^2}$$

$$T_2 \quad 500 \cdot K$$

$$T_2 = root\left(\int_{T_1}^{T_2} \frac{C_{pm}(T)}{T} \, dT + R \cdot \ln\left(\frac{T_1 \cdot V_2}{T_2 \cdot V_1} \right), T_2 \right) \qquad\qquad T_2 = 323.027 \cdot K$$

We do something similar for the RK gas. The entropy imperfection for an RK gas is given on Table 3.3.

$$a = 1.551 \cdot \frac{Pa \cdot m^6 \cdot \sqrt{K}}{mole} \qquad b = 26.74 \cdot 10^{-6} \cdot \frac{m^3}{mole}$$

$$S_i(V,T) = \frac{a}{2 \cdot b \cdot T^{1.5}} \cdot \ln\left(\frac{V}{V + b} \right) - R \cdot \ln\left(\frac{V}{V - b} \right)$$

$$T_2 = root\left(\int_{T_1}^{T_2} \frac{C_{pm}(T)}{T} \, dT + R \cdot \ln\left(\frac{T_1 \cdot V_2}{T_2 \cdot V_1} \right) + S_i(V_2, T_2) - S_i(V_1, T_1), T_2 \right)$$

$$T_2 = 318.213$$

SECTION 3.8

3.59

Equation (3.42) gives us the definition of work, and we plug eq. (3.45) into this expression to obtain work in terms of stress.

$$w = \int_{l_o}^{l} f\, dl \qquad\qquad \sigma = \frac{f}{A_o} \qquad\qquad w = \int_{l_o}^{l} \sigma A_o\, dl$$

$$\frac{w}{V} = \int_{l_o}^{l} \frac{\sigma A_o}{l_o A_o}\, dl = \int_{l_o}^{l} \frac{\sigma}{l_o}\, dl$$

Equation (3.46) gives us the relationship between length and strain. We can substitute this into the integral:

$$\varepsilon = \frac{l - l_o}{l_o} \qquad d\varepsilon = \frac{dl}{l_o} \qquad \varepsilon(l_o) = \frac{l_o - l_o}{l_o} = 0 \qquad \varepsilon(l) = \frac{l - l_o}{l_o} = \varepsilon$$

$$\frac{w}{V} = \int_0^{\varepsilon} \sigma\, d\varepsilon$$

The stress is given by:

$$\sigma = \frac{\rho RT}{zM}\left(\frac{l}{l_o} - \left(\frac{l_o}{l}\right)^2\right) \qquad \frac{l}{l_o} = \frac{l}{l_o} - \frac{l_o}{l_o} + \frac{l_o}{l_o} \qquad \frac{l}{l_o} = \frac{l - l_o}{l_o} + 1 = \varepsilon + 1$$

$$\sigma = \frac{\rho RT}{zM}\left(\varepsilon + 1 - \left(\frac{1}{\varepsilon + 1}\right)^2\right)$$

We can perform the integration to obtain:

$$\frac{w}{V} = \int_0^\varepsilon \frac{\rho RT}{zM}\left(\varepsilon + 1 - \left(\frac{1}{\varepsilon + 1}\right)^2\right) d\varepsilon \qquad\qquad \text{Let } u = \varepsilon + 1.$$

$$\frac{w}{V} = \int_1^{\varepsilon+1} \frac{\rho RT}{zM}\left(u - \frac{1}{u^2}\right) du = \frac{\rho RT}{zM}\left[\frac{u^2}{2} + \frac{1}{u}\right]_1^{\varepsilon+1} = \frac{\rho RT}{zM}\left[\frac{(\varepsilon + 1)^2}{2} + \frac{1}{\varepsilon + 1} - \left[\frac{1^2}{2} + \frac{1}{1}\right]\right]$$

$$\frac{w}{V} = \frac{\rho RT}{zM}\left[\frac{(\varepsilon + 1)^3 + 2 - 3(\varepsilon + 1)}{2(\varepsilon + 1)}\right] = \frac{\rho RT}{zM}\left[\frac{\varepsilon^3 + 3\varepsilon^2 + 3\varepsilon + 1 + 1 - 3\varepsilon - 3}{2(\varepsilon + 1)}\right]$$

$$\frac{w}{V} = \frac{\rho RT}{zM}\left[\frac{\varepsilon^2(\varepsilon + 3)}{2(\varepsilon + 1)}\right]$$

For isoprene, we have:

$T := 298.15 \ K \qquad\qquad M := 0.068 \ kg \qquad\qquad \rho := 0.970 \ g/cm^3$

$l_o := 10 \ cm \qquad\qquad l := 50 \ cm \qquad\qquad A_o := 0.05 \ cm^2 \qquad\qquad V := A_o \ l_o$

$R := 8.31451 \ \dfrac{joule}{mole \ K} \qquad\qquad \varepsilon := \dfrac{l - l_o}{l_o}$

The percent cross-linking is 1.25, so 1.25 out of 100 monomer units is cross-linked. Thus, the number of monomer units in a chain is 100/1.25.

$\%\text{cross-link} := 1.25 \qquad\qquad z := 100/1.25 \qquad\qquad z = 80$

$$w := \frac{\rho VRT}{zM}\left[\frac{\varepsilon^2(\varepsilon + 3)}{2(\varepsilon + 1)}\right]$$

$w = 2.47 \ joule$

3.61

The stress is given by:

$$\sigma = \frac{\rho RT}{zM}\left(\frac{1}{l_0} - \left(\frac{l_0}{l}\right)^2\right) \qquad \frac{l}{l_0} = \frac{l}{l_0} - \frac{l_0}{l_0} + \frac{l_0}{l_0} \qquad \frac{l}{l_0} = \frac{l - l_0}{l_0} + 1 = \varepsilon + 1$$

$$\sigma = \frac{\rho RT}{zM}\left(\varepsilon + 1 - \left(\frac{1}{\varepsilon + 1}\right)^2\right)$$

We differentiate with respect to the strain to yield:

$$\left(\frac{\partial \sigma}{\partial \varepsilon}\right) = \frac{\rho RT}{zM}\left(1 + \frac{2}{(\varepsilon + 1)^3}\right)$$

To obtain the initial slope, we plug in $l = l_0$. This is equivalent to substituting $\varepsilon = 0$.

$$\left(\frac{\partial \sigma}{\partial \varepsilon}\right)_{l_0} = \frac{\rho RT}{zM}\left(1 + \frac{2}{(0 + 1)^3}\right) \qquad \left(\frac{\partial \sigma}{\partial \varepsilon}\right)_{l_0} = \frac{3\rho RT}{zM}$$

From Table 3.4:

$$T := 298\ K \qquad\qquad M := 0.086\ kg \qquad\qquad \rho := 1.320\ g/cm^3$$

$$R := 8.31451\ \frac{joule}{mole\ K}$$

$$initslope := 1.6\ 10^6\ Pa$$

$$z := \frac{3\rho RT}{initslope\ M}$$

$$z = 71.3$$

The number of monomer units in a chain is 71.3, so the number of molecules cross-linked will be 100/71.3.

$$\%cross\text{-}link := 100/z \qquad\qquad \%cross\text{-}link = 1.40\ \%$$

3.63

By analogy with the coefficient of thermal expansion, we define the coefficient of linear expansion to be:

$$\alpha_L = \frac{1}{L} \cdot \left(\frac{dL}{dT}\right)_P$$

We will take the coefficient of linear expansion to be 1/3 the coefficient of thermal expansion. Approximating the derivative as a difference gives us:

$$\Delta L = \frac{1}{3} \cdot \alpha \cdot L \cdot \Delta T$$

We use a typical value from Table 3.4 to approximate the change in L.

$$L = 15 \cdot cm \qquad \Delta T = 100 \cdot K \qquad \alpha = 6.6 \cdot 10^{-4} \cdot \frac{1}{K} \qquad \Delta L = \frac{1}{3} \cdot \alpha \cdot L \cdot \Delta T \qquad \Delta L = 0.33 \cdot cm$$

3.65

From eq. (3.48),

$$dU = TdS + fdl$$

The total differential with respect to temperature and elongation is:

$$dU = \left(\frac{\partial U}{\partial T}\right)_l dT + \left(\frac{\partial U}{\partial l}\right)_T dl$$

Comparing these two equations, we obtain

$$TdS + fdl = \left(\frac{\partial U}{\partial T}\right)_l dT + \left(\frac{\partial U}{\partial l}\right)_T dl$$

Since we are concerned with a change that occurs at constant length, the dl terms on both sides of the equation will be zero, leaving us with:

$$TdS = \left(\frac{\partial U}{\partial T}\right)_l dT \qquad \text{Now we define } \left(\frac{\partial U}{\partial T}\right)_l = C_l \text{ by analogy with } C_v. \text{ This gives us:}$$

$$TdS = C_l dT$$

$$\left(\frac{\partial S}{\partial T}\right)_l = \frac{C_l}{T}$$

3.67

(a) An isoprene unit looks like this:

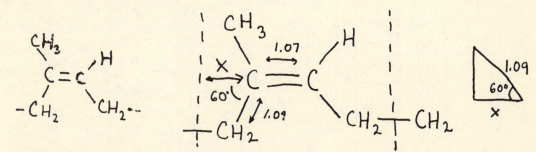

(I don't know how you were supposed to know that. I didn't know it when I took PChem either). From trignometry, we find that

$$x = 1.09 \cdot 10^{-8} \cdot cm \cdot cos\left(\frac{\pi}{3}\right) \qquad x = 5.45 \cdot 10^{-9} \cdot cm$$

which gives us a total length of:

$$l = 1.07 \cdot 10^{-8} \cdot cm + 2 \cdot x + 1.09 \cdot 10^{-8} \cdot cm \qquad l = 3.25 \cdot 10^{-8} \cdot cm$$

(b) The mean length is denoted as Imean, and is calculated using the equation given in the problem:

$$N = 80 \qquad l_{mean} = \sqrt{N \cdot l} \qquad l_{mean} = 2.907 \cdot 10^{-7} \cdot cm$$

(c) First we calculate the volume of the cell:

$$V = l_{mean}^3 \qquad V = 2.456 \cdot 10^{-20} \cdot cm^3$$

Now, each of the 12 sides of a cell is shared by four other cells (again, refer to the excellent artwork below to make this clearer). Thus, the weight of one cell is 1/4 of the weight of the 12 chains which make up the sides.

$$M = 68 \cdot \frac{gm}{mole} \qquad L = 6.022 \cdot 10^{23} \cdot \frac{1}{mole} \qquad W = \frac{12}{4} \cdot N \cdot \frac{M}{L} \qquad W = 2.71 \cdot 10^{-20} \cdot gm$$

$$\rho = \frac{W}{V} \qquad \rho = 1.103 \cdot \frac{gm}{cm^3}$$

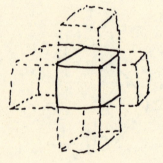

CHAPTER 4 *Equilibrium in Pure Substances*

SECTION 4.1

4.1
_____

We can parallel a similar derivation given in section 4.1 of the text. First, we will start with the definition of the Helmholtz free energy,

$$A = U - TS$$

$$\frac{A}{T} = \frac{U}{T} - S$$

Taking the partial derivative of this equation with respect to temperature gives

$$\left(\frac{\partial (A/T)}{\partial T}\right)_V = \frac{1}{T}\left(\frac{\partial U}{\partial T}\right)_V - \frac{U}{T^2} - \left(\frac{\partial S}{\partial T}\right)_V$$

From Table 3.1, the following relations show that the first and third terms on the right add to zero.

$$C_v = \left(\frac{\partial U}{\partial T}\right)_V = T\left(\frac{\partial S}{\partial T}\right)_V$$

Therefore, our expression for the temperature dependence of the Helmholtz free energy simplifies to

$$\left(\frac{\partial A/T}{\partial T}\right)_V = -\frac{U}{T^2}$$

This can further be simplified by use of the chain rule:

$$\frac{d(A/T)}{dT} = \frac{d(A/T)}{d(1/T)}\frac{d(1/T)}{dT} = -\frac{1}{T^2}\frac{d(A/T)}{d(1/T)}$$

Combining the previous two results gives

$$-\frac{1}{T^2}\left(\frac{\partial(A/T)}{\partial(1/T)}\right)_V = -\frac{U}{T^2}$$

$$\left(\frac{\partial(A/T)}{\partial(1/T)}\right)_V = U$$

4.3

The dependence of the chemical potential on pressure is given by eq. 4.12

$$\left(\frac{\partial\mu}{\partial P}\right)_T = V_m$$

For constant temperature,

$$d\mu = V_m dP$$

Inserting the given equation of state and integrating gives

$$\int_{\mu_1}^{\mu_2} d\mu = \int_{P_1}^{P_2}\left[\frac{RT}{P} + B(T)\right] dP$$

$$\mu_2 - \mu_1 = RT\ln\left(\frac{P_2}{P_1}\right) + B(P_2 - P_1)$$

SECTION 4.2

4.5

The activity is defined by eq. 4.15, where the standard chemical potential is the chemical potential at 1 bar.

$$a \equiv \exp\left[\frac{\mu - \mu^\theta}{RT}\right]$$

The change in the chemical potential with pressure is given by eq. 4.12. We will be able to use this equation to calculate the difference in chemical potential from 1 bar to 100 atm.

$$\left(\frac{\partial \mu}{\partial P}\right)_T = V_m$$

At constant temperature this may be integrated to give,

$$\int_{\mu^\theta}^{\mu} d\mu = V_m \int_{P^\theta}^{100 \; atm} dP$$

$$\mu - \mu^\theta = V_m (100 \; atm - P^\theta)$$

We can now get an expression for the activity of water at 100 atm by inserting the above expression into eq. 4.15, the definition of activity.

$$a = \exp\left[\frac{V_m (100 \; atm - P^\theta)}{RT}\right]$$

With

$$V_m = 18.069 \times 10^{-6} \; \frac{m^3}{mol} \quad , \quad 100 \; atm = 10132500 \; Pa \quad , \quad P^\theta = 10^5 \; Pa \quad , \quad R = 8.31451 \; \frac{J}{mol \; K} \quad , \quad T = 298.15 \; K$$

$$a = 1.076$$

4.7

The activity is defined by eq. 4.15

$$a \equiv \exp\left(\frac{\mu - \mu^\theta}{RT}\right)$$

The ratio of the activities of two phases, alpha and beta, is thus given by

$$\frac{a_\beta}{a_\alpha} = \exp\left(\frac{\mu_\beta - \mu_\beta^\theta}{RT} - \frac{\mu_\alpha - \mu_\alpha^\theta}{RT}\right)$$

The requirement for equilibrium between the two phases is that the chemical potentials are equal. Therefore the following must be true

$$\mu_\alpha = \mu_\beta$$

Using this to simplify the ratio of activities expression gives

$$\frac{a_\beta}{a_\alpha} = \exp\left(\frac{\mu_\alpha^\theta - \mu_\beta^\theta}{RT}\right)$$

Recall that the chemical potential is just the partial molar Gibbs free energy. In addition, for a pure component, it is also equal to the molar Gibbs free energy. In the problem text, the following relation was defined.

$$\Delta G^\theta = G_\beta^\theta - G_\alpha^\theta$$

In this equation, the quantities on the right are the standard molar Gibbs free energies for their respective phases. They are equal to the standard chemical potentials for the two phases as well because we are dealing with a pure component. If we were dealing with a multicomponent system, we would have to use the partial molar Gibbs free energy for each component, which generally is not equal to the pure component molar Gibbs free energy.

Thus we may write the ratio of activities as

$$\frac{a_\beta}{a_\alpha} = \exp\left(\frac{-\Delta G^\theta}{RT}\right)$$

SECTION 4.3

4.9

If we can assume that the heat of vaporization is constant, we can use eq. 4.22, the integrated form of the Clausius-Clapeyron equation, to find the new boiling point at the reduced pressure.

$$\ln\left(\frac{P_2}{P_1}\right) = -\frac{\Delta vH}{R} \cdot \left(\frac{1}{T_2} - \frac{1}{T_1}\right)$$

This equation gives the equilibrium relation between the temperature and vapor pressure at two points. If we take the normal boiling point of water of 373.15 K (and the corresponding vapor pressure of 1 atmosphere), we can estimate the vapor pressure at any other temperature or the temperature at any other vapor pressure. In this case, the latter will be more useful since we know the pressure but not the boiling temperature at the top of Mount Everest.

$$P_1 = 1 \cdot atm \qquad\qquad T_1 = 373.15 \cdot K$$

$$P_2 = \frac{1}{3} \cdot atm \qquad\qquad R = 8.31451 \cdot \frac{joule}{mole \cdot K}$$

From Table 4.2, we find that $\quad \Delta vH = 40.66 \cdot 10^3 \cdot \frac{joule}{mole}$

$$T_2 = -\Delta vH \cdot \frac{T_1}{\left(\ln\left(\frac{P_2}{P_1}\right) \cdot R \cdot T_1 - \Delta vH\right)}$$

$$T_2 = 344 \cdot K$$

4.11

The data from Table 4.2 give the normal boiling point of carbon tetrachloride and the heat of vaporization at the normal boiling point.

$$T_1 = 349.9 \cdot K \qquad P_1 = 1 \cdot atm \qquad R = 8.31451 \cdot \frac{joule}{mole \cdot K}$$

$$\Delta vH = 30.0 \cdot \frac{kJ}{mole} \qquad kJ \equiv 1000 \cdot joule$$

The Clausius-Clapeyron equation (eq. 4.20) gives the dependence of the vapor pressure-temperature relation on the heat of vaporization. For a constant heat of vaporization, this equation can be integrated to give eq. 4.22. This will allow us to predict the vapor pressure of carbon tetrachloride at any temperature.

$$\ln\left(\frac{P_2}{P_1}\right) = -\frac{\Delta vH}{R} \cdot \left(\frac{1}{T_2} - \frac{1}{T_1}\right)$$

$$T_2 = 400 \cdot K$$

$$P_2 = \exp\left[\frac{-(\Delta vH \cdot T_1 - \Delta vH \cdot T_2)}{[R \cdot (T_2 \cdot T_1)]}\right] \cdot P_1$$

$$P_2 = 3.64 \cdot atm$$

4.13

We are given two vapor pressure-temperature pairs. From these data, we can estimate the enthalpy of vaporization using the Clausius-Clapeyron equation (eq. 4.20). The form that is most useful in this case is the integrated form at constant enthalpy of vaporization, eq. 4.22. In that equation, the only unknown is the enthalpy of vaporization itself.

$$\ln\left(\frac{P_2}{P_1}\right) = -\frac{\Delta vH}{R} \cdot \left(\frac{1}{T_2} - \frac{1}{T_1}\right)$$

Given

$$T_1 = (1284 - 273.15) \cdot K \qquad T_2 = (1487 - 273.15) \cdot K$$

$$P_1 = 1 \cdot torr \qquad P_2 = 10 \cdot torr$$

$$R = 8.31451 \cdot \frac{joule}{mole \cdot K}$$

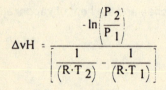

$$\Delta vH = \frac{-\ln\left(\frac{P_2}{P_1}\right)}{\left[\frac{1}{(R \cdot T_2)} - \frac{1}{(R \cdot T_1)}\right]}$$

$$\Delta vH = 258 \cdot \frac{kJ}{mole} \qquad kJ \equiv 1000 \cdot joule$$

Now we can use the calculated heat of vaporization and one of the given data points in eq. 4.22 to solve for the temperature for which the vapor pressure is 5 torr.

$$\ln\left(\frac{P_2}{P_1}\right) = -\frac{\Delta vH}{R} \cdot \left(\frac{1}{T_2} - \frac{1}{T_1}\right)$$

$P_1 = 1 \cdot torr \qquad P_2 = 5 \cdot torr$

$T_1 = (1284 + 273.15) \cdot K$

$$T_2 = -\Delta vH \cdot \frac{T_1}{\left(\ln\left(\frac{P_2}{P_1}\right) \cdot R \cdot T_1 - \Delta vH\right)}$$

$T_2 = 1693.68 \cdot K \qquad\qquad C = K$

$T_2 - 273.15 \cdot K = 1420.53 \cdot C$

4.15

The data from Table 4.2 give the normal boiling point of n-butane and the heat of vaporization at the normal boiling point.

$T_1 = 272.65 \cdot K \qquad P_1 = 1 \cdot atm \qquad R = 8.31451 \cdot \frac{joule}{mole \cdot K}$

$\Delta vH = 22.39 \cdot \frac{kJ}{mole} \qquad kJ = 1000 \cdot joule$

The Clausius-Clapeyron equation (eq. 4.20) gives the dependence of the vapor pressure-temperature relation on the heat of vaporization. For a constant heat of vaporization, this equation can be integrated to give eq. 4.22. This will allow us to predict the vapor pressure of n-butane at any temperature.

$$\ln\left(\frac{P_2}{P_1}\right) = -\frac{\Delta vH}{R} \cdot \left(\frac{1}{T_2} - \frac{1}{T_1}\right)$$

$T_2 = 300 \cdot K$

$$P_2 = \exp\left[\frac{-(\Delta vH \cdot T_1 - \Delta vH \cdot T_2)}{R \cdot (T_2 \cdot T_1)}\right] \cdot P_1$$

$P_2 = 2.46 \cdot atm$

4.17

The Clausius-Clapeyron equation, eq. 4.20, tells us that the slope of ln(P) versus 1/T is equal to $-\Delta_v H/R$.

points = 3   i = 0 .. points − 1

T_i = P_i =

273.15
277.60
283.15

1.268
1.490
1.810

$x_i = \dfrac{1}{T_i}$ $y_i = \ln(P_i \cdot 101325)$

We are now ready to perform the regression. Notice that the pressures have been multiplied by a factor that converts atm to the SI unit of pascals. Actually, since we are interested only in the slope of the regression and not the intercept, conversion of pressure units is unnecessary, but it is included here for clarity and rigor.

m = slope(x,y) $m = -2.753 \cdot 10^3$ m is the slope of the ln P vs. 1/T graph

r = corr(x,y)   r = −0.9999998   r is the correlation factor related to the goodness of fit

b = intercept(x,y)   b = 21.841   b is the intercept of the ln P vs 1/T graph

The standard deviation of the slope may be calculated using eq. AI.25 of the Appendix.

$$\sigma_m = \frac{m}{r} \cdot \sqrt{\frac{1 - r^2}{\text{points} - 2}}$$ $\sigma_m = 1.783$ R = 8.31451

kJ = 1000·joule

$$\Delta v H = -m \cdot R \cdot \frac{\text{joule}}{\text{mole}}$$ $\Delta v H = 22.887 \cdot \dfrac{kJ}{\text{mole}}$

$$\sigma_{\Delta v H} = \sigma_m \cdot R \cdot \frac{\text{joule}}{\text{mole}}$$ $\sigma_{\Delta v H} = 0.015 \cdot \dfrac{kJ}{\text{mole}}$

It may be helpful to see a plot of the data and regressed fit on a ln P vs 1/T graph.

$\ln P(x) = m \cdot x + b$

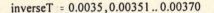

inverseT $= 0.0035, 0.00351 .. 0.00370$

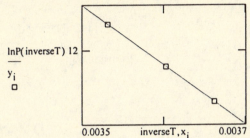

This graph shows the fit of the ln P vs 1/T data. The slope of this line is $-\Delta_v H/R$ according to the Clausius-Clapeyron equation.

Summary

$$\Delta vH = 22.887 \cdot \frac{kJ}{mole} \qquad \sigma_{\Delta vH} = 0.015 \cdot \frac{kJ}{mole}$$

4.19

The Clausius-Clapeyron equation, eq. 4.20, tells us that the slope of ln(P) vs 1/T is equal to $-\Delta_v H/R$.

points 8   i 0 .. points 1

T_i	P_i
187.45	5
195.35	10
204.25	20
214.05	40
220.35	60
228.95	100
241.95	200
256.85	400

$$x_i \quad \frac{1}{T_i} \qquad y_i \quad \ln P_i \cdot 133.332$$

We are now ready to perform the regression. Notice that the pressures have been multiplied by a factor that converts torr to the SI unit of pascals. Actually, since we are interested only in the slope of the regression and not the intercept, conversion of pressure units is unnecessary, but it is included here for clarity and rigor.

m   slope(x,y)   m = $3.04 \cdot 10^3$ m is the slope of the ln P vs. 1/T graph

r   corr(x,y)   r = -0.999847   r is the correlation factor related to the goodness of fit

b   intercept(x,y)   b = 22.759   b is the intercept of the ln P vs 1/T graph

The standard deviation of the slope may be calculated using eq. AI.25 of the Appendix.

$$\sigma_m \quad \frac{m}{r} \cdot \sqrt{\frac{1 \cdot r^2}{\text{points} \quad 2}} \qquad \sigma_m = 21.738 \qquad R \quad 8.31451$$

$$\Delta vH \quad m \cdot R \cdot \frac{\text{joule}}{\text{mole}} \qquad \Delta vH = 25.28 \cdot \frac{kJ}{\text{mole}} \qquad kJ \quad 1000 \cdot \text{joule}$$

$$\sigma_{\Delta vH} \quad \sigma_m \cdot R \cdot \frac{\text{joule}}{\text{mole}} \qquad \sigma_{\Delta vH} = 0.18 \cdot \frac{kJ}{\text{mole}}$$

It may be helpful to see a plot of the data and regressed fit on a ln P vs 1/T graph.

lnP(x)   m·x · b

inverseT   0.0038,0.0039..0.0054

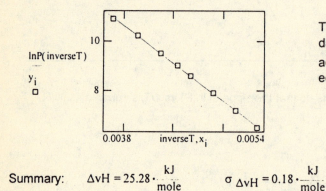

This graph shows the fit of the ln P vs 1/T data. The slope of this line is $-\Delta_V H/R$ according to the Clausius-Clapeyron equation.

Summary: $\Delta vH = 25.28 \cdot \frac{kJ}{\text{mole}}$ $\sigma_{\Delta vH} = 0.18 \cdot \frac{kJ}{\text{mole}}$

4.21

$$\ln\left(\frac{P}{\text{torr}}\right) = 25.7735 - \frac{7769.32}{T} - 1.05576 \cdot \ln(T)$$

part a)

$$P(T) = (T)^{-1.05576} \cdot \exp\left(25.7735 - \frac{7769.32}{(T)}\right) \cdot \text{torr}$$

$$P(25 - 273.15) = 1.84 \cdot 10^{-3} \cdot \text{torr}$$

part b)

Tguess $= 500$

Tboil $= \text{root}\left(P(\text{Tguess}) - 10^5 \cdot \text{Pa}, \text{Tguess}\right) \cdot \text{K}$

Tboil $= 629.11 \cdot \text{K}$

part c)

From the Clausius-Clapeyron equation, we know that the derivative of ln P with respect to 1/T has a slope of $-\Delta_v\text{H}/\text{R}$. That derivative is referred to here as dlnP.

dlnP(T) $= -7769.32 + 1.05576 \cdot \text{T}$

The original vapor pressure expression is given as ln(P/torr). The difference between ln(P/torr) and ln(P/Pa) is only a constant. When we take the derivative with respect to 1/T, the constant term vanishes. Thus, for calculating the enthalpy of vaporization, it doesn't matter what pressure units are.

$\Delta v\text{H}(\text{T}) = -\text{R} \cdot \text{dlnP}(\text{T})$ $\qquad \text{R} \equiv 8.31451 \cdot \dfrac{\text{joule}}{\text{mole} \cdot \text{K}} \qquad$ $\text{kJ} \equiv 1000 \cdot \text{joule}$

$\Delta v\text{H}(273) \cdot \text{K} = 62.2 \cdot \dfrac{\text{kJ}}{\text{mole}}$

$\Delta v\text{H}(473) \cdot \text{K} = 60.4 \cdot \dfrac{\text{kJ}}{\text{mole}}$

$\Delta v\text{H}(673) \cdot \text{K} = 58.7 \cdot \dfrac{\text{kJ}}{\text{mole}}$

4.23

a) The Clausius-Clapeyron equation can be integrated to give an expression for the vapor pressure of N_2 as a function of temperature.

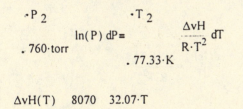

$$\int_{760 \cdot \text{torr}}^{P_2} \ln(P) \, dP = \int_{77.33 \cdot \text{K}}^{T_2} \dfrac{\Delta v\text{H}}{\text{R} \cdot \text{T}^2} \, dT \qquad\qquad \text{R} = 8.31451$$

$\Delta v\text{H}(\text{T}) \quad 8070 \quad 32.07 \cdot \text{T}$

$$\ln\!\left(\frac{P_2}{760}\right) = \int_{77.33}^{T} \frac{8070 - 32.07 \cdot T}{R \cdot T^2}\, dT$$

$$P(T) = 760 \cdot \exp\!\left(-\frac{8070}{R \cdot T} + \frac{32.07 \cdot \ln(T)}{R} + \frac{8070}{R \cdot 77.33} - \frac{32.07 \cdot \ln(77.33)}{R}\right)$$

This can also be expressed as follows.

$$\ln\!\left(\frac{P}{torr}\right) = 35.9557 - \frac{970.5924}{T} - 3.8571 \cdot \ln(T)$$

b) The boiling temperature at 100 torr is the temperature at which the vapor pressure equals the ambient pressure.

$$Tguess = 150$$

$$T_{boil} = root(P(Tguess) - 100, Tguess)$$

$$T_{boil} = 63.20$$

4.25

With heat capacity data, we no longer have to assume a constant heat of vaporization. We can use equation 4.24 to calculate the dependence of $\Delta_V H$ on temperature. With $\Delta_V H(T)$, we can integrate the Clausius-Clapeyron equation to obtain a more accurate description of the temperature dependence of vapor pressure.

From Table 4.2: $\Delta vH = 30.0 \cdot \dfrac{kJ}{mole}$ $T_b = 349.9 \cdot K$ $kJ = 1000 \cdot joule$

$$C_{pgas} = 83.30 \cdot \frac{joule}{mole \cdot K} \qquad C_{pliq} = 131.74 \cdot \frac{joule}{mole \cdot K} \qquad\qquad R = 8.31451 \cdot \frac{joule}{mole \cdot K}$$

$$\Delta vH(T) = \Delta vH - \left(C_{pgas} - C_{pliq}\right) \cdot (T - 349.9 \cdot K)$$

We can now use this in the Clausius Clapeyron equation: $\dfrac{d(\ln P)}{dT} = \dfrac{\Delta vH(T)}{R \cdot T^2}$

Integrating from P=1 atm and T=349.9 K to any temperature T and pressure P gives

$$\ln\left(\frac{P}{1\cdot atm}\right) = \int_{349.9\cdot K}^{T} \frac{\Delta vH(T)}{R\cdot T^2}\, dT$$

$$P(T) = 1\cdot atm\cdot exp\left(\int_{349.9\cdot K}^{T} \frac{\Delta vH(T)}{R\cdot T^2}\, dT\right)$$

$$P(400\cdot K) = 3.46\cdot atm$$

4.27

Part a)

The vapor pressure expression may be rearranged to the following form.

$$P(T) = MPa\cdot exp\left[45.8006 - \frac{2681.5}{\left(\frac{T}{K}\right)} + .012366\cdot\frac{T}{K} - 6.8688\cdot\ln\left(\frac{T}{K}\right)\right] \qquad MPa = 10^6\cdot Pa$$

$$Tguess = 150\cdot K$$

$$Tboil = root(P(Tguess) - 101325\cdot Pa, Tguess)$$

$$Tboil = 184.52\cdot K$$

Part b)

The Clausius-Clapeyron equation tells us that the heat of vaporization is proportional to the derivative of ln(P) with respect to temperature.

$$\frac{d\ln(P)}{dT} = \frac{\Delta vH}{R\cdot T^2} \qquad\qquad R = 8.31451\cdot\frac{joule}{mole\cdot K} \qquad kJ = 1000\cdot joule$$

We can easily obtain an expression for the left side by taking the derivative of the vapor pressure equation given in the problem text.

$$\ln P(T) = 45.8006 - \frac{2681.5}{T} + 0.012366\cdot T - 6.8688\cdot\ln(T)$$

$$\text{deriv}(T) = \left(\frac{2681.5}{T^2} + 0.012366 - \frac{6.8688}{T} \right) \cdot \frac{1}{K}$$

$$\Delta vH(T) = \text{deriv}(T) \cdot R \cdot (T \cdot K)^2$$

$$\Delta vH(125) = 16.76 \cdot \frac{kJ}{mole}$$

$$\Delta vH(150) = 16.04 \cdot \frac{kJ}{mole}$$

$$\Delta vH(175) = 15.45 \cdot \frac{kJ}{mole}$$

SECTION 4.4

4.29
___

The work required can be calculated from equation 4.25. This tells us that the work is equal to the product of the surface tension and the change in surface area.

$$\gamma = 72.0 \cdot 10^{-3} \cdot \frac{newton}{m}$$

$$r = 25 \cdot nm \qquad\qquad nm = 10^{-9} \cdot m$$

$$areadroplet = 4 \cdot \pi \cdot r^2$$

$$volumedroplet = \frac{4}{3} \cdot \pi \cdot r^3$$

$$N_{droplets} = \frac{16 \cdot cm^3}{volumedroplet}$$

$$totalsurface = N_{droplets} \cdot areadroplet \qquad\qquad totalsurface = 1920 \cdot m^2$$

$$work = \gamma \cdot totalsurface$$

Strictly, only the surface area change should be included here, but the initial surface area is negligible compared to the final surface area.

$$work = 138 \cdot joule$$

4.31
___

The surface free energy, surface internal energy, and surface entropy are defined by eqs. 4.28, 4.30, and 4.29, respectively. We can approximate these partial derivatives using data from Table 4.3.

$$\gamma_1 = 10.5 \cdot 10^{-3} \cdot \frac{newton}{m} \qquad\qquad T_1 = 203 - 273. \ K$$

$$\gamma_2 = 8.3 \cdot 10^{-3} \cdot \frac{newton}{m} \qquad\qquad T_2 = 193 - 273. \ K$$

$$\gamma_3 = 6.2 \cdot 10^{-3} \cdot \frac{newton}{m} \qquad\qquad T_3 = 183 - 273. \ K$$

Equation 4.28 tells us that the surface free energy is equal to the surface tension. Therefore:

$$A_s = \gamma_2$$

$$A_s = 0.0083 \cdot \frac{newton}{m}$$

Equation 4.30 shows that we need to evaluate the partial derivative of surface tension with respect to temperature, here denoted dγdt.

$$dgdt = \frac{g_3 - g_1}{T_3 - T_1} \qquad dgdt = -2.15 \cdot 10^{-4} \cdot \frac{newton}{m \cdot K}$$

$$U_s = g_2 - T_2 \cdot dgdt$$

$$U_s = 0.0255 \cdot \frac{joule}{m^2}$$

Equation 4.29 tells us that the surface entropy is equal to the negative of dγdt.

$$S_s = -dgdt$$

$$S_s = 2.15 \cdot 10^{-4} \cdot \frac{joule}{K \cdot m^2}$$

4.33

The effect of surface tension on vapor pressure is given by eq. 4.36. This effect occurs because surface tension affects the liquid phase free energy, which in turn affects where the minimum of system free energy occurs, ie. equilibrium.

$$\ln\left(\frac{P}{P_o}\right) = \frac{2 \cdot g \cdot M}{r \cdot r \cdot R \cdot T}$$

$$P = \exp\left(2 \cdot g \cdot \frac{M}{(r \cdot r \cdot R \cdot T)}\right) \cdot P_o$$

$$P = 769 \cdot torr$$

Data:

$$P_o = 760 \cdot torr$$

$$T = (64.7 + 273.15) \cdot K$$

$$M = 32.04 \cdot \frac{gm}{mole}$$

$$R = 8.31451 \cdot \frac{joule}{mole \cdot K}$$

$$\rho = 0.7510 \cdot \frac{gm}{cm^3}$$

$$r = 50 \cdot 10^{-9} \cdot m$$

$$g = 20.2 \cdot 10^{-3} \cdot \frac{newton}{m}$$

Surface tension data from Table 4.3 at 50 C.

4.35

The effect of surface tension on vapor pressure is given by eq. 4.36. This effect occurs because surface tension affects the liquid phase free energy, which in turn affects where the minimum of system free energy occurs, ie. equilibrium.

$$\ln \frac{P}{P_o} = \frac{2 \cdot \gamma \cdot M}{\rho \cdot r \cdot R \cdot T}$$

$$P = \exp\left(2 \cdot \gamma \cdot \frac{M}{(\rho \cdot r \cdot R \cdot T)}\right) \cdot P_o$$

$P = 1.285 \cdot \text{millitorr}$

Data:

$P_o = 1.201 \cdot \text{millitorr}$

$T = (20 + 273.15) \cdot K$

$M = 200.59 \cdot \frac{\text{gm}}{\text{mole}}$

$R = 8.31451 \cdot \frac{\text{joule}}{\text{mole} \cdot K}$

$\text{millitorr} = 10^{-3} \cdot \text{torr}$

$\rho = 13.546 \cdot \frac{\text{gm}}{\text{cm}^3}$

$r = 85 \cdot 10^{-9} \cdot m$

$\gamma = 472 \cdot 10^{-3} \cdot \frac{\text{newton}}{m}$

Surface tension data from Table 4.3 at 20 C.

SECTION 4.5

4.37

The enthalpy change of this process consists of 5 steps. Three of the steps are the enthalpy changes of H_2O with temperature corresponding to the solid, liquid, and gas phases. The simplest way to calculate this contribution is to use an average heat capacity for each phase. The enthalpy change then becomes the average heat capacity multiplied by the temperature change. (Remember from chapter 2 that, at constant pressure, the enthalpy change is the integral of C_p with respect to temperature. This is the same as finding the area under the heat capacity lines on Figure 2.3.) Due to the phase changes that occur, the heats of fusion and vaporization must also be added to account for the total enthalpy change.

$$\Delta H_{solid} \quad 1.4 \cdot \frac{joule}{K \cdot gm} \cdot (273.15 \cdot K \quad 73.15 \cdot K)$$

$$\Delta H_{liquid} \quad 4.2 \cdot \frac{joule}{K \cdot gm} \cdot (373.15 \cdot K \quad 273.15 \cdot K)$$

$$\Delta H_{gas} \quad 2.0 \cdot \frac{joule}{K \cdot gm} \cdot (573.15 \cdot K \quad 373.15 \cdot K)$$

$$\Delta H_{fusion} \quad 6010 \cdot \frac{joule}{mole} \cdot \frac{mole}{18.015 \cdot gm}$$

$$\Delta H_{vaporization} \quad 40660 \cdot \frac{joule}{mole} \cdot \frac{mole}{18.015 \cdot gm}$$

$$\Delta H_{total} \quad \Delta H_{solid} \cdot \Delta H_{liquid} \cdot \Delta H_{gas} \cdot \Delta H_{fusion} \cdot \Delta H_{vaporization}$$

$$\Delta H_{total} = 3.7 \cdot 10^3 \cdot \frac{joule}{gm}$$

4.39

If we may assume that the heat of fusion and molar volumes of liquid and solid lead are constant, we can use the integrated form of the Clapeyron equation (eq. 4.19) given by eq. 4.39.

$$\ln \frac{T_f}{T_{fo}} = \frac{(P - P_o) \cdot \Delta fV}{\Delta fH}$$

Given

$$\rho_{sol} = 11.23 \cdot \frac{gm}{cm^3} \qquad \rho_{liq} = 10.51 \cdot \frac{gm}{cm^3}$$

$$T_{fo} = (327.5 + 273.15) \cdot K$$

$$P_o = 1 \cdot atm \qquad \Delta fH = 5100 \cdot \frac{joule}{mole} \cdot \frac{mole}{200.59 \cdot gm}$$

$$P = 1000 \cdot atm$$

Heat of fusion from Table 4.2

$$\Delta fV = \frac{1}{\rho_{liq}} - \frac{1}{\rho_{sol}}$$

$$T_f = \exp\left(\frac{\Delta fV \cdot P - \Delta fV \cdot P_o}{\Delta fH}\right) \cdot T_{fo}$$

$$T_f = 615.4 \cdot K \qquad \text{This corresponds to 342 degrees Celsius.}$$

$$\Delta T_f = T_f - T_{fo}$$

$$\Delta T_f = 14.8 \cdot K$$

4.41

Recall that the equilibrium condition is that the chemical potentials are equal. The problem text gives us the difference between the chemical potentials at the standard state (P=1 bar) and 25 C. However, as the pressure changes, so do the chemical potentials. Therefore, we must find the pressure at which the chemical potentials become equal at 25 C. The partial derivative of a pure component chemical potential with respect to pressure at constant temperature is the volume, which was derived in chapter 3. Therefore, we can use the volume to predict the variation of the chemical potential with pressure. Since the phases are virtually incompressible, we can use the equilibrium condition of eq. 4.37 to find the pressure at which the two phases are in equilibrium.

$$\mu o_{dia} + Vm_{dia} \cdot (P - Po) = \mu o_{gra} + Vm_{gra} \cdot (P - Po)$$

Given:

$$\Delta Go = 2900 \cdot \frac{joule}{mole} \qquad Po = 10^5 \cdot Pa$$

$$\rho_{dia} = 3.51 \cdot \frac{gm}{cm^3} \qquad \rho_{gra} = 2.26 \cdot \frac{gm}{cm^3}$$

$$\mu o_{dia} - \mu o_{gra} = \Delta Go = (Vm_{gra} - Vm_{dia}) \cdot (P - Po)$$

$$MW\_C = 12.011 \cdot \frac{gm}{mole}$$

$$P = Po + \frac{\Delta Go}{Vm_{gra} - Vm_{dia}}$$

$$Vm_{dia} = \frac{MW\_C}{\rho_{dia}} \qquad Vm_{gra} = \frac{MW\_C}{\rho_{gra}}$$

$$P = 1.53 \cdot GPa$$

$$GPa = 10^9 \cdot Pa$$

4.43

Part a)

Since the difference in standard free energy at 25 C is positive for the transition of calcite to aragonite, the calcite form is more stable under those conditions. Recall that the equilibrium form is the one which minimizes free energy.

Part b)

The equilibrium condition is that the chemical potentials are equal. The problem text gives us the difference between the chemical potentials (ΔGo) at the standard state (P=1 bar) and 25 C. However, as the pressure changes, so do the chemical potentials. Therefore, we must find the pressure at which the chemical potentials become equal at 25 C. Since the phases are virtually incompressible, we can use the equilibrium condition of eq. 4.37 to find the pressure at which the two phases are in equilibrium.

Given

$$\mu o_{ara} \cdot Vm_{ara} \cdot (P \quad Po) = \mu o_{cal} \cdot Vm_{cal} \cdot (P - Po)$$

$$\rho_{cal} \quad 2.71 \cdot \frac{gm}{cm^3} \qquad \rho_{ara} \quad 2.93 \cdot \frac{gm}{cm^3}$$

$$MW\_CaCO3 \quad 100.0872 \cdot \frac{gm}{mole}$$

$$\mu o_{ara} \quad \mu o_{cal} = \Delta Go = (Vm_{cal} \quad Vm_{ara}) \cdot (P \quad Po)$$

$$MPa \quad 10^6 \cdot Pa \qquad Po \quad 10^5 \cdot Pa$$

$$\Delta Go \quad 1040 \cdot \frac{joule}{mole}$$

$$P \quad Po \cdot \frac{\Delta Go}{(Vm_{cal} \quad Vm_{ara})}$$

$$Vm_{cal} \quad \frac{MW\_CaCO3}{\rho_{cal}} \qquad Vm_{ara} \quad \frac{MW\_CaCO3}{\rho_{ara}}$$

$$P = 375 \cdot MPa$$

Part c)

If we can assume that ΔHo and ΔSo are invariant with temperature, we can calculated ΔGo at 1000 K using G=H-TS.

$$kJ \quad 1000 \cdot joule$$

$$\Delta Ho \quad 0.21 \cdot \frac{kJ}{mole} \qquad \Delta So = 4.2 \cdot \frac{joule}{K \cdot mole}$$

$$\Delta Go(T) = \Delta Ho \quad T \cdot \Delta So$$

$$P(T) = Po \cdot \frac{\Delta Go(T)}{(Vm_{cal} \quad Vm_{ara})}$$

$$P(1000 \cdot K) = 1439 \cdot MPa$$

SECTION 4.6

4.45

The Clausius-Clapeyron equation relates the vapor and sublimation pressures of a substance to the heats of vaporization and sublimation, respectively.

$$\frac{d \cdot \ln P_{vap}}{d \cdot \frac{1}{T}} = \frac{\Delta vH}{R} \quad \text{and} \quad \frac{d \cdot \ln P_{sub}}{d \cdot \frac{1}{T}} = \frac{\Delta sH}{R} \qquad R = 8.31451 \cdot \frac{joule}{mole \cdot K} \qquad kJ \quad 1000 \cdot joule$$

From Figure 4.2 we can estimate the derivatives in the above equations. The slopes of the figure are related to the needed derivatives. If we multiply the derivatives by -R, that will give us $\Delta_v H$ and $\Delta_s H$.

$$slope_{vap} = \frac{1.6 - 4.2}{4.6 - 3.3} \qquad\qquad slope_{sub} = \frac{1.6 - 1.6}{5.6 - 4.6}$$

$$deriv_{vap} = slope_{vap} \cdot 1000 \cdot K \qquad\qquad deriv_{sub} = slope_{sub} \cdot 1000 \cdot K$$

$$\Delta vH \quad deriv_{vap} \cdot (R) \qquad\qquad \Delta sH \quad deriv_{sub} \cdot (R)$$

$$\Delta vH = 16.6 \cdot \frac{kJ}{mole} \qquad\qquad \Delta sH = 26.6 \cdot \frac{kJ}{mole}$$

The enthalpy of fusion can be estimated as the difference between the enthalpies of vaporization and sublimation. (See example 4.7 in the text.)

$$\Delta fH \quad \Delta sH \cdot \Delta vH$$

$$\Delta fH = 10 \cdot \frac{kJ}{mole}$$

4.47

The Clausius-Clapeyron equation relates the vapor and sublimation pressures of a substance to the heats of vaporization and sublimation, respectively.

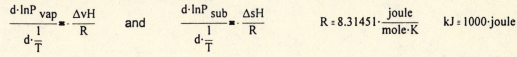

$$\frac{d \cdot \ln P_{vap}}{d \cdot \frac{1}{T}} = \frac{\Delta vH}{R} \quad \text{and} \quad \frac{d \cdot \ln P_{sub}}{d \cdot \frac{1}{T}} = \frac{\Delta sH}{R} \qquad R \equiv 8.31451 \cdot \frac{joule}{mole \cdot K} \qquad kJ \equiv 1000 \cdot joule$$

From the given data we can estimate the derivatives in the above equations. They will be referred to as $deriv_{vap}$ and $deriv_{sub}$. If we multiply the derivatives by -R, that will give us Δ_vH and Δ_sH.

$i = 0..1$

$Tsolid_i =$ $Psolid_i =$ $Tliquid_i =$ $Pliquid_i =$

$-62.7 + 273.15$	40
$-51.8 + 273.15$	100

$-33 + 273.15$	400
$-21 + 273.15$	760

$lnPsolid_i = ln(Psolid_i)$

$inverseTsolid_i = \dfrac{1}{Tsolid_i}$

$lnPliquid_i = ln(Pliquid_i)$

$inverseTliquid_i = \dfrac{1}{Tliquid_i}$

$deriv_{sub} = slope(inverseTsolid, lnPsolid) \cdot K$

$intercept_{sub} = intercept(inverseTsolid, lnPsolid)$

$\Delta sH = -R \cdot deriv_{sub}$

$\Delta sH = 32.6 \cdot \dfrac{kJ}{mole}$

$deriv_{vap} = slope(inverseTliquid, lnPliquid) \cdot K$

$intercept_{vap} = intercept(inverseTliquid, lnPliquid)$

$\Delta vH = -R \cdot deriv_{vap}$

$\Delta vH = 26.9 \cdot \dfrac{kJ}{mole}$

The enthalpy of fusion can be estimated as the difference between the enthalpies of vaporization and sublimation. (See example 4.7 in the text.)

$\Delta fH = \Delta sH - \Delta vH$

$\Delta fH = 5.6 \cdot \dfrac{kJ}{mole}$

The triple point occurs where the sublimation pressure equals the vapor pressure.

$lnP_{sub}(T) = intercept_{sub} + \dfrac{deriv_{sub}}{T}$ $lnP_{vap}(T) = intercept_{vap} + \dfrac{deriv_{vap}}{T}$

$Tguess = 300 \cdot K$

$T_c = root(lnP_{sub}(Tguess) - lnP_{vap}(Tguess), Tguess)$

$T_c = 240.3 \cdot K$

$P_c = exp(lnP_{vap}(T_c)) \cdot torr$

$P_c = 402.5 \cdot torr$

SECTION 4.7

4.49

For a second order phase transition, the entropy is continuous at the phase transition. Therefore, the entropies of phases a and b must be equal.

$$S_a = S_b$$

In order for the system to remain at equilibrium if the conditions are changed (such as temperature and pressure), the entropy change of each phase must also be equal.

$$dS_a = dS_b$$

We may also write the differential change in entropy using the slope formula for partial derivatives

$$dS = \left(\frac{\partial S}{\partial T}\right)_P dT + \left(\frac{\partial S}{\partial P}\right)_T dP$$

From Table 3.1, we can easily identify the slopes in the above equation as follows

$$\left(\frac{\partial S}{\partial T}\right)_P = \frac{C_P}{T}$$

$$\left(\frac{\partial S}{\partial P}\right)_T = -\left(\frac{\partial V}{\partial T}\right)_P = -V\alpha$$

Thus the slope formula becomes

$$dS = \frac{C_P}{T} dT - V\alpha \, dP$$

Now if we equate the differential entropy changes using the above equation for phases a and b, we get

$$\frac{C_{P_a}}{T} dT - V_a \alpha_a \, dP = \frac{C_{P_b}}{T} dT - V_b \alpha_b \, dP$$

Since the volume change at the phase transition is zero, the volumes of phases a and b must be th[e] same. Thus we may drop the volume subscripts. Collecting terms in the above equation gives

$$V(\alpha_a - \alpha_b)dP = \frac{C_{P_a} - C_{P_b}}{T}dT$$

Further rearrangement gives

$$\frac{dP}{dT} = \frac{\Delta C_p}{TV\Delta \alpha}$$

SECTION 4.9

4.51

Gas Constant and Units:	R $\quad$ 8.31451	cm $\quad$ 0.01	MPa $\quad$ 10^6	atm $\quad$ 101325

Constants for Nitrogen: $\quad T_c \quad 126 \qquad P_c \quad 3.394 \cdot MPa$

van der Waals constants: $\quad a \quad \dfrac{27 \cdot R^2 \cdot T_c^2}{64 \cdot P_c} \qquad b \quad \dfrac{R \cdot T_c}{8 \cdot P_c}$

T $\quad$ 77.33

$$P(V_m) \quad \frac{R \cdot T}{V_m \quad b} \quad \frac{a}{V_m^2}$$

volume $\quad 50 \cdot cm^3, 51 \cdot cm^3 .. 1000 \cdot cm^3$

This plot can be used to obtain an
initial guess for the vapor pressure.
The estimate should be chosen in
the two phase region where the
"areas" bounded by the curve above
and below the estimate are about equal.

Vapor Pressure Estimate
$\quad p_0 \quad 5 \cdot atm$

The solution can be found by solving eq. 4.40 when $\Delta_v G = 0$. It will be useful to define functions for calculating the liquid volume (V_1), the gas volume (V_2), and the indefinite integral of $P(V_m)dV_m$. Then, it will be possible to iterate to find the vapor pressure.

	Initial Guess	Volume Function
Liquid	$v_b \quad 1.01 \cdot b$	$V_1(px) \quad root(px \quad P(v_b), v_b)$
Vapor	$v_a \quad \dfrac{R \cdot T}{p_0}$	$V_2(px) \quad root(px \quad P(v_a), v_a)$

These initial guesses should converge on the liquid and vapor volumes. Recall that there are three real roots in the critical region. The liquid volume initial guess was chosen to be slightly larger than the van der Waals b constant because this term accounts for the volume occupied by the molecules. Since the liquid phase is compact, this volume is a good initial guess for the liquid volume. (A factor of 1.01 was used so division by 0 is not encounted in the equation solver.) The initial guess for the vapor volume is just the ideal gas volume.

Integral of $P(V_m)dV_m$ $\quad pi(V_m) = R \cdot T \cdot \ln(V_m - b) - \dfrac{a}{V_m}$

Now we can iterate for the vapor pressure. Eq. 4.40 can be rearranged into the function below when $\Delta_V G = 0$. This iteration uses the pressures and volumes from the previous iteration to calculate a new guess for the pressure.

$$\text{iterations} \quad 5 \qquad k \quad 1 .. \text{iterations}$$

$$P_k = \frac{pi(V_2(P_{k-1})) - pi(V_1(P_{k-1}))}{V_2(P_{k-1}) - V_1(P_{k-1})}$$

$$p = \begin{array}{l} 5 \\ 2.835 \\ 3.271 \\ 3.311 \\ 3.312 \\ 3.312 \end{array} \cdot \text{atm}$$

The iteration has converged.

$P_{vap} \quad P_{iterations}$

$P_{vap} = 3.3 \cdot \text{atm}$

4.53

To calculate the enthalpy of vaporization, we need to estimate the enthalpy imperfections for the liquid and vapor phases.

Given: $\quad V_{liq} \quad 0.0490855 \cdot \dfrac{\text{liter}}{\text{mole}} \qquad V_{gas} \quad 19.8698 \cdot \dfrac{\text{liter}}{\text{mole}}$

Gas Constant and Units: $\quad R \quad 8.31451 \cdot \dfrac{\text{joule}}{\text{mole} \cdot \text{K}} \qquad MPa \quad 10^6 \cdot Pa \qquad kJ \quad 1000 \cdot \text{joule}$

Constants for SO_2: $\quad T_c \quad 430 \cdot K \qquad P_c \quad 7.873 \cdot MPa$

Redlich-Kwong constants: $\quad a \quad \dfrac{0.42748 \cdot R^2 \cdot T_c^{\frac{5}{2}}}{P_c} \qquad b \quad \dfrac{0.086640 \cdot R \cdot T_c}{P_c}$

$T \quad 247.184 \cdot K$

$$P(V_m) \quad \frac{R \cdot T}{V_m - b} \quad \frac{a}{\sqrt{T} \cdot V_m \cdot (V_m + b)}$$

From Table 3.3

$$U_i(V_m) \quad \frac{3 \cdot a}{2 \cdot b \cdot \sqrt{T}} \cdot \ln\left(\frac{V_m}{V_m + b}\right) \qquad H_i(V_m) \quad U_i(V_m) + P(V_m) \cdot V_m \quad R \cdot T$$

$$H_i(V_{gas}) = 110.938 \cdot \frac{joule}{mole} \qquad H_i(V_{liq}) = -2.259 \cdot 10^4 \cdot \frac{joule}{mole}$$

The enthalpy imperfections represent deviations from the standard state. Since the standard state enthalpies for both the liquid and vapor phases are the same, the enthalpy of vaporization is just the difference between the liquid and gas phase enthalpy imperfections.

$$\Delta vH \quad H_i(V_{gas}) - H_i(V_{liq})$$

$$\Delta vH = 22.48 \cdot \frac{kJ}{mole}$$

The observed value given in Table 4.2 is 24.92 kJ/mole, so the value calculated here is very close.

4.55

Gas Constant and Units: $\quad R = 8.31451 \qquad cm = 0.01 \qquad MPa = 10^6 \qquad atm = 101325$

Constants for Ethane: $\quad T_c = 305.33 \qquad P_c = 4.871 \cdot MPa$

Redlich-Kwong constants: $\qquad \frac{0.42748 \cdot R^2 \cdot T_c^{\frac{5}{2}}}{P_c} \qquad \qquad \frac{0.086640 \cdot R \cdot T_c}{P_c}$
$\qquad\qquad\qquad\qquad\qquad\qquad a \qquad\qquad\qquad\qquad\qquad b$

T $\quad$ 184.57

$$P(V_m) \quad \frac{R \cdot T}{V_m - b} - \frac{a}{\sqrt{T} \cdot V_m \cdot (V_m + b)}$$

$\qquad\qquad\qquad\qquad\qquad\qquad\qquad$ volume $\quad 50 \cdot cm^3, 51 \cdot cm^3 .. 2000 \cdot cm^3$

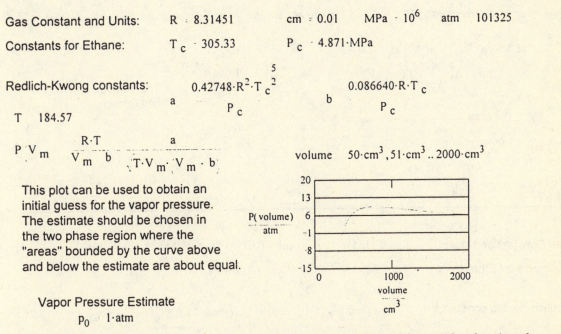

This plot can be used to obtain an initial guess for the vapor pressure. The estimate should be chosen in the two phase region where the "areas" bounded by the curve above and below the estimate are about equal.

Vapor Pressure Estimate
$\qquad p_0 \quad 1 \cdot atm$

The solution can be found by solving eq. 4.40 when $\Delta_v G = 0$. It will be useful to define functions for calculating the liquid volume (V_1), the gas volume (V_2), and the indefinite integral of $P(V_m)dV_m$. Then, it will be possible to iterate to find the vapor pressure.

	Initial Guess	Volume Function
Liquid	v_b $1.01 \cdot b$	$V_1(px)$ $root(px - P(v_b), v_b)$
Vapor	v_a $\dfrac{R \cdot T}{P_0}$	$V_2(px)$ $root(px - P(v_a), v_a)$

These initial guesses should converge on the liquid and vapor volumes. Recall that there are three real roots in the critical region. The liquid volume initial guess was chosen to be slightly larger than the Redlich-Kwong b constant because this term accounts for the volume occupied by the molecules. Since the liquid phase is compact, this volume is a good initial guess for the liquid volume. (A factor of 1.01 was used so division by 0 is not encounted in the equation solver.) The initial guess for the vapor volume is just the ideal gas volume.

Integral of $P(V_m)dV_m$ $\qquad pi(V_m) = R \cdot T \cdot \ln(V_m - b) - \dfrac{a}{\sqrt{T} \cdot b} \cdot \ln\left(\dfrac{V_m - b}{V_m}\right)$

Now we can iterate for the vapor pressure. Eq. 4.40 can be rearranged into the function below when $\Delta_v G = 0$. This iteration uses the pressures and volumes from the previous iteration to calculate a new guess for the pressure.

iterations   3   k   1 .. iterations

$$p_k = \frac{pi(V_2(p_{k-1})) - pi(V_1(p_{k-1}))}{V_2(p_{k-1}) - V_1(p_{k-1})} \qquad\qquad p = \begin{matrix} 1 \\ 1.062 \\ 1.064 \\ 1.064 \end{matrix} \cdot atm$$

The iteration has converged.

P_{vap} $P_{iterations}$

$P_{vap} = 1.06 \cdot atm$

4.57

Gas Constant and Units: $R = 8.31451$ $cm = 0.01$ $MPa = 10^6$ $atm = 101325$

Constants for Ethane: $T_c = 305.33$ $P_c = 4.871 \cdot MPa$

Redlich-Kwong constants: $a = \dfrac{0.42748 \cdot R^2 \cdot T_c^{\frac{5}{2}}}{P_c}$ $b = \dfrac{0.086640 \cdot R \cdot T_c}{P_c}$

T 184.57

$$P(V_m) = \frac{R \cdot T}{V_m - b} - \frac{a}{\sqrt{T} \cdot V_m \cdot (V_m + b)}$$

volume $= 50 \cdot cm^3, 51 \cdot cm^3 .. 2000 \cdot cm^3$

This plot can be used to obtain an initial guess for the vapor pressure. The estimate should be chosen in the two phase region where the "areas" bounded by the curve above and below the estimate are about equal.

$\frac{P(volume)}{atm}$

Vapor Pressure Estimate
$$p_0 = 1 \cdot atm$$

The solution can be found by solving eq. 4.40 when $\Delta_v G = 0$. It will be useful to define functions for calculating the liquid volume (V_1), the gas volume (V_2), and the indefinite integral of $P(V_m) dV_m$. Then, it will be possible to iterate to find the vapor pressure.

	Initial Guess	Volume Function
Liquid	$v_b = 1.01 \cdot b$	$V_1(px) = root(px - P(v_b), v_b)$
Vapor	$v_a = \frac{R \cdot T}{p_0}$	$V_2(px) = root(px - P(v_a), v_a)$

These initial guesses should converge on the liquid and vapor volumes. Recall that there are three real roots in the critical region. The liquid volume initial guess was chosen to be slightly larger than the Redlich-Kwong b constant because this term accounts for the volume occupied by the molecules. Since the liquid phase is compact, this volume is a good initial guess for the liquid volume. (A factor of 1.01 was used so division by 0 is not encounted in the equation solver.) The initial guess for the vapor volume is just the ideal gas volume.

Integral of $P(V_m) dV_m$ $\qquad$ $pi(V_m) = R \cdot T \cdot \ln(V_m - b) + \frac{a}{\sqrt{T} \cdot b} \cdot \ln\left(\frac{V_m + b}{V_m}\right)$

Now we can iterate for the vapor pressure. Eq. 4.40 can be rearranged into the function below when $\Delta_v G = 0$. This iteration uses the pressures and volumes from the previous iteration to calculate a new guess for the pressure.

iterations $= 3$ $\qquad$ $k = 1 ..$ iterations

$$p_k = \frac{pi(V_2(p_{k-1})) - pi(V_1(p_{k-1}))}{V_2(p_{k-1}) - V_1(p_{k-1})}$$

$$p = \begin{bmatrix} 1 \\ 1.062 \\ 1.064 \\ 1.064 \end{bmatrix} \cdot atm$$

The iteration has converged.

$P_{vap} = P_{iterations}$

$P_{vap} = 1.06 \cdot atm$

121

CHAPTER 5 *Statistical Thermodynamics*

SECTION 5.1

5.1

By direct expansion, we obtain:

$$ans = (1 - x)^4 \qquad ans = 1 - 4 \cdot x - 6 \cdot x^2 - 4 \cdot x^3 - x^4$$

Using eq. (5.2) gives:

$$C(N,p) \quad \frac{N!}{p! \cdot (N - p)!}$$

$$C(4,0) = 1 \qquad C(4,2) = 6 \qquad C(4,4) = 1$$

$$C(4,1) = 4 \qquad C(4,3) = 4$$

We get the same results. Computing the sum:

$$sum \quad 1 - 4 - 6 - 4 - 1 \qquad sum = 16$$

$$sum \quad 2^4 \qquad sum = 16$$

Again, we obtain identical results.

5.3

To find the probability, we use eq. (5.3). N is the total number of flips, and p is the number of heads.

$$P(N,p) \quad \frac{1}{2^N} \cdot \frac{N!}{p! \cdot (N - p)!}$$

$$P(5,0) = 0.03125 \qquad \text{0 heads, 5 tails}$$

$$P(5,1) = 0.15625 \qquad \text{1 head, 4 tails}$$

$$P(5,2) = 0.3125 \qquad \text{2 heads, 3 tails}$$

$$P(5,3) = 0.3125 \qquad \text{3 heads, 2 tails}$$

$P(5,4) = 0.15625$ 4 heads, 1 tail

$P(5,5) = 0.03125$ 5 heads, 0 tails

sum $\cdot$ 0.03125 $\cdot$ 0.15625 $\cdot$ 0.3125 $\cdot$ 0.3125 $\cdot$ 0.15625 $\cdot$ 0.03125

sum $= 1$ Not surprisingly, the sum of the probabilities of all the outcomes is one.

5.5

This is similar to example 5.4. First, we calculate the mean number of molecules using the ideal gas law. Then, we calculate the deviation using the expression given in the problem statement.

$$R = 8.31451 \cdot \frac{joule}{mole \cdot K} \qquad V = 1 \cdot cm^3 \qquad T = 300 \cdot K \qquad L = 6.02 \cdot 10^{23} \cdot \frac{1}{mole} \qquad P_{av} = 0.133 \cdot Pa$$

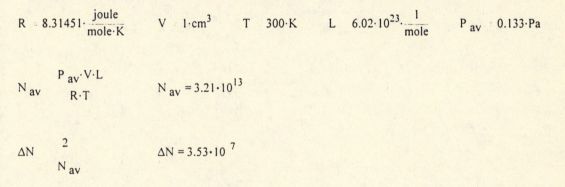

$$N_{av} = \frac{P_{av} \cdot V \cdot L}{R \cdot T} \qquad N_{av} = 3.21 \cdot 10^{13}$$

$$\Delta N = \frac{2}{N_{av}} \qquad \Delta N = 3.53 \cdot 10^{-7}$$

The fluctuation in pressure will be proportional to this:

$$\Delta P = P_{av} \cdot \Delta N \qquad \Delta P = 4.695 \cdot 10^{-8} \cdot Pa$$

5.7

In example 5.5, it is derived:

$$\Delta S = C_p \cdot \frac{\Delta T^2}{T}$$

Eq. (5.7) gives us the probability:

$$P = e^{\frac{\Delta S}{k}}$$

The heat capacity can be found on Table 2.1:

$$C_p \quad 20.88 \cdot \frac{joule}{mole \cdot K} \qquad k \quad 1.38066 \cdot 10^{23} \cdot \frac{joule}{K} \qquad R \quad 8.31451 \cdot \frac{joule}{mole \cdot K}$$

$$V \quad 2 \cdot cm^3 \qquad P \quad 1 \cdot Pa \qquad \Delta T \quad 1 \cdot 10^6 \cdot K \qquad T \quad 10 \cdot K$$

$$n \quad \frac{P \cdot V}{R \cdot T} \qquad \Delta S \quad n \cdot C_p \cdot \frac{\Delta T^2}{T} \qquad P \quad e^{\frac{\Delta S}{k}}$$

$$P = 1.03 \cdot 10^{158}$$

5.9

This is similar to example 5.8. We use eq. (5.8). There are a total of 9 letters in the word, but since there are four E's, two N's, and two S's, we have:

$$N = 9 \qquad N_E = 4 \qquad N_N \quad 2 \qquad N_S \quad 2$$

$$W \quad \frac{N!}{N_E! \cdot N_N! \cdot N_S!} \qquad W = 3780$$

5.11

(a) We use eq. (5.8). If the words are kept separately, then for the word PEEWEE, we have:

$$N = 6 \qquad N_E = 4 \qquad W = \frac{N!}{N_E!} \qquad W = 30$$

For the word REFEREE, we have:

$$N = 7 \qquad N_R = 2 \qquad N_E = 4 \qquad W = \frac{N!}{N_E! \cdot N_R!} \qquad W = 105$$

The number of distinct sequences possible if the letters are not mixed is the product of these:

$$W_{unmixed} = 30 \cdot 105 \qquad W_{unmixed} = 3150$$

(b) When the words are mixed, we obtain:

$$N = 13 \qquad N_E = 8 \qquad N_R = 2 \qquad W_{mixed} = \frac{N!}{N_E! \cdot N_R!}$$

$$W_{mixed} = 77220$$

5.13

This is similar to example 5.9. We use eq. (5.10 to calculate the entropy.

$$i = 1..6$$

$$X_1 = 0.0035 \qquad \text{mass } 78$$

$$X_2 = 0.0227 \qquad \text{mass } 80$$

$$X_3 = 0.1156 \qquad \text{mass } 82$$

$$X_4 = 0.1155 \qquad \text{mass } 83$$

$$X_5 = 0.569 \qquad \text{mass } 84$$

$$X_6 = 0.1737 \qquad \text{mass } 86$$

sum $\sum_i X_i$ sum = 1   Check to see if fractions sum to unity.

Now, we use eq. (5.10) to compute the entropy:

R $8.31451 \cdot \dfrac{joule}{mole \cdot K}$ ΔS_{mix} $R \cdot \sum_i X_i \cdot \ln X_i$

$\Delta S_{mix} = 10.221 \cdot \dfrac{joule}{mole \cdot K}$

5.15

There are a total of five species: CH4, CH3D, CH2D2, CHD3, and CD4. The fractions of each of these species is given by:

i   1..5

X_1 0.15^4 $X_1 = 0.00050625$ CH4

X_2 $4 \cdot 0.15^3 \cdot (0.85)$ $X_2 = 0.011475$ CH3D

X_3 $6 \cdot 0.15^2 \cdot 0.85^2$ $X_3 = 0.097538$ CH2D2

X_4 $4 \cdot (0.15) \cdot 0.85^3$ $X_4 = 0.368475$ CHD3

X_5 0.85^4 $X_5 = 0.522006$ CD4

Remember, there is more than one way to form some of these species, which is why we multiply by 4 and 6. The coefficients correspond to those of the binomial expansion.

sum $\sum_i X_i$ sum = 1   Check to see if fractions sum to unity.

Now, we use eq. (5.10) to compute the entropy:

R $8.31451 \cdot \dfrac{joule}{mole \cdot K}$ ΔS_{mix} $R \cdot \sum_i X_i \cdot \ln X_i$

$\Delta S_{mix} = 8.226 \cdot \dfrac{joule}{mole \cdot K}$

5.17

The easiest way to do this problem is to work out problems 5.15 and 5.16 and see the pattern developing. If f is the fraction of enriched molecules, and n is the number of available positions, then the fraction of each species, p, is given by:

$$X_p = C(n,p) \cdot e^p \cdot e^{(n-p)}$$

The entropy of mixing is then calculated using

$$S = -R \cdot \sum_p X_p \cdot \ln(X_p)$$

Using the binomial coefficients works because we have only two possibilities for each site: an enriched molecule and a regular molecule. Thus, the problem is analogous to the problem of flipping a coin: there are only two possible outcomes. However, if we had more than two different molecules which could occupy a site, then this approach would not work.

5.19

It may be helpful to reread the subsection entitled "Third Law Anomalies." Since NNO can crystallize in two different orientations, the contribution to the entropy will be:

$$R \quad 8.31451 \cdot \frac{joule}{mole \cdot K}$$

$$\Delta S \; = \; R \cdot \ln(2) \qquad \Delta S = 5.763 \cdot \frac{joule}{mole \cdot K}$$

Since phosgene can crystallize in three orientations, the contribution to the entropy will be:

$$\Delta S \quad R \cdot \ln(3) \qquad \Delta S = 9.134 \cdot \frac{joule}{mole \cdot K}$$

5.21

(a) From eq. (5.1),

$$S = k \cdot \ln(W)$$

$$S = k \cdot \ln\left(A \cdot L^2 \cdot e^{\frac{L^2}{N \cdot \lambda^2}}\right)$$

$$S = k \cdot \left(\ln A \cdot L^2 + \frac{L^2}{N \cdot \lambda^2}\right)$$

To find the minimum, we set the derivative equal to zero and solve for L:

$$\frac{dS}{dL} = \frac{2 \cdot A \cdot L}{A \cdot L^2} + \frac{2 \cdot L}{N \cdot \lambda^2}$$

$$0 = \frac{2}{L} + \frac{2 \cdot L}{N \cdot \lambda^2} \qquad 0 = \frac{1}{L} + \frac{L}{N \cdot \lambda^2} \qquad 0 = 1 + \frac{L^2}{N \cdot \lambda^2}$$

$$L^2 = N \cdot \lambda^2$$

$$L = \lambda \cdot \sqrt{N}$$

(b) Let Lo be the original length. The change in entropy will be:

$$\Delta S = k \cdot \left[\ln \left(A \cdot (1.1 \cdot L_o)^2 \right) \cdot \frac{(1.1 \cdot L_o)^2}{N \cdot \lambda^2} - \ln \left(A \cdot L_o^2 \right) \cdot \frac{L_o^2}{N \cdot \lambda^2} \right]$$

$$\Delta S = k \cdot \left[\ln \left(\frac{A \cdot (1.1 \cdot L_o)^2}{A \cdot L_o^2} \right) \cdot \frac{(1.1 \cdot L_o)^2 - L_o^2}{N \cdot \lambda^2} \right]$$

$$\Delta S = k \cdot \left[\ln \left(1.1^2 \right) \cdot \frac{(1.1 \cdot L_o)^2 - L_o^2}{N \cdot \lambda^2} \right] \qquad \Delta S = k \cdot \left[\ln \left(1.1^2 \right) \cdot \frac{L_o^2 \cdot (1.1^2 - 1)}{N \cdot \lambda^2} \right]$$

Substituting in for Lo:

$$L_o = \lambda \cdot N \qquad \Delta S = k \cdot \left[\ln \left(1.1^2 \right) \cdot \frac{N \cdot \lambda^2 \cdot (1.1^2 - 1)}{N \cdot \lambda^2} \right]$$

For one mole of chains:

$$k \quad 1.38066 \cdot 10^{-23} \cdot \frac{joule}{mole \cdot K} \qquad L \quad 6.02 \cdot 10^{23}$$

$$\Delta S \quad k \cdot L \cdot \left[\ln \left(1.1^2 \right) \cdot \left(1.1^2 - 1 \right) \right] \qquad \Delta S = 0.161 \cdot \frac{joule}{mole \cdot K}$$

SECTION 5.2

5.23

To find the total energy, we use eq. (5.11):

i   0..3

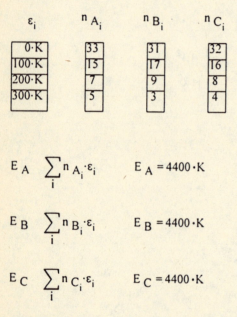

	ε_i	n_{A_i}	n_{B_i}	n_{C_i}
$0 \cdot K$		33	31	32
$100 \cdot K$		15	17	16
$200 \cdot K$		7	9	8
$300 \cdot K$		5	3	4

$$E_A \quad \sum_i n_{A_i} \cdot \varepsilon_i \qquad E_A = 4400 \cdot K$$

$$E_B \quad \sum_i n_{B_i} \cdot \varepsilon_i \qquad E_B = 4400 \cdot K$$

$$E_C \quad \sum_i n_{C_i} \cdot \varepsilon_i \qquad E_C = 4400 \cdot K$$

Thus, all three have the same energy. To calculate the number of configurations, we use eq. (5.13):

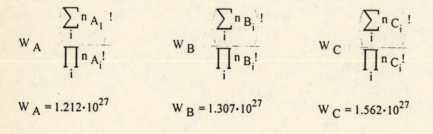

$$W_A \quad \frac{\sum_i n_{A_i}!}{\prod_i n_{A_i}!} \qquad W_B \quad \frac{\sum_i n_{B_i}!}{\prod_i n_{B_i}!} \qquad W_C \quad \frac{\sum_i n_{C_i}!}{\prod_i n_{C_i}!}$$

$$W_A = 1.212 \cdot 10^{27} \qquad W_B = 1.307 \cdot 10^{27} \qquad W_C = 1.562 \cdot 10^{27}$$

For a Boltzmann distribution, the ration of populations is given by:

$$\frac{n_i}{n_j} = \exp\left(\frac{\Delta\varepsilon}{k \cdot T}\right)$$

For equally spaced levels, this means that:

$$\frac{P_2}{P_1} = \frac{P_3}{P_2}$$

The only set of populations which mets this condition is set C, so set C is a Boltzmann distribution and is more probable than A or B.

5.25

The Boltzmann distribution is given by eq. (5.21)

$i \quad 0..2$

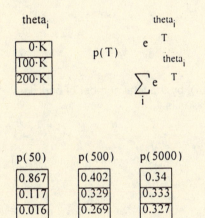

$$\text{theta}_i$$

$$p(T) \quad \frac{e^{\frac{\text{theta}_i}{T}}}{\sum_i e^{\frac{\text{theta}_i}{T}}}$$

	0·K
	100·K
	200·K

p(50)	p(500)	p(5000)
0.867	0.402	0.34
0.117	0.329	0.333
0.016	0.269	0.327

At very high temperatures, all of the energy levels will be equally accessible. The high temperature limit is to have 1/3 of the molecules in each level.

5.27

For systems with degeneracy, we use eq. (5.24). The partition function is defined by eq. (5.23).

$i \quad 0..1 \qquad T \quad 987 \cdot K$

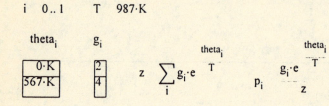

theta$_i$ g$_i$

0·K
567·K

2
4

$$z \quad \sum_i g_i \cdot e^{\frac{theta_i}{T}} \qquad p_i \quad \frac{g_i \cdot e^{\frac{theta_i}{T}}}{z}$$

p$_i$

0.4704
0.5296

The average energy is obtained by dividing the total energy, eq. (5.11), by the total number of particles. Since we are working with population ratios, the total number of particles is 1.

$$E \quad \sum_i theta_i \cdot p_i \qquad\qquad E = 300.3 \cdot K$$

We can get this in terms of J/mole by multiplying by R:

$$R \quad 8.31451 \cdot \frac{joule}{mole \cdot K} \qquad\qquad E \cdot R = 2497 \cdot \frac{joule}{mole}$$

SECTION 5.3

5.29

The partition function is defined by eq. (5.22). For the system of problem 5.25:

i   0..2

theta$_i$

0·K
100·K
200·K

$$z(T) \quad \sum_i e^{\dfrac{-theta_i}{T}}$$

$z(50) = 1.154$

$z(500) = 2.489$

$z(5000) = 2.941$

The high temperature limit is z = 3.

5.31

The heat capacity at constant colume is defined as:

$$C_v = \left(\frac{\partial U}{\partial T}\right)_v$$

The relationship between the partition function, z, and the internal energy, U, is given by eq. (5.26). We take the derivative with respect to temperature to obtain Cv:

$$U - U_o = RT^2 \left(\frac{\partial \ln z}{\partial T} \right)_V$$

$$C_v = \frac{\partial}{\partial T} \left[RT^2 \left(\frac{\partial \ln z}{\partial T} \right)_V + U_o \right]$$

$$C_v = 2RT \left(\frac{\partial \ln z}{\partial T} \right)_V + RT^2 \left(\frac{\partial^2 \ln z}{\partial T^2} \right)_V$$

$$R = Nk$$

$$C_v = 2NkT \left(\frac{\partial \ln z}{\partial T} \right)_V + NkT^2 \left(\frac{\partial^2 \ln z}{\partial T^2} \right)_V$$

SECTION 5.4

5.33

Equation (5.31) gives ztr:

$$z_{tr} = \frac{(2 \cdot \pi \cdot m \cdot k \cdot T)^{\frac{3}{2}} \cdot V}{h^3}$$

The SI units are:

m   kg   k $\frac{joule}{K}$ T   K   V m^3 h   joule·sec

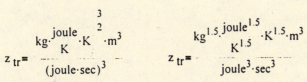

$$z_{tr} = \frac{kg \cdot \frac{joule}{K}^{\frac{3}{2}} \cdot K^2 \cdot m^3}{(joule \cdot sec)^3}$$

$$z_{tr} = \frac{kg^{1.5} \cdot \frac{joule^{1.5}}{K^{1.5}} \cdot K^{1.5} \cdot m^3}{joule^3 \cdot sec^3}$$

$$z_{tr} = \frac{kg^{1.5} \cdot joule^{1.5} \cdot m^3}{joule^3 \cdot sec^3}$$

$$z_{tr} = \frac{kg^{1.5} \cdot m^3}{joule^{1.5} \cdot sec^3}$$

A joule is defined as $joule = \frac{kg \cdot m^2}{sec^3}$ Substituting this in:

$$z_{tr} = \frac{kg^{1.5} \cdot m^3 \cdot sec^3}{kg^{1.5} \cdot m^3 \cdot sec^3} = 1$$

Thus, ztr is unitless.

5.35

Following example 5.16 in the text, we use the Sackur-Tetrode equation, eq. (5.35), with the constants substituted in. This gives us:

$$R \quad 8.31451 \cdot \frac{joule}{mole \cdot K} \qquad T \quad 298.15 \cdot K \qquad M \quad 4.022602 \cdot \frac{gm}{mole}$$

$$S \quad 1.5 \cdot R \cdot \ln \frac{M}{1 \cdot gm \cdot mole^{1}} \cdot 2.5 \cdot R \cdot \ln \frac{T}{1 \cdot K} \quad 1.15167 \cdot R$$

$$S = 126.216 \cdot \frac{joule}{mole \cdot K}$$

5.37

Following example 5.16 in the text, we use the Sackur-Tetrode equation, eq. (5.35), with the constants substituted in. This gives us:

$$R = 8.31451 \cdot \frac{\text{joule}}{\text{mole} \cdot \text{K}} \qquad m = 9.109390 \cdot 10^{-28} \cdot \text{gm} \qquad L = 6.022137 \cdot 10^{23} \cdot \frac{1}{\text{mole}} \qquad T = 1000 \cdot \text{K}$$

$$S = 1.5 \cdot R \cdot \ln \left[\frac{m \cdot L}{1 \cdot \left(\text{gm} \cdot \text{mole}^{-1} \right)} \right] + 2.5 \cdot R \cdot \ln \left(\frac{T}{1 \cdot K} \right) - 1.15167 \cdot R \qquad S = 40.371 \cdot \frac{\text{joule}}{\text{mole} \cdot \text{K}}$$

SECTION 5.5

5.39

This problem is similar to example 5.19. We use eq. (5.39) to calculate the vibrational partition function. The constant for vibration for HCl can be found on Table 5.1. Boltzmann's law gives the fraction of the molecules in the lower state.

$\theta_v = 4301.6 \qquad n = 0$

$T = 300 \cdot K \qquad z_{vib} = \cfrac{1}{1 - \exp\left(-\cfrac{\theta_v}{T}\right)} \qquad p = \cfrac{\exp\left(-n \cdot \cfrac{\theta_v}{T}\right)}{z_{vib}}$

$z_{vib} = 1.000000593 \qquad\qquad p = 0.999999407$

$T = 3000 \cdot K \qquad z_{vib} = \cfrac{1}{1 - \exp\left(-\cfrac{\theta_v}{T}\right)} \qquad p = \cfrac{\exp\left(-n \cdot \cfrac{\theta_v}{T}\right)}{z_{vib}}$

$z_{vib} = 1.313 \qquad\qquad p = 0.7616$

5.41

As T goes to infinity, the expression becomes indeterminate. We can either use L'Hopital's rule or expand the function in a Taylor series. Here, we will expand in a series for the exponents:

$$C_v = \frac{R\left(\dfrac{\theta}{T}\right)^2 e^{\theta T}}{\left(e^{\theta T} - 1\right)^2} = \frac{R\left(\dfrac{\theta}{T}\right)^2 e^{\theta T}}{e^{2\theta T} - 2e^{\theta T} + 1}$$

$$C_v = \frac{R \left(\frac{\theta}{T}\right)^2 \left(1 + \frac{\theta}{T} + \frac{\theta^2}{2T^2} + \frac{\theta^3}{6T^3} + \ldots\right)}{\left(1 + \frac{2\theta}{T} + \frac{4\theta^2}{2T^2} + \frac{8\theta^3}{6T^3} + \ldots\right) - \left(2 + \frac{2\theta}{T} + \frac{2\theta^2}{2T^2} + \frac{2\theta^3}{6T^3} + \ldots\right) + 1}$$

We combine terms in the denominator to obtain:

$$C_v = \frac{R \left(\frac{\theta}{T}\right)^2 \left(1 + \frac{\theta}{T} + \frac{\theta^2}{2T^2} + \frac{\theta^3}{6T^3} + \ldots\right)}{\frac{2\theta^2}{2T^2} + \frac{4\theta^3}{6T^3} + \ldots}$$

Multiplying top and bottom by $(T/\theta)^2$:

$$C_v = \frac{R \left(1 + \frac{\theta}{T} + \frac{\theta^2}{2T^2} + \frac{\theta^3}{6T^3} + \ldots\right)}{1 + \frac{4\theta}{6T} + \ldots}$$

Now, if we take the limit as T goes to infinity, all of the terms in the sum in the denominator will be zero except for the leading 1, and all the terms in the denominator will be zero except for the leading one. This gives us:

$$C_v = R$$

5.43

Equation (5.43) is the high temperature approximation of the rotational partition function. The molecular constants for oxygen can be found on Table 5.1. For oxygen, a linear, symmetric molecule, the symmetry number is 2:

$\theta_r \quad 2.08 \qquad \sigma \quad 2 \qquad T \quad 20.8 \cdot K$

$z_{rot} \quad \dfrac{T}{\sigma \cdot \theta_r} \qquad z_{rot} = 5$

By direct summation of eq. (5.42):

$$n = 0..10$$

$$J_n = 2 \cdot n + 1$$

$$z_{rot} = \frac{1}{\sigma} \cdot \left[\sum_n (2 \cdot J_n + 1) \cdot exp\left[-J_n \cdot (J_n + 1) \cdot \frac{\theta_r}{T} \right] \right] \qquad z_{rot} = 2.585$$

5.45

Equation (5.43) is the high temperature approximation of the rotational partition function. The molecular constants for HF can be found on Table 5.1. For HCl, a linear, unsymmetric molecule, the symmetry number is 1.

$$\theta_r = 30.127 \qquad \sigma = 1 \qquad T = 300 \cdot K$$

$$z_{rot} = \frac{T}{\sigma \cdot \theta_r} \qquad z_{rot} = 9.958$$

The percent population is given by Boltzmann's law (see example 5.24)::

$$J = 0..7 \qquad p_J = \frac{(2 \cdot J + 1) \cdot exp\left[-J \cdot (J + 1) \cdot \frac{\theta_r}{T} \right]}{z_{rot}}$$

J	p_J
0	0.1
1	0.246
2	0.275
3	0.211
4	0.121
5	0.054
6	0.019
7	0.005

5.47

This problem is similar to exmaple 5.27 in the text. We must write out the partition function explicitly. The electronic energy levels are listed in Table 5.2.

$T = 10000 \cdot K$ $i = 0..3$ $R = 8.31451 \cdot \dfrac{joule}{mole \cdot K}$ $kJ = 1000 \cdot joule$

$g_i =$ θ_{e_i}

g_i	θ_{e_i}
4	0
6	27658.7
4	27671.7
6	41492.4

$z_{elec} \quad \displaystyle\sum_i g_i \cdot \exp\left(\dfrac{\theta_{e_i}}{T}\right)$ $p_i \quad \dfrac{g_i \cdot \exp\left(\dfrac{\theta_{e_i}}{T}\right)}{z_{elec}}$

$z_{elec} = 4.723545$

i	p_i
0	0.846822
1	0.079925
2	0.053214
3	0.02004

$U_{elec} \quad R \cdot \displaystyle\sum_i p_i \cdot \theta_{e_i}$ $U_{elec} = 37.537 \cdot \dfrac{kJ}{mole}$

5.49

As in example 5.27, eq. (5.27) can be used to calculate the entropy. For a species in which only the ground state contributes, the electronic entropy will be equal to R ln (g0).

$$R \quad 8.31451 \cdot \frac{joule}{mole \cdot K} \qquad T = 1500 \cdot K \qquad M = 22.98977$$

$$S_{elec} \quad R \cdot \ln(2) \qquad S_{elec} = 5.763 \cdot \frac{joule}{mole \cdot K}$$

The entropy also has a contribution from translational motion:

$$S_{trans} \quad 1.5 \cdot R \cdot \ln(M) \cdot 2.5 \cdot R \cdot \ln(T) \quad 1.15167 \cdot R \qquad S_{trans} = 181.539 \cdot \frac{joule}{mole \cdot K}$$

Since we are dealing with an atomic species, there are not contributions from vibration or rotation. The total entropy is:

$$S \quad S_{elec} \cdot S_{trans} \qquad S = 187.302 \cdot \frac{joule}{mole \cdot K}$$

5.51

Since there is no degeneracy, the partition function is just a sum over all the energy levels:

$$z = \sum_m e^{-m^2 \theta_x / T}$$

At high temperatures, we would expect many levels to be populated, so the sum can be approximated as an integral up to infinity. This integral can be computed using an integral table:

$$z = \int_0^\infty e^{-m^2 \theta_x / T} \, dm \quad = \frac{1}{2} \sqrt{\frac{\pi T}{\theta_x}}$$

Equation (5.26) relates the partition function to the internal energy:

$$U = RT^2 \left(\frac{\partial \ln z}{\partial T} \right)_V + U_o$$

$$\ln z = \ln \left(\frac{1}{2} \right) + \frac{1}{2} \left[\ln \pi + \ln T - \ln \theta_x \right]$$

$$\left(\frac{\partial \ln z}{\partial T} \right)_V = \frac{1}{2T} \qquad\qquad U = RT^2 \left(\frac{1}{2T} \right) + U_o \qquad\qquad U = \frac{RT}{2} + U_o$$

The heat capacity at constant volume is defined as:

$$C_V = \left(\frac{\partial U}{\partial T} \right)_V$$

$$C_V = \left(\frac{\partial U}{\partial T} \right)_V = \frac{R}{2}$$

5.53

The partition function for this system is:

$$z = \sum_m m^2 e^{-m \theta_x / T}$$

At high temperatures, we would expect many levels to be populated, so the sum can be approximated as an integral up to infinity. This integral can be computed using an integral table:

$$z = \int_0^\infty m^2 e^{-m \theta_x / T} \, dm \qquad\qquad \int_0^\infty x^n e^{-ax} dx = \frac{n!}{a^{n+1}}$$

$$z = \int_0^\infty m^2 e^{-m \theta_x / T} \, dm = \frac{2}{a^3}$$

Equation (5.26) relates the partition function to the internal energy:

$$U = RT^2 \left(\frac{\partial \ln z}{\partial T} \right)_V + U_o$$

$$\ln z = \ln 2 + 3 \ln \left(\frac{T}{\theta_x} \right)$$

$$\left(\frac{\partial \ln z}{\partial T} \right)_V = \frac{3}{T} \qquad\qquad U = RT^2 \left(\frac{3}{T} \right) + U_o \qquad\qquad U = 3RT + U_o$$

The heat capacity at constant volume is defined as:

$$C_v = \left(\frac{\partial U}{\partial T} \right)_V$$

$$C_v = \left(\frac{\partial U}{\partial T} \right)_V = 3R$$

SECTION 5.6

5.55
_____

We use the formulas on Table 5.3 for a linear molecule.

$$R \quad 8.31451 \cdot \frac{joule}{mole \cdot K} \qquad T = 298.15 \cdot K \qquad \theta_v = 3323 \cdot K \qquad kJ = 1000 \cdot joule$$

$$U \quad 1.5 \cdot R \cdot T \cdot R \cdot T \cdot \frac{R \cdot \theta_v}{\exp\left(\frac{\theta_v}{T}\right) 1} \qquad U = 6.199 \cdot \frac{kJ}{mole}$$

To compute the enthaply, we use the definition, H = U + PV. For an ideal gas, PV = RT, so we have H = U + RT.

$$H \quad U + R \cdot T \qquad H = 8.677 \cdot \frac{kJ}{mole}$$

5.57
_____

We use the formulas on Table 5.3 for a nonlinear molecule.

$$i : 0..3 \qquad \theta_{v_i} \qquad g_i = \qquad R = 8.31451 \cdot \frac{joule}{mole \cdot K} \qquad T = 500 \cdot K$$

θ_{v_i}	g_i
1151·K	1
374·K	2
1470·K	3
604·K	3

$$u_i \quad \frac{\theta_{v_i}}{T}$$

145

$$C_{vm} = 1.5 \cdot R + 1.5 \cdot R + \sum_i \frac{g_i \cdot R \cdot (u_i)^2 \cdot e^{u_i}}{\left(e^{u_i} - 1\right)^2} \qquad\qquad C_{vm} = 81.087 \cdot \frac{joule}{mole \cdot K}$$

For an ideal gas,

$$C_{pm} = C_{vm} + R \qquad\qquad C_{pm} = 89.402 \cdot \frac{joule}{mole \cdot K}$$

5.59

We use the formulas on Table 5.3 for a nonlinear molecule. Also, because Table 5.1 indicates that NO2 has a ground electronic state of 2, we must add on a term of R ln(g0) to the entropy.

$$R = 8.31451 \cdot \frac{joule}{mole \cdot K} \qquad M = 46.0067 \qquad T = 298.15 \cdot K \qquad \sigma = 2 \qquad i = 0..2$$

$$\theta a \theta b \theta c = 4.24244 \cdot K^3$$

$$\theta_{v_i} =$$

$1953.55 \cdot K$
$1088.85 \cdot K$
$2396.25 \cdot K$

$$z_r = \frac{\sqrt{\pi \cdot T^{1.5}}}{\sigma \cdot \sqrt{\theta a \theta b \theta c}}$$

$$S = 1.5 \cdot R \cdot \ln(M) + 2.5 \cdot R \cdot \ln(T) - 1.15167 \cdot R + 1.5 \cdot R + R \cdot \ln(z_r) \dots$$

$$+ \sum_i \left[\frac{R \cdot \frac{\theta_{v_i}}{T}}{\exp\left(\frac{\theta_{v_i}}{T}\right) - 1} - R \cdot \ln\left(1 - \exp\left(\frac{\theta_{v_i}}{T}\right)\right) \right] + R \cdot \ln(2)$$

$$S = 240.031 \cdot \frac{joule}{mole \cdot K} \qquad\qquad \text{This agrees well with the literature value of 239.9 J/moleK.}$$

SECTION 5.7

5.61

Since we only have a table of energy levels, we must use direct sums to compute the desired quantities. We use eqs. (5.47) and (5.48).

$i := 0 .. 9$ $\qquad$ $T := 600 \cdot K$

$\theta_i :=$
$0 \cdot K$
$155 \cdot K$
$310 \cdot K$
$465 \cdot K$
$620 \cdot K$
$775 \cdot K$
$930 \cdot K$
$1085 \cdot K$
$1240 \cdot K$
$1395 \cdot K$

$R := 8.31451 \cdot \dfrac{joule}{mole \cdot K}$ $\qquad u_i := \dfrac{\theta_i}{T}$ $\qquad z := \sum_i \exp\left(-u_i\right)$ $\qquad p_i := \dfrac{\exp\left(-u_i\right)}{z}$

$u_{av} := \sum_i u_i \cdot p_i$ $\qquad u_{rms} := \sum_i \left(u_i\right)^2 \cdot p_i$

$C_v := R \cdot \left(u_{rms} - u_{av}^2\right)$ $\qquad C_v = 3.365 \cdot \dfrac{joule}{mole \cdot K}$

$\dfrac{C_v}{R} = 0.405$

SECTION 5.8

5.63

Because we are concerned with an atomic species, we will only have contributions from translational motion and electrical energy. The formulas for the free energy function can be found on Table 5.3. Equation (5.53) gives the electronic contribution to the free energy function.

$$i \quad 0..3 \qquad R = 8.31451 \cdot \frac{joule}{mole \cdot K} \qquad M \quad 15.9994$$

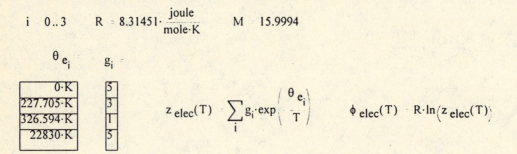

$$\theta_{e_i} \qquad g_i =$$

θ_{e_i}
$0 \cdot K$
$227.705 \cdot K$
$326.594 \cdot K$
$22830 \cdot K$

g_i
5
3
1
5

$$z_{elec}(T) \quad \sum_i g_i \cdot exp\left(\frac{\theta_{e_i}}{T}\right) \qquad \phi_{elec}(T) \quad R \cdot ln\left(z_{elec}(T)\right)$$

$$\phi(T) \quad 1.5 \cdot R \cdot ln(M) \quad 2.5 \cdot R \cdot ln(T) \quad 3.65167 \cdot R \quad \phi_{elec}(T)$$

$$\phi(298.15 \cdot K) = 138.504 \cdot \frac{joule}{mole \cdot K} \qquad \phi(2000 \cdot K) = 180.031 \cdot \frac{joule}{mole \cdot K}$$

$$\phi(500 \cdot K) = 150.062 \cdot \frac{joule}{mole \cdot K}$$

5.65

We use the equations on Table 5.3 for a linear molecule.

$$R = 8.31451 \cdot \frac{joule}{mole \cdot K} \qquad M = 2 \cdot 14.0067 \qquad \theta_v = 3395 \cdot K \qquad \theta_r = 2.89 \cdot K \qquad \sigma = 2$$

$$z_r(T) = \frac{T}{\sigma \cdot \theta_r}$$

$$\phi(T) = 1.5 \cdot R \cdot \ln(M) + 2.5 \cdot R \cdot \ln(T) - 3.65167 \cdot R + R \cdot \ln(z_r(T)) - R \cdot \ln\left(1 - \exp\left(\frac{\theta_v}{T}\right)\right)$$

$$\phi(298.15 \cdot K) = 162.42 \cdot \frac{joule}{mole \cdot K} \qquad \phi(1500 \cdot K) = 210.349 \cdot \frac{joule}{mole \cdot K}$$

5.67

We use the equations on Table 5.3 for a linear molecule. For oxygen, we must include the electronic contribution, eq. (5.53).

$$R = 8.31451 \cdot \frac{joule}{mole \cdot K} \qquad M = 2 \cdot 15.9994 \qquad \theta_v = 2274 \cdot K \qquad \theta_r = 2.08 \cdot K \qquad \sigma = 2$$

$$i = 0 .. 2 \qquad\qquad z_r(T) = \frac{T}{\sigma \cdot \theta_r}$$

$\theta_{e_i} = \qquad g_i =$

θ_{e_i}
$0 \cdot K$
$11392 \cdot K$
$18984 \cdot K$

g_i
3
2
1

$$z_{elec}(T) = \sum_i g_i \cdot \exp\left(-\frac{\theta_{e_i}}{T}\right) \qquad \phi_{elec}(T) = R \cdot \ln(z_{elec}(T))$$

$$\phi(T) \quad 1.5 \cdot R \cdot \ln(M) \quad 2.5 \cdot R \cdot \ln(T) \quad 3.65167 \cdot R \quad R \cdot \ln\, z_r(T) \quad R \cdot \ln\left[1 \quad \exp\left(\frac{\theta_v}{T}\right)\right] \quad \phi_{elec}(T)$$

$$\phi(298.15 \cdot K) = 175.952 \cdot \frac{joule}{mole \cdot K} \qquad \phi(1500 \cdot K) = 225.028 \cdot \frac{joule}{mole \cdot K}$$

$$\phi(1000 \cdot K) = 212.068 \cdot \frac{joule}{mole \cdot K}$$

SECTION 5.9

5.69
_____

At low temperatures, the state of the system is dominated by energy considerations. Thus, at low temperatures, a graph of the order parameter versus the configurational free energy would have a minimum in the middle at J = 0, since in this state there are mainly A-B interactions. This is exactly the opposite of the low-temperature graph in Figure 5.11. At higher temperatures, entropy begins to play a role. The entropy will want to drive the system towards a disordered state, where J = 0, so the shape of the high temperature graph in Figure 5.11 will stay the same. However, a disordered state is already favorable from an energetics point of view. So we will probably not see an order-disorder transition if A-B interactions are higher in energy than A-A and B-B interactions. If this seems confusing, reread section 5.9.

5.71
_____

The order parameter is given by eq. (5.61):

$$T \cdot \ln \left(\frac{1 + J}{1 + J} \right) = \frac{4 \cdot J \cdot (\varepsilon_{AA} + \varepsilon_{BB} - 2 \cdot \varepsilon_{AB})}{k}$$

Solving for T, we obtain:

$$J = 0.5 \qquad \theta_{AA} = 100 \cdot K \qquad \theta_{BB} = 200 \cdot K \qquad \theta_{AB} = 50 \cdot K$$

$$T = \frac{4 \cdot J \cdot (\theta_{AA} + \theta_{BB} - 2 \cdot \theta_{AB})}{\ln \left(\frac{1 + J}{1 - J} \right)} \qquad T = 364 \cdot K$$

CHAPTER 6 *Chemical Reactions*
SECTION 6.2

6.1

Formation reactions of compounds are written using the component elements in their most stabl forms at 1 bar and a given (usually room) temperature.

$$C(s,graphite)+2H_2(g)+\frac{1}{2}O_2(g)=CH_3OH(liq)$$

$$C(s,graphite)+2H_2(g)+\frac{1}{2}O_2(g)=CH_3OH(g)$$

$$H_2(g)+S(s,rhombic)+2O_2(g)=H_2SO_4(liq)$$

$$2Na(s)+C(s,graphite)+\frac{3}{2}O_2(g)=Na_2CO_3(s)$$

$$Na(s)+I(s)+\frac{3}{2}O_2(g)=NaIO_3(s)$$

6.3

Combustion products are usually a combination of CO_2, SO_2, and H_2O (liq).

$$C_{12}H_{22}O_{11}+12O_2=12CO_2+11H_2O$$

$$C_4H_4S+6O_2=4CO_2+2H_2O+SO_2$$

6.5

The standard enthalpy change is the heat of reaction (at standard conditions), and thus can be calculated from eq. 6.4 in the text. Stoichiometric coefficients are positive for products and negative for reactants. Heats of formation are given in Table 6.1 at 298.15 K.

$$\Delta fH_{N2H4gas} = 95.40 \cdot \frac{kJ}{mole} \quad \Delta fH_{O2} = 0 \cdot \frac{kJ}{mole} \qquad \qquad kJ \equiv 1000 \cdot joule$$

$$\Delta fH_{NO2} = 33.18 \cdot \frac{kJ}{mole} \quad \Delta fH_{H2Ogas} = -241.818 \cdot \frac{kJ}{mole}$$

$$\Delta rxnH = 2 \cdot \Delta fH_{NO2} - 2 \cdot \Delta fH_{H2Ogas} - \Delta fH_{N2H4gas} - 3 \cdot \Delta fH_{O2}$$

$$\Delta rxnH = -512.68 \cdot \frac{kJ}{mole}$$

6.7

The formation reaction of TiN can be obtained by reversing the second equation and adding it to the first equation. The heat of the formation reaction can be obtained using Hess's law of heat summation (remembering to reverse the sign of the enthalpy change for the second equation because we flipped it). The result is

$$Ti(s) + \frac{1}{2}N_2 = TiN \qquad \Delta_f H^\theta = -333.0 \frac{kJ}{mole}$$

6.9

The only components that can undergo combustion are CO and H_2. The heats of combustion of these do not appear in Table 6.2, but they can be calculated using heats of formation given in Table 6.1. (Heats of formation of elements in their most stable form at standard conditions are defined to be zero.)

$$CO(g) + \frac{1}{2} \cdot O_2(g) = CO_2(g)$$

$$kJ \equiv 1000 \cdot joule$$

$$MJ \equiv 10^6 \cdot joule$$

$$\Delta fH_{CO2} = -393.509 \cdot \frac{kJ}{mole} \qquad \Delta fH_{CO} = -110.525 \cdot \frac{kJ}{mole}$$

$$\Delta cH_{CO} = \Delta fH_{CO2} - \Delta fH_{CO}$$

$$H_2(g) + \frac{1}{2} \cdot O_2(g) = H2O(liq) \qquad \text{Remember that } H_2O(liq) \text{ is used for defining}$$
combustion reactions.

$$\Delta fH_{H2O} = -285.830 \cdot \frac{kJ}{mole}$$

$$\Delta cH_{H2} - \Delta fH_{H2O}$$

For 1 kg of water gas,

$$moles_{CO} = \frac{0.40 \cdot (1 \cdot kg)}{28.0104 \cdot \frac{gm}{mole}} \qquad moles_{H2} = \frac{0.52 \cdot (1 \cdot kg)}{2.01588 \cdot \frac{gm}{mole}}$$

$$\Delta cH_{watergas} = moles_{CO} \cdot \Delta cH_{CO} + moles_{H2} \cdot \Delta cH_{H2}$$

$$\Delta cH_{watergas} = -78 \cdot MJ$$

6.11

Table 6.2 gives the heat of combustion of thiophene

$$C_4H_4S + 6O_2(g) = 4CO_2(g) + 2H_2O(liq) + SO_2(g) \qquad \Delta_c H^\theta = -2805 \frac{kJ}{mole}$$

This can be converted to the formation reaction of thiophene using the following formation reactio (Table 6.1)

$$C(s,graphite)+O_2(g)=CO_2(g) \qquad \Delta_f H^\theta = -393.509\frac{kJ}{mole} \qquad (2$$

$$H_2(g)+\frac{1}{2}O_2(g)=H_2O(liq) \qquad \Delta_f H^\theta = -285.830\frac{kJ}{mole} \qquad (3$$

$$S(s,rhombic)+O_2(g)=SO_2(g) \qquad \Delta_f H^\theta = -296.830\frac{kJ}{mole} \qquad (4$$

The formation reaction of thiophene can be obtained by adding 4 times eq.(2), 2 times eq.(3), eq.(4) and the reverse of eq.(1). The result is

$$4C(s,graphite)+2H_2(g)+S(s,rhombic)=C_4H_4S \qquad \Delta_f H^\theta = 362.474\frac{kJ}{mole}$$

6.13

Eq. 6.8 in the text gives a formula to calculate the temperature dependence of the heat of reaction If $\Delta_{rxn}C_p$ (defined by eq. 6.6) is constant, this equation simplifies to

$$\Delta_{rxn}H(T_2)=\Delta_{rxn}H(T_1)+\Delta_{rxn}C_p(T_2-T_1)$$

It will be most convenient to take $T_1 = 298.15$ K because that is where we can most easily calculate the heat of reaction (using heats of formation from Table 6.1).

$$\Delta_{rxn}H(298.15)=\Delta_f H^\theta(CO,g)+\Delta_f H^\theta(Cl_2,g)-\Delta_f H^\theta(COCl_2,g)$$

$$\Delta_{rxn}H(298.15)=-110.525+0-(-218.8)=108.275\frac{kJ}{mole}$$

Similarly, we can use the heat capacities in Table 6.1 to determine $\Delta_{rxn}C_p$

$$\Delta_{rxn}C_p=C_{pm}(CO,g)+C_{pm}(Cl_2,g)-C_{pm}(COCl_2,g)$$

$$\Delta_{rxn}C_p=29.142+33.907-57.66=5.389\frac{J}{mole\cdot K}$$

155

Student's Solutions Manual

Thus

$$\Delta_{rxn}H^\theta(T) = 108275 + (5.389)(T - 298.15)$$

$$\Delta_{rxn}H^\theta(T) = (106668 + 5.389T)\frac{J}{mole}$$

6.15

We can use eq. 6.8 in the text to derive an accurate formula for the enthalpy of the reaction. It is easiest to take $T_1 = 298.15$ K because then we can use Table 6.1 to calculate the heat of reaction at T_1.

$$\Delta rxnH_{298.15} = -241.818 \cdot \frac{kJ}{mole} \qquad kJ \equiv 1000 \cdot joule$$

Table 2.2 gives the heat capacities needed to calculate $\Delta_{rxn}C_p$.

$$Cpm_{H2O}(T) = \left(26.06 + 17.7 \cdot 10^{-3} \cdot T - 2.63 \cdot 10^{-6} \cdot T^2 + \frac{2.20 \cdot 10^5}{T^2}\right) \cdot \frac{joule}{mole \cdot K}$$

$$Cpm_{H2}(T) = \left(26.36 + 4.35 \cdot 10^{-3} \cdot T - 0.245 \cdot 10^{-6} \cdot T^2 + \frac{1.15 \cdot 10^5}{T^2}\right) \cdot \frac{joule}{mole \cdot K}$$

$$Cpm_{O2}(T) = \left(29.30 + 6.14 \cdot 10^{-3} \cdot T - 0.88 \cdot 10^{-6} \cdot T^2 - \frac{1.59 \cdot 10^5}{T^2}\right) \cdot \frac{joule}{mole \cdot K}$$

$$\Delta rxnCp(T) = Cpm_{H2O}(T) - Cpm_{H2}(T) - \frac{1}{2} \cdot Cpm_{O2}(T)$$

$$\Delta rxnH(T) = \Delta rxnH_{298.15} + \int_{298.15}^{T} \Delta rxnCp(T) \, dT$$

$$\Delta rxnH(T) = \left(-2.37182 \cdot 10^5 - 14.95 \cdot T + 0.00514 \cdot T^2 - 6.483 \cdot 10^{-7} \cdot T^3 - \frac{1.845 \cdot 10^5}{T}\right) \cdot \frac{joule}{mole}$$

SECTION 6.3

6.17

The adiabatic flame temperature is the temperature at which the reaction products have absorbed all of the heat generated by the reaction. However, in these calculations, the "reaction products" include any species that are present after the reaction has taken place, including inerts and unreacted reactants as well as the "new" products. The adiabatic flame temperature can be found by finding the temperature that satisfies eq. 6.9.

The overall reaction can be written as

$$C(s, graphite) + O_2(g) + 4 \cdot N_2(g) = CO_2(g) + 4 \cdot N_2(g)$$

$$\Delta rxnH_\theta = -393509 \text{ joule/mole} \qquad \text{from Table 6.1}$$

Heat capacities are given by Table 2.2 as follows

$$Cpm_{CO2}(T) = 41.58 + 15.6 \cdot 10^{-3} \cdot T - 2.95 \cdot 10^{-6} \cdot T^2 - \frac{7.97 \cdot 10^5}{T^2}$$

$$Cpm_{N2}(T) = 25.79 + 8.09 \cdot 10^{-3} \cdot T - 1.46 \cdot 10^{-6} \cdot T^2 + \frac{0.88 \cdot 10^5}{T^2}$$

$$Cp_{prod}(T) = Cpm_{CO2}(T) + 4 \cdot Cpm_{N2}(T)$$

Solving eq. 6.9:

$$T_{guess} = 2000$$

$$T_{flame} = root\left(\Delta rxnH_\theta + \int_{298.15}^{T_{guess}} Cp_{prod}(T) \, dT, T_{guess}\right)$$

$$T_{flame} = 2376 \text{Kelvin}$$

6.19

The adiabatic flame temperature is the temperature at which the reaction products have absorbed all of the heat generated by the reaction. However, in these calculations, the "reaction products" include any species that are present after the reaction has taken place, including inerts and unreacted reactants as well as the "new" products. The adiabatic flame temperature can be found by finding the temperature that satisfies eq. 6.9.

The overall reaction can be written as

$$CH_3 \cdot OH(g) + 1.5 \cdot O_2(g) + 6 \cdot N_2(g) = CO_2(g) + 2 \cdot H_2O(g) + 6 \cdot N_2(g)$$

$$\Delta rxnH_\theta = (-393509) + 2 \cdot (-241818) - (-200660) \quad \text{joule/mole} \qquad \text{from Table 6.1}$$

Heat capacities are given by Table 2.2 as follows

$$Cpm_{CO2}(T) = 41.58 + 15.6 \cdot 10^{-3} \cdot T - 2.95 \cdot 10^{-6} \cdot T^2 - \frac{7.97 \cdot 10^5}{T^2}$$

$$Cpm_{H2O}(T) = 26.06 + 17.7 \cdot 10^{-3} \cdot T - 2.63 \cdot 10^{-6} \cdot T^2 + \frac{2.20 \cdot 10^5}{T^2}$$

$$Cpm_{N2}(T) = 25.79 + 8.09 \cdot 10^{-3} \cdot T - 1.46 \cdot 10^{-6} \cdot T^2 + \frac{0.88 \cdot 10^5}{T^2}$$

$$Cp_{prod}(T) = Cpm_{CO2}(T) + 2 \cdot Cpm_{H2O}(T) + 6 \cdot Cpm_{N2}(T)$$

Solving eq. 6.9:

$$T_{guess} = 2000$$

$$T_{flame} = root\left(\Delta rxnH_\theta + \int_{298.15}^{T_{guess}} Cp_{prod}(T) \, dT, T_{guess}\right)$$

$$T_{flame} = 2270 \text{Kelvin}$$

6.21

Again, eq. 6.9 in the text is the basis for the solution. From Table 6.1 we have the following data

$$\Delta fH_{NO} = 90250 \qquad \Delta fH_{SO3} = -395720 \qquad \Delta fH_{NO2} = 33180 \qquad \Delta fH_{SO2} = -296830$$

All are in joule/mole. Thus, the standard heat of reaction is

$$\Delta rxnH_{\theta} = \Delta fH_{NO} - \Delta fH_{SO3} - \Delta fH_{NO2} - \Delta fH_{SO2}$$

$$\Delta rxnH_{\theta} = -4.182 \cdot 10^4$$

Heat capacities are given by Table 6.1 as follows

$$Cpm_{NO} = 29.844 \qquad Cpm_{SO3} = 50.67 \qquad \text{Both are in joule/mole/K}$$

$$Cp_{prod}(T) = Cpm_{NO} - Cpm_{SO3}$$

Solving eq. 6.9:

$$T_{guess} = 1000$$

$$T_{max} = root\left(\Delta rxnH_{\theta} - \int_{373}^{T_{guess}} Cp_{prod}(T)\, dT, T_{guess} \right)$$

$$T_{max} = 892 \quad \text{Kelvin}$$

SECTION 6.4

6.23

Given: $P = 1 \cdot atm$ $P_\theta = 10^5 \cdot Pa$

$X_{CS2} = 0.85$ $X_{S2} = 1 - X_{CS2}$ There is no graphite in the gas phase.

The equilibrium constant is defined by eq. 6.13 and 6.15 in the text.

$$K_a = \frac{a_{CS2}}{a_C \cdot a_{S2}}$$

The activity of any solid can be approximated as one at reasonable pressures. If we assume ideal gas behavior, then the activity of each gas is its partial pressure divided by the standard pressure. Dalton's law can be used to express the partial pressures as the mole fraction times the total pressure.

$$a_c = 1$$

$$K_a = \frac{\dfrac{X_{CS2} \cdot P}{P_\theta}}{a_c \cdot \left(\dfrac{X_{S2} \cdot P}{P_\theta}\right)} \qquad K_a = \frac{X_{CS2}}{X_{S2}} \qquad K_a = 5.67$$

We can see that with the ideal gas assumption and Dalton's law of partial pressures, the resulting expression for K_a does not depend on the reaction pressure. Therefore, the percentage of CS_2 at 10 atm will still be 85%. The reason is there is no change in the number of moles of gas during the reaction. You may recall from Le Chatelier's Principle that increasing the pressure drives equilibrium toward the direction of fewer moles of gas. Since the number of moles of gas does not change, the equilibrium does not change either (with the assumptions made).

6.25

(a) Given: $\quad P = 0.5523 \cdot bar \quad P_\theta = 1 \cdot bar \qquad bar = 10^5 \cdot Pa$

$X_{H2O} = 0.5 \qquad X_{CO2} = 0.5 \qquad$ Water vapor and carbon dioxide are formed in equimolar amounts.

The equilibrium constant is defined by eq. 6.13 and 6.15 in the text.

$$K_a = \frac{a_{Na2CO3} \cdot a_{H2O} \cdot a_{CO2}}{a_{NaHCO3}^2}$$

The activity of any solid can be approximated as one at reasonable pressures. If we assume ideal gas behavior, then the activity of each gas is its partial pressure divided by the standard pressure. Dalton's law can be used to express the partial pressures as the mole fraction times the total pressure.

$$a_{Na2CO3} = 1 \qquad a_{NaHCO3} = 1$$

$$K_a = \frac{a_{Na2CO3} \cdot \left(\frac{X_{H2O} \cdot P}{P_\theta}\right) \cdot \left(\frac{X_{CO2} \cdot P}{P_\theta}\right)}{a_{NaHCO3}^2} \qquad K_a = 0.076259$$

(b) $\quad P_{CO2}(\xi) = 1.6500 \cdot bar + \xi \qquad P_{H2O}(\xi) = \xi$

The equilibrium constant can be expressed in terms of the partial pressures

$$K_a = \frac{P_{CO2} \cdot P_{H2O}}{P_\theta^2}$$

$\xi_{guess} = 0.5 \cdot bar$

$$\xi = root\left(\frac{P_{CO2}(\xi_{guess}) \cdot P_{H2O}(\xi_{guess})}{P_\theta^2} - K_a, \xi_{guess}\right)$$

$P_{H2O}(\xi) = 0.0450 \cdot bar$

$P_{total} = P_{CO2}(\xi) + P_{H2O}(\xi)$

$P_{total} = 1.7400 \cdot bar$

SECTION 6.5

6.27
___

From Table 6.1 we can calculate the free energy change for the reaction, $CCl_4(liq) = CCl_4(g)$

$$\Delta fG\ CCl4liq = -65.21 \cdot \frac{kJ}{mole} \qquad \Delta fG\ CCl4gas = -60.59 \cdot \frac{kJ}{mole}$$

$$kJ = 1000 \cdot joule$$

$$R = 8.31451 \cdot \frac{joule}{mole \cdot K}$$

$$\Delta rxnG_\theta = \Delta fG\ CCl4gas - \Delta fG\ CCl4liq$$

$$T = 298.15 \cdot K$$

We can use eq. 6.15 in the text to calculate the equilibrium constant.

$$K_a = exp\left(\frac{\Delta rxnG_\theta}{-R \cdot T}\right) \qquad K_a = 0.155$$

In this reaction, K_a is equal to the activity of the vapor divided by the activity of the liquid. Because liquids are largely incompressible, their activities can be assumed to be one at reasonable pressures. The activity of the vapor can be estimated as the vapor pressure divided by the standard pressure (assuming an ideal gas).

$$K_a = \frac{P_{vap}}{P_\theta} \qquad P_{vap} = K_a \cdot P_\theta \qquad\qquad\qquad P_\theta = 10^5 \cdot Pa$$

$$P_{vap} = 116.3 \cdot torr$$

The methods of chapter 4 can also be used to estimate the vapor pressure. The data from Table 4.2 give the normal boiling point of carbon tetrachloride and the heat of vaporization at the normal boiling point.

$$T_1 = 349.9 \cdot K \qquad P_1 = 1 \cdot atm \qquad \Delta vH = 30.0 \cdot \frac{kJ}{mole}$$

The Clausius-Clapeyron equation (eq. 4.20) gives the dependence of the vapor pressure-temperature relation on the heat of vaporization. For a constant heat of vaporization, this equation can be integrated to give eq. 4.22. This will allow us to predict the vapor pressure of carbon tetrachloride at 25 C.

$$\ln\left(\frac{P_2}{P_1}\right) = \frac{\Delta vH}{R}\cdot\left(\frac{1}{T_2} - \frac{1}{T_1}\right)$$

$$T_2 = 298.15\cdot K$$

$$P_2 = \exp\left[\frac{-(\Delta vH\cdot T_1 - \Delta vH\cdot T_2)}{R\cdot(T_2\cdot T_1)}\right]\cdot P_1$$

$$P_2 = 126.9\cdot torr$$

In order to tell which answer is more accurate, we have to look at the assumptions that accompany both solutions. Using the methods of chapter 6, we assumed that the liquid activity was one and that the vapor could be treated as an ideal gas. The first is a very good assumption, and the second is reasonable. Using the methods of chapter 4, we assumed that the heat of vaporization did not vary with temperature over approximately a 50 K range. This assumption turns out to cause the greatest error. Therefore, the first answer is most accurate. The actual vapor pressure is about 115 torr.

6.31

Eq. 6.25 can be used to calculate the equilibrium constant at 1000 K. In order to use this it is necessary the standard free energy of reaction and the standard enthalpy of reaction (both at 298.15 K) using data from Table 6.1

$$\Delta fG_{ClO} = 98.11\cdot\frac{kJ}{mole} \qquad \Delta fG_{CO2} = -394.359\cdot\frac{kJ}{mole} \qquad \Delta fG_{CCl4} = -60.59\cdot\frac{kJ}{mole} \qquad \Delta fG_{O2} = 0\cdot\frac{kJ}{mole}$$

$$\Delta rxnG_{ref} = 4\cdot\Delta fG_{ClO} + \Delta fG_{CO2} - \Delta fG_{CCl4} - 3\cdot\Delta fG_{O2}$$

$$\Delta fH_{ClO} = 101.84\cdot\frac{kJ}{mole} \qquad \Delta fH_{CO2} = -393.509\cdot\frac{kJ}{mole} \qquad \Delta fH_{CCl4} = -102.9\cdot\frac{kJ}{mole} \qquad \Delta fH_{O2} = 0\cdot\frac{kJ}{mole}$$

$$\Delta rxnH_{ref} = 4\cdot\Delta fH_{ClO} + \Delta fH_{CO2} - \Delta fH_{CCl4} - 3\cdot\Delta fH_{O2}$$

We also need to consider the temperature dependence of the enthalpy

$$Cpm_{ClO} = 31.46\cdot\frac{joule}{K\cdot mole} \qquad Cpm_{CO2} = 37.11\cdot\frac{joule}{K\cdot mole}$$

$$Cpm_{CCl4} = 83.30\cdot\frac{joule}{K\cdot mole} \qquad Cpm_{O2} = 29.355\cdot\frac{joule}{K\cdot mole}$$

$$\Delta rxnCp = 4\cdot Cpm_{ClO} + Cpm_{CO2} - Cpm_{CCl4} - 3\cdot Cpm_{O2}$$

$$kJ = 1000\cdot joule$$

$$R = 8.31451\cdot\frac{joule}{mole\cdot K}$$

$$T_{ref} = 298.15\cdot K$$

$$\Delta rxnH(T) = \Delta rxnH_{ref} + \int_{T_{ref}}^{T} \Delta rxnCp\, dT$$

$$\Delta rxnH(1000 \cdot K) = 111 \cdot \frac{kJ}{mole}$$

Now we can use eq. 6.25 to calculate the equilibrium constant at 1000 K.

$$K_a(T) = \exp\left(\frac{-\Delta rxnG_{ref}}{R \cdot T_{ref}} + \int_{T_{ref}}^{T} \frac{\Delta rxnH(T)}{R \cdot T^2}\, dT \right)$$

$$K_a(1000 \cdot K) = 7.1 \cdot 10^3$$

6.29

We will use the free energy function based on 298.15 K because Table 6.5 contains the data needed for this problem. The equilibrium constant can then be determined by solving eq. 6.21. Necessary data are

$$\phi_{TiN} = 68.3 \cdot \frac{joule}{K} \qquad \phi_{Ti} = 54.2 \cdot \frac{joule}{K} \qquad \phi_{N2} = 216.2 \cdot \frac{joule}{K}$$

$$\Delta\phi = \phi_{TiN} - \phi_{Ti} - \frac{1}{2} \cdot \phi_{N2} \qquad\qquad R = 8.31451 \cdot \frac{joule}{mole \cdot K} \qquad kJ = 1000 \cdot joule$$

$$\Delta fH_{TiN} = -338 \cdot \frac{kJ}{mole} \qquad \Delta fH_{Ti} = 0 \cdot \frac{kJ}{mole} \qquad \Delta fH_{N2} = 0 \cdot \frac{kJ}{mole}$$

$$\Delta H_\theta = \Delta fH_{TiN} - \Delta fH_{Ti} - \frac{1}{2} \cdot \Delta fH_{N2}$$

$$T = 1500 \cdot K$$

$$K_a = \exp\left(\frac{\Delta\phi}{R} - \frac{\Delta H_\theta}{R \cdot T} \right) \qquad \text{eq. 6.21}$$

$$K_a = 7.2 \cdot 10^6$$

6.33

The equilibrium constant for this reaction can be expressed as follows if we assume the ideal gas law applies and the activity of solids is one.

$$K_a = \left(\frac{P}{P_\theta}\right)^{10} \qquad P = 0.249 \cdot bar \qquad P_\theta = 1 \cdot bar \qquad bar \equiv 10^5 \cdot Pa \qquad kJ \equiv 1000 \cdot joule$$

$$K_a = 9.16 \cdot 10^{-7}$$

The standard free energy of reaction can be calculated using eq. 6.15

$$T = 298.15 \cdot K \qquad\qquad R = 8.31451 \cdot \frac{joule}{mole \cdot K}$$

$$\Delta rxnG_\theta = -R \cdot T \cdot \ln(K_a) \qquad \Delta rxnG_\theta = 34.5 \cdot \frac{kJ}{mole}$$

6.35

The equilibrium constant can then be determined by solving eq. 6.21. Data at 773 K can be interpolated from the data given by assuming a linear relationship.

$$\phi_{NH3} = 49.2456 \cdot \frac{cal}{K} \qquad\qquad \phi_{N2} = 48.14856 \cdot \frac{cal}{K} \qquad\qquad \phi_{H2} = 33.56326 \cdot \frac{cal}{K}$$

$$\Delta\phi = \phi_{NH3} - \frac{1}{2} \cdot \phi_{N2} - \frac{3}{2} \cdot \phi_{H2}$$

$$\Delta fH_{NH3} = -10.97 \cdot \frac{kcal}{mole} \qquad \Delta fH_{N2} = 0 \cdot \frac{kcal}{mole} \qquad \Delta fH_{H2} = 0 \cdot \frac{kcal}{mole}$$

$$\Delta H_\theta = \Delta fH_{NH3} - \frac{1}{2} \cdot \Delta fH_{N2} - \frac{3}{2} \cdot \Delta fH_{H2}$$

$$T = 773 \cdot K \qquad\qquad\qquad R = 8.31451 \cdot \frac{joule}{mole \cdot K} \qquad kcal \equiv 1000 \cdot cal$$

$$K_a = \exp\left(\frac{\Delta\phi}{R} - \frac{\Delta H_\theta}{R \cdot T}\right) \qquad \text{eq. 6.21}$$

$$K_a = 4.0 \cdot 10^{-3}$$

6.37

(a) In order to tell which form is more stable at 25 C and 1 bar, all we have to do is examine the standard free energies of formation in Table 6.1

$$\Delta fG_{PbOred} = -188.93 \cdot \frac{kJ}{mole} \qquad \Delta fG_{PbOyel} = -187.89 \cdot \frac{kJ}{mole} \qquad kJ \equiv 1000 \cdot joule$$

Since red PbO has a lower standard free energy at 25 C, it is the thermodynamically more stable form under these conditions.

(b) Consider the reaction PbO(red) = PbO(yel). The standard free energy of reaction is just the difference between the standard free energies of formation. The temperature dependence of $\Delta_{rxn}G_\theta$ can be determined assuming $\Delta_{rxn}H_\theta$ and $\Delta_{rxn}S_\theta$ are constant. The temperature at which $\Delta_{rxn}G_\theta$ equals zero represents the point at which the standard free energies of formation are equal, and the temperature above which the yellow form will presumably be more stable.

$$\Delta fH_{PbOred} = -218.99 \cdot \frac{kJ}{mole} \qquad \Delta fH_{PbOyel} = -217.32 \cdot \frac{kJ}{mole} \qquad \Delta rxnH_\theta = \Delta fH_{PbOyel} - \Delta fH_{PbOred}$$

$$S_{PbOred} = 66.5 \cdot \frac{joule}{mole \cdot K} \qquad S_{PbOyel} = 68.70 \cdot \frac{joule}{mole \cdot K} \qquad \Delta rxnS_\theta = S_{PbOyel} - S_{PbOred}$$

$$\Delta rxnG_\theta(T) = \Delta rxnH_\theta - T \cdot \Delta rxnS_\theta$$

$$T_{guess} = 500 \cdot K$$

$$T = root\left(\Delta rxnG_\theta(T_{guess}), T_{guess}\right)$$

$$T = 759 \cdot K$$

Above this temperature, the yellow form of PbO is predicted to be more stable.

Thus far we have only dealt with the standard free energy of reaction, which does not by itself predict whether a reaction is spontaneous or not. Spontaneity is determined by $\Delta_{rxn}G$, not $\Delta_{rxn}G_\theta$. Eq. 6.14 in the text relates the two quantities. In this example because we are dealing only with solids, all activities will be equal to one and the distinction is not significant.

6.39

Solving eq. 6.24 in the text is the easiest way to estimate $\Delta rxnH_\theta$.

$$\ln(K_2) = \ln(K_1) - \frac{\Delta rxnH_\theta}{R} \cdot \left(\frac{1}{T_2} - \frac{1}{T_1}\right)$$

$$R = 8.31451 \cdot \frac{joule}{mole \cdot K}$$

$$kJ = 1000 \cdot joule$$

$$K_1 = 5.85 \cdot 10^{-4} \qquad T_1 = 1000 \cdot K$$

$$K_2 = 2.04 \cdot 10^{-5} \qquad T_2 = 2000 \cdot K$$

$$\Delta rxnH_\theta = \frac{(\ln(K_2) - \ln(K_1))}{\frac{1}{(R \cdot T_2)} - \frac{1}{(R \cdot T_1)}}$$

$$\Delta rxnH_\theta = -55.8 \cdot \frac{kJ}{mole}$$

This value represents the mean reaction enthalpy over the temperature interval 1000 K to 2000 K. Data in Table 6.1 are for 298.15 K. The value given in Table 6.1 is -46.11 kJ/mole. Remember that $\Delta rxnH_\theta$ is a function of temperature through eq. 6.8.

6.41

Eq. 6.23 tells us that the slope of $\ln(K_a)$ versus $1/T$ is equal to $-\Delta rxnH/R$. For this reaction $K_p = K_a$ since the activity of solids is one and there is no net change in the total moles of gas.

points = 5   i = 0 .. points − 1

$T_i =$	$Ka_i =$
831	0.0154
857	0.0170
878	0.0182
906	0.0202
918	0.0215

$$x_i = \frac{1}{T_i} \qquad y_i = \ln(Ka_i)$$

We are now ready to perform the regression.

$m = slope(x,y)$ $m = -2.861 \cdot 10^3$ m is the slope of the ln K_a vs. 1/T graph

$r = corr(x,y)$ $r = -0.997168$ r is the correlation factor related to the goodness of fit

$b = intercept(x,y)$ $b = -0.737$ b is the intercept of the ln K_a vs 1/T graph

The standard deviation of the slope may be calculated using eq. AI.25 of the Appendix.

$$\sigma_m = \frac{m}{r} \cdot \sqrt{\frac{1-r^2}{points - 2}}$$ $\sigma_m = 124.56$ $R = 8.31451$

$$\Delta rxnH_\theta = -m \cdot R \cdot \frac{joule}{mole}$$ $\Delta rxnH_\theta = 23.786 \cdot \frac{kJ}{mole}$ $kJ = 1000 \cdot joule$

$$\sigma_{\Delta rxnH\theta} = \sigma_m \cdot R \cdot \frac{joule}{mole}$$ $\sigma_{\Delta rxnH\theta} = 1.036 \cdot \frac{kJ}{mole}$

The confidence interval is defined by eq. AI.27 of the Appendix. In this problem there are 5 points and 2 parameters, which means there are 3 degrees of freedom. For 90% confidence, $t_c = 2.35$. Therefore, the confidence interval, λ, is

$t_c = 2.35$ $\lambda_{90\%} = t_c \cdot \sigma_{\Delta rxnH\theta}$ $\lambda_{90\%} = 2.434 \cdot \frac{kJ}{mole}$

It may be helpful to see a plot of the data and regressed fit on a ln K_a vs 1/T graph.

$lnKa(x) = m \cdot x + b$

$inverseT = 0.00100, 0.00101 .. 0.00130$

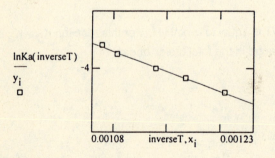

$$\frac{lnKa(inverseT)}{y_i}$$

This graph shows the fit of the ln K_a vs 1/T data. The slope of this line is $-\Delta rxnH_\theta/R$ according to eq. 6.23

Summary

$\Delta rxnH_\theta = 24 \cdot \frac{kJ}{mole}$ $\lambda_{90\%} = 2 \cdot \frac{kJ}{mole}$ $\sigma_{\Delta rxnH\theta} = 1 \cdot \frac{kJ}{mole}$

SECTION 6.6

6.43

Since the given equation of state is explicit in volume, we can use eq. 6.29 to calculate the fugacity coefficient, and hence the fugacity.

$$\ln\phi = \frac{1}{RT}\int_0^P\left(V_m - \frac{RT}{P}\right)dP$$

Substituting the given equation of state for V_m gives

$$\ln\phi = \frac{1}{RT}\int_0^P(\beta + \gamma P)dP$$

$$\ln\phi = \frac{1}{RT}\left(\beta P + \frac{\gamma P^2}{2}\right)\Big|_0^P$$

$$\phi = \exp\left(\frac{\beta P}{RT} + \frac{\gamma P^2}{2RT}\right)$$

$$f = P\exp\left(\frac{\beta P}{RT} + \frac{\gamma P^2}{2RT}\right)$$

6.45

Eq. 6.30 in the text can be used to calculate the fugacity coefficient. A cubic spline will be used to fit the data.

$i = 0 .. 20$

$p_i =$	$z_i =$
$1 \cdot atm$	0.994
$10 \cdot atm$	0.937
$20 \cdot atm$	0.8683
$30 \cdot atm$	0.7928
$40 \cdot atm$	0.7034
$50 \cdot atm$	0.5936
$60 \cdot atm$	0.4515
$80 \cdot atm$	0.3429
$100 \cdot atm$	0.3767
$120 \cdot atm$	0.4259
$140 \cdot atm$	0.4753
$160 \cdot atm$	0.5252
$180 \cdot atm$	0.5752
$200 \cdot atm$	0.6246
$250 \cdot atm$	0.7468
$300 \cdot atm$	0.8663
$400 \cdot atm$	1.098
$500 \cdot atm$	1.3236
$600 \cdot atm$	1.5409
$800 \cdot atm$	1.9626
$1000 \cdot atm$	2.3684

$fit = cspline(p, z)$ $atm = 101325$

$Z(P) = interp(fit, p, z, P)$

$P = 0 \cdot atm, 1 \cdot atm .. 1000 \cdot atm$

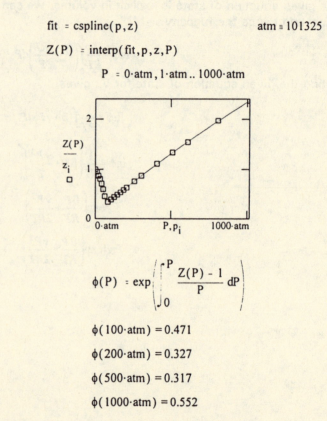

$$\phi(P) = exp\left(\int_0^P \frac{Z(P) - 1}{P} \, dP \right)$$

$\phi(100 \cdot atm) = 0.471$

$\phi(200 \cdot atm) = 0.327$

$\phi(500 \cdot atm) = 0.317$

$\phi(1000 \cdot atm) = 0.552$

6.47

Eq. 6.30 in the text can be used to calculate the fugacity coefficient. A cubic spline will be used to fit the data.

$i = 0 .. 24$

$p_i =$	$z_i =$
$0.1 \cdot MPa$	0.9943
$1 \cdot MPa$	0.9410
$2 \cdot MPa$	0.8768
$3 \cdot MPa$	0.8055
$4 \cdot MPa$	0.7241
$5 \cdot MPa$	0.6266
$6 \cdot MPa$	0.5013
$7 \cdot MPa$	0.3701
$8 \cdot MPa$	0.3338
$9 \cdot MPa$	0.3398
$10 \cdot MPa$	0.3560
$12 \cdot MPa$	0.3974
$14 \cdot MPa$	0.4424
$16 \cdot MPa$	0.4884
$20 \cdot MPa$	0.5803
$24 \cdot MPa$	0.6709
$28 \cdot MPa$	0.7600
$30 \cdot MPa$	0.8040
$35 \cdot MPa$	0.9123
$40 \cdot MPa$	1.0186
$45 \cdot MPa$	1.1231
$50 \cdot MPa$	1.226
$60 \cdot MPa$	1.4278
$70 \cdot MPa$	1.6251
$80 \cdot MPa$	1.8188

$fit = cspline(p, z)$ $MPa \equiv 10^6$

$Z(P) := interp(fit, p, z, P)$

$P = 0.1 \cdot MPa, 0.2 \cdot MPa .. 82 \cdot MPa$

$$\phi(P) = \exp\left(\int_0^P \frac{Z(P) - 1}{P} \, dP \right)$$

$\phi(5 \cdot MPa) = 0.726$

$\phi(10 \cdot MPa) = 0.485$

$\phi(20 \cdot MPa) = 0.329$

$\phi(40 \cdot MPa) = 0.270$

$\phi(80 \cdot MPa) = 0.394$

6.49

(a) For an incompressible solid, we can use eq. 6.31 to calculate the activity

$P = 1000 \cdot atm$ $P_\theta = 1 \cdot bar$ $R = 8.31451 \cdot \dfrac{joule}{mole \cdot K}$ $bar = 10^5 \cdot Pa$

$T = 298.15 \cdot K$ $V_m = 7.1 \cdot cm^3$

$\Gamma = \exp\left[\dfrac{(P - P_\theta) \cdot V_m}{R \cdot T}\right]$ $\Gamma = 1.336$

(b) If we no longer assume the molar volume is constant, then we have to integrate the following equation

$R \cdot T \cdot d(\ln(\Gamma)) = V_m(P) \cdot dP$

The dependence of the molar volume on pressure is given by eq. 1.51

$V_m(P) = V_m \cdot [1 - \kappa_T \cdot (P - P_\theta)]$ $\kappa_T = \dfrac{1.5 \cdot 10^{-7}}{atm}$

$\displaystyle \int_1^\Gamma 1 \, d\ln(\Gamma) = \int_{P_\theta}^{1000 \cdot atm} \dfrac{V_m(P)}{R \cdot T} \, dP$

$\Gamma = \exp\left(\displaystyle\int_{P_\theta}^{1000 \cdot atm} \dfrac{V_m(P)}{R \cdot T} \, dP\right)$ $\Gamma = 1.336$

6.51

K_p is defined by the following equation

$K_p = \dfrac{K_a \cdot P_\theta^{\Delta v_g}}{K_\phi}$

where Δv_g represents the change in moles of gas for the reaction and K_ϕ represents the collection of fugacity coefficients (and activity coefficients for solids).

$$K_a = 0.46 \qquad \Delta v_g = -1 \qquad P = 1000 \cdot atm \qquad P_\theta = 1 \cdot bar \qquad T = 873 \cdot K$$

$$bar = 10^5 \cdot Pa$$

$$R = 8.31451 \cdot \frac{joule}{mole \cdot K}$$

$$K_\phi = \frac{\phi_{CH4}}{\Gamma_C \cdot \phi_{H2}^2}$$

$$MPa = 10^6 \cdot Pa$$

Estimation of fugacity (or activity) coefficients:

Carbon

$$\rho = 2.26 \cdot \frac{gm}{cm^3} \qquad MW = 12.011 \cdot \frac{gm}{mole} \qquad V_m = \frac{MW}{\rho}$$

$$\Gamma_C = \exp\left[\frac{(P - P_\theta) \cdot V_m}{R \cdot T}\right] \qquad \Gamma_C = 1.077$$

Hydrogen

$$Tc_{H2} = 33.2 \cdot K \qquad Pc_{H2} = 1.297 \cdot MPa$$

$$Tr_{H2} = \frac{T}{Tc_{H2}} \qquad Pr_{H2} = \frac{P}{Pc_{H2}}$$

$$Tr_{H2} = 26.295 \qquad Pr_{H2} = 78.123$$

$$\phi_{H2} = 1.36 \quad \text{from Figure 6.8}$$

Methane

$$Tc_{CH4} = 190.6 \cdot K \qquad Pc_{CH4} = 4.641 \cdot MPa$$

$$Tr_{CH4} = \frac{T}{Tc_{CH4}} \qquad Pr_{CH4} = \frac{P}{Pc_{CH4}}$$

$$Tr_{CH4} = 4.58 \qquad Pr_{CH4} = 21.833$$

$$\phi_{CH4} = 1.43 \qquad \text{from Figure 6.8}$$

Therefore

$$K_\phi = \frac{\phi_{CH4}}{\Gamma_C \cdot \phi_{H2}^2} \qquad K_\phi = 0.718$$

$$K_p = \frac{K_a \cdot P_\theta^{\Delta v_g}}{K_\phi} \qquad K_p = 0.649 \cdot atm^1$$

We can easily solve for mole fraction of methane using K_p.

$$K_p = \frac{P_{CH4}}{P_{H2}^2} \qquad P_{CH4}(X_{CH4}) = X_{CH4} \cdot P \qquad P_{H2}(X_{CH4}) = (1 - X_{CH4}) \cdot P$$

$$X_{guess} = 0.5$$

$$X_{CH4} = \text{root}\left(P_{CH4}(X_{guess}) - K_p \cdot P_{H2}(X_{guess})^2, X_{guess}\right)$$

$$X_{CH4} = 96 \cdot \%$$

SECTION 6.7

6.53

The most systematic way to approach equilibrium problems is to set up a mass balance table. With no loss of generality we may assume that there is one mole of CO and two moles of H_2 initially. ξ is the extent of reaction.

	moles initial	moles at equilibrium	
CO	1	$1 - \xi$	
H_2	2	$2 \cdot (1 - \xi)$	
CH3OH	0	ξ	$totalmoles(\xi) = 3 - 2 \cdot \xi$

$$X_{CO}(\xi) = \frac{1 - \xi}{totalmoles(\xi)} \qquad X_{H2}(\xi) = \frac{2 \cdot (1 - \xi)}{totalmoles(\xi)} \qquad X_{CH3OH}(\xi) = \frac{\xi}{totalmoles(\xi)}$$

The equilibrium constant may be expressed as follows using Dalton's Law of Partial Pressures.

$$K_p(\xi) = \frac{X_{CH3OH}(\xi) \cdot P}{\left(X_{CO}(\xi) \cdot P\right) \cdot \left(X_{H2}(\xi) \cdot P\right)^2}$$

$P = 56 \cdot atm$

$TOL = 0.001$

$K_{given} = 0.00608 \cdot atm^{-2}$

Adjust tolerance until the calculated value of K_p matches the given one. It may also be necessary to refine the initial guess if convergence problems are encountered.

Solving for ξ and X_{CH3OH}:

$\xi_{guess} = 0.9$

$$\xi = root\left[X_{CH3OH}(\xi_{guess}) \cdot P - \left(X_{CO}(\xi_{guess}) \cdot P\right) \cdot \left(X_{H2}(\xi_{guess}) \cdot P\right)^2 \cdot K_{given}, \xi_{guess}\right]$$

$\xi = 0.715$

$X_{CH3OH}(\xi) = 0.456$

Recalculate K_p to check: $\qquad K_p(\xi) = 0.00608 \cdot atm^{-2}$

6.55

The equilibrium constant can easily be calculated with data from Table 6.1 and eq. 6.15 in the text.

$$\Delta fG_{NO2} = 51.31 \cdot \frac{kJ}{mole} \qquad \Delta fG_{N2O4} = 97.89 \cdot \frac{kJ}{mole} \qquad\qquad kJ = 1000 \cdot joule \qquad R = 8.31451 \cdot \frac{joule}{mole \cdot K}$$

$$\Delta rxnG_\theta = 2 \cdot \Delta fG_{NO2} - \Delta fG_{N2O4} \qquad\qquad T = 298.15 \cdot K$$

$$Ka_{calc} = exp\left(-\frac{\Delta rxnG_\theta}{R \cdot T}\right) \qquad Ka_{calc} = 0.148$$

The most systematic way to solve equilibrium problems is to set up a mass balance table. We may assume that there is one mole of N_2O_4 initially. α is the degree of dissociation.

	moles initial	moles at equilibrium
N_2O_4	1	$1 - \alpha$
NO_2	0	$2 \cdot \alpha$ $\qquad$ totalmoles$(\alpha) = 1 + \alpha$

$$X_{N2O4}(\alpha) = \frac{1 - \alpha}{totalmoles(\alpha)} \qquad X_{NO2}(\alpha) = \frac{2 \cdot \alpha}{totalmoles(\alpha)}$$

If we neglect non-idealities, the activities and equilibrium constant may be expressed as follows

$$a_{N2O4}(\alpha) = \frac{X_{N2O4}(\alpha) \cdot P}{P_\theta} \qquad a_{NO2}(\alpha) = \frac{X_{NO2}(\alpha) \cdot P}{P_\theta} \qquad\qquad P = 1 \cdot bar \qquad bar = 10^5 \cdot Pa$$

$$\qquad\qquad\qquad P_\theta = 1 \cdot bar$$

$$K_a(\alpha) = \frac{a_{NO2}(\alpha)^2}{a_{N2O4}(\alpha)}$$

Solving for α:

$$\alpha_{guess} = 0.5 \qquad\qquad\qquad TOL = 10^{-4}$$

Adjust tolerance until the calculated value of K_a matches the given one. It

$$\alpha = root\left(a_{NO2}(\alpha_{guess})^2 - Ka_{calc} \cdot a_{N2O4}(\alpha_{guess}), \alpha_{guess}\right)$$

may also be necessary to refine the initial guess if convergence problems

$$\alpha = 0.189$$

are encountered.

Recalculate K_a to check: $\qquad K_a(\alpha) = 0.148$

6.57

The most systematic way to approach equilibrium problems is to set up a mass balance table. With no loss of generality we may assume that there is one mole each of NO_2 and SO_2 initially. ξ is the extent of reaction, which is also the conversion of SO_2 to SO_3.

	moles initial	moles at equilibrium
NO_2	1	$1 - \xi$
SO_2	1	$1 - \xi$
NO	0	ξ
SO_3	0	ξ totalmoles 2

$$X_{NO2}(\xi) = \frac{1 - \xi}{totalmoles} \quad X_{SO2}(\xi) = \frac{1 - \xi}{totalmoles} \quad X_{NO}(\xi) = \frac{\xi}{totalmoles} \quad X_{SO3}(\xi) = \frac{\xi}{totalmoles}$$

The equilibrium constant may be expressed as follows. Notice that because there is no net change in the moles of gas, the pressure does not appear in K_p.

$$K_p(\xi) = \frac{X_{NO}(\xi) \cdot X_{SO3}(\xi)}{X_{NO2}(\xi) \cdot X_{SO2}(\xi)}$$

TOL $= 0.001$

$$K_{given} = 15.8 \cdot 10^3$$

Adjust tolerance until the calculated value of K_p matches the given one. It may also be necessary to refine the initial guess if convergence problems are encountered.

Solving for ξ:

$$\xi_{guess} = 0.9$$

$$\xi = root\left(X_{NO}(\xi_{guess}) \cdot X_{SO3}(\xi_{guess}) - X_{NO2}(\xi_{guess}) \cdot X_{SO2}(\xi_{guess}) \cdot K_{given}, \xi_{guess}\right)$$

$$\xi = 99.2 \cdot \%$$

Recalculate K_p to check: $\quad K_p(\xi) = 1.58 \cdot 10^4$

6.59

The most systematic way to approach equilibrium problems is to set up a mass balance table. ξ is the extent of reaction.

	moles initial	moles at equilibrium
HCl	4	$4 \cdot (1 - \xi)$
O_2	1	$1 - \xi$
Cl_2	0	$2 \cdot \xi$
$H2O$	0	$2 \cdot \xi$ totalmoles(ξ) $5 - \xi$

$$X_{HCl}(\xi) = \frac{4 \cdot (1 - \xi)}{totalmoles(\xi)} \qquad X_{O2}(\xi) = \frac{1 - \xi}{totalmoles(\xi)}$$

$$X_{Cl2}(\xi) = \frac{2 \cdot \xi}{totalmoles(\xi)} \qquad X_{H2O}(\xi) = \frac{2 \cdot \xi}{totalmoles(\xi)}$$

The equilibrium constant may be expressed as follows using Dalton's Law.

$$K_p(\xi) = \frac{\left(X_{Cl2}(\xi) \cdot P\right)^2 \cdot \left(X_{H2O}(\xi) \cdot P\right)^2}{\left(X_{HCl}(\xi) \cdot P\right)^4 \cdot \left(X_{O2}(\xi) \cdot P\right)} \qquad\qquad P = 5 \cdot atm$$

$$TOL = 0.001$$

$$K_{given} = 23.14 \cdot atm^{-1}$$

Solving for ξ:

$$\xi_{gss} = 0.9$$

$$\xi = root\left[\left(X_{Cl2}(\xi_{gss}) \cdot P\right)^2 \cdot \left(X_{H2O}(\xi_{gss}) \cdot P\right)^2 - \left(X_{HCl}(\xi_{gss}) \cdot P\right)^4 \cdot \left(X_{O2}(\xi_{gss}) \cdot P\right) \cdot K_{given}, \xi_{gss}\right]$$

$$\xi = 0.762$$

$$P_{Cl2}(\xi) = X_{Cl2}(\xi) \cdot P$$

$$P_{Cl2}(\xi) = 1.80 \cdot atm$$

Recalculate K_p to check: $\qquad K_p(\xi) = 23.14 \cdot atm^{-1}$

6.61

The equilibrium constant for the reaction $4 S_2 = S_8$ can be calculated with free energy function data from Table 6.5 and eq. 6.21 in the text.

$$\phi S2_{1000K} = 245.7 \cdot \frac{joule}{mole \cdot K} \qquad \phi S2_{1500K} = 256.7 \cdot \frac{joule}{mole \cdot K} \qquad \Delta fH_{S2} = 129.0 \cdot \frac{kJ}{mole} \qquad R = 8.31451 \cdot \frac{joule}{mole \cdot K}$$

$$\phi S8_{1000K} = 516.1 \cdot \frac{joule}{mole \cdot K} \qquad \phi S8_{1500K} = 569.8 \cdot \frac{joule}{mole \cdot K} \qquad \Delta fH_{S8} = 101.3 \cdot \frac{kJ}{mole} \qquad kJ = 1000 \cdot joule$$

$$\Delta\phi_{1000K} = \phi S8_{1000K} - 4 \cdot \phi S2_{1000K} \qquad\qquad \Delta H = \Delta fH_{S8} - 4 \cdot \Delta fH_{S2}$$

$$\Delta\phi_{1500K} = \phi S8_{1500K} - 4 \cdot \phi S2_{1500K}$$

$$Ka_{1000K} = exp\left(\frac{\Delta\phi_{1000K}}{R} - \frac{\Delta H}{R \cdot 1000 \cdot K}\right) \qquad Ka_{1000K} = 1.923 \cdot 10^{-3}$$

$$Ka_{1500K} = exp\left(\frac{\Delta\phi_{1500K}}{R} - \frac{\Delta H}{R \cdot 1500 \cdot K}\right) \qquad Ka_{1500K} = 3.717 \cdot 10^{-10}$$

The most systematic way to solve equilibrium problems is to set up a gas phase mass balance table. We may assume that there are 4 moles of S_2 initially. ξ is the extent of reaction.

	moles initial	moles at equilibrium
S_2	4	$4 \cdot (1 - \xi)$
S_8	0	ξ

$$totalmoles(\xi) = 4 - 3 \cdot \xi$$

$$X_{S2}(\xi) = \frac{4 \cdot (1 - \xi)}{totalmoles(\xi)} \qquad X_{S8}(\xi) = \frac{\xi}{totalmoles(\xi)}$$

If we neglect non-idealities, the activities and equilibrium constant may be expressed as follows

$$a_{S2}(\xi) = \frac{X_{S2}(\xi) \cdot P}{P_\theta} \qquad a_{S8}(\xi) = \frac{X_{S8}(\xi) \cdot P}{P_\theta} \qquad\qquad P = 3 \cdot atm \qquad bar = 10^5 \cdot Pa$$

$$P_\theta = 1 \cdot bar$$

$$K_a(\xi) = \frac{a_{S8}(\xi)}{a_{S2}(\xi)^4}$$

Solving for ξ at 1000K:

$\xi_{guess} := 0.5$

$\xi_{1000K} := root\left(a_{S8}\left(\xi_{guess}\right) - a_{S2}\left(\xi_{guess}\right)^4 \cdot Ka_{1000K}, \xi_{guess}\right)$ TOL $:= 10^{-9}$

$\xi_{1000K} = 0.158$

$X_{S8}\left(\xi_{1000K}\right) = 0.045$

Adjust tolerance until the calculated value of K_a matches the given one. It may also be necessary to refine the initial guess if convergence problems are encountered.

Solving for ξ at 1500K:

$\xi_{guess} := 0.5$

$\xi_{1500K} := root\left(a_{S8}\left(\xi_{guess}\right) - a_{S2}\left(\xi_{guess}\right)^4 \cdot Ka_{1500K}, \xi_{guess}\right)$

$\xi_{1500K} = 4.176 \cdot 10^{-8}$

$X_{S8}\left(\xi_{1500K}\right) = 1.0 \cdot 10^{-8}$

Recalculate K_a to check: $K_a\left(\xi_{1000K}\right) = 1.923 \cdot 10^{-3}$ $K_a\left(\xi_{1500K}\right) = 3.717 \cdot 10^{-10}$

6.63

The most systematic way to approach equilibrium problems is to set up a mass balance table. ξ is the extent of reaction, which is also the fractional conversion of A to B.

	moles initial	moles at equilibrium
A	1	$1 - \xi$
I_2	1	$1 - \xi$
HI	0	$2 \cdot \xi$
B	0	ξ totalmoles$(\xi) := 2 - \xi$

$$X_A(\xi) := \frac{1 - \xi}{totalmoles(\xi)} \quad X_{I2}(\xi) := \frac{1 - \xi}{totalmoles(\xi)} \quad X_{HI}(\xi) := \frac{2 \cdot \xi}{totalmoles(\xi)} \quad X_B(\xi) := \frac{\xi}{totalmoles(\xi)}$$

The equilibrium constant may be expressed as follows using Dalton's Law of Partial Pressures.

$$K_p(\xi) = \frac{\left(X_{HI}(\xi) \cdot P\right)^2 \cdot \left(X_B(\xi) \cdot P\right)}{\left(X_A(\xi) \cdot P\right) \cdot \left(X_{I2}(\xi) \cdot P\right)}$$

$P = 2 \cdot atm$

$K_{given} = 0.30 \cdot atm$

$TOL = 0.001$

Solving for ξ:

$\xi_{guess} = 0.9$

Adjust tolerance until the calculated value of K_p matches the given one. It may also be necessary to refine the initial guess if convergence problems are encountered.

$$\xi = root\left[\left(X_{HI}(\xi_{guess}) \cdot P\right)^2 \cdot \left(X_B(\xi_{guess}) \cdot P\right) - \left(X_A(\xi_{guess}) \cdot P\right) \cdot \left(X_{I2}(\xi_{guess}) \cdot P\right) \cdot K_{given}, \xi_{guess}\right]$$

$\xi = 34 \cdot \%$

Recalculate K_p to check: $K_p(\xi) = 0.30 \cdot atm$

6.65

We can determine the equilibrium constant from free energy function data in Tables 6.4 and 6.5 for 1500 K.

$\phi_C = 17.5 \cdot \dfrac{joule}{mole \cdot K} + \dfrac{1.050 \cdot \dfrac{kJ}{mole}}{1500 \cdot K}$ Equation 5.27 was used to convert ϕ based on 0 K to ϕ based on 298.15 K.

$\phi_{Ti} = 54.2 \cdot \dfrac{joule}{mole \cdot K}$ $\phi_{TiC} = 60.8 \cdot \dfrac{joule}{mole \cdot K}$

$kJ \equiv 1000 \cdot joule$

$R = 8.31451 \cdot \dfrac{joule}{mole \cdot K}$

$\Delta fH_{Ti} = 0 \cdot \dfrac{kJ}{mole}$ $\Delta fH_{TiC} = -185 \cdot \dfrac{kJ}{mole}$ $\Delta fH_C = 0 \cdot \dfrac{kJ}{mole}$

$T = 1500 \cdot K$

$\Delta\phi = \phi_{TiC} - \phi_C - \phi_{Ti}$ $\Delta H = \Delta fH_{TiC} - \Delta fH_C - \Delta fH_{Ti}$

$K_a = exp\left(\dfrac{\Delta\phi}{R} - \dfrac{\Delta H}{R \cdot T}\right)$ eq. 6.21

$K_a = 6.858 \cdot 10^5$

We can calculate $\Delta rxnG_\theta$ using equation 6.15

$$\Delta rxnG_\theta = -R \cdot T \cdot \ln(K_a) \qquad \Delta rxnG_\theta = -167.6 \cdot \frac{kJ}{mole}$$

However, it is $\Delta rxnG$, not $\Delta rxnG_\theta$, that determines spontaneity. The two are related by eq. 6.14. Since we are dealing only with solids, the activity quotient is one (and ln Q = 0). Thus, for this case the two are equal.

$$Q = 1$$

$$\Delta rxnG = \Delta rxnG_\theta + R \cdot T \cdot \ln(Q) \qquad \Delta rxnG = -167.6 \cdot \frac{kJ}{mole}$$

Since $\Delta rxnG < 0$, the reaction is spontaneous under these conditions.

6.67

The most stable hydrate is the one which minimizes free energy. The general reaction for hydrate formation is given below. We will first calculate $\Delta rxnG_\theta$, and then use eq. 6.14 to calculate $\Delta rxnG$. (Lone subscripts refer to the degree of hydration of $MgCl_2$.)

$$MgCl_2 - n \cdot H2O = MgCl2 \cdot n \cdot H2O$$

$$kJ = 1000 \cdot joule$$

$$\Delta fG_0 \quad 592.33 \cdot \frac{kJ}{mole} \quad \Delta fG_1 \quad 862.36 \cdot \frac{kJ}{mole} \quad \Delta fG_2 \quad 1118.5 \cdot \frac{kJ}{mole} \quad \Delta fG_4 \quad 1633.8 \cdot \frac{kJ}{mole}$$

$$\Delta fG_6 \quad 1278.8 \cdot \frac{kJ}{mole} \quad \Delta fG_{H2O} \quad 228.572 \cdot \frac{kJ}{mole}$$

$$\Delta rxnG_{\theta1} \quad \Delta fG_1 \quad \Delta fG_0 \quad \Delta fG_{H2O} \qquad \Delta rxnG_{\theta1} = -41.458 \cdot \frac{kJ}{mole}$$

$$\Delta rxnG_{\theta2} \quad \Delta fG_2 \quad \Delta fG_0 \quad 2 \cdot \Delta fG_{H2O} \qquad \Delta rxnG_{\theta2} = -69.026 \cdot \frac{kJ}{mole}$$

$$\Delta rxnG_{\theta4} \quad \Delta fG_4 \quad \Delta fG_0 \quad 4 \cdot \Delta fG_{H2O} \qquad \Delta rxnG_{\theta4} = -127.182 \cdot \frac{kJ}{mole}$$

$$\Delta rxnG_{\theta6} \quad \Delta fG_6 \quad \Delta fG_0 \quad 6 \cdot \Delta fG_{H2O} \qquad \Delta rxnG_{\theta6} = 684.962 \cdot \frac{kJ}{mole}$$

Because the only non-condensed phase in this reaction is H_2O (g), the activity quotient found in eq. 6.14 simplifies to $(P_\theta/P_{H2O})^n$.

$$T \quad 298.15 \cdot K \qquad P_{H2O} \quad 0.8 \cdot (23.76 \cdot torr) \quad P_\theta \quad 1 \cdot atm \quad R \cdot 8.31451 \cdot \frac{joule}{mole \cdot K}$$

$$\Delta rxnG_1 \quad \Delta rxnG_{\theta 1} \cdot R \cdot T \cdot \ln\left(\frac{P_\theta}{P_{H2O}}\right) \qquad \Delta rxnG_1 = 32.314 \cdot \frac{kJ}{mole}$$

$$\Delta rxnG_2 \quad \Delta rxnG_{\theta 2} \cdot R \cdot T \cdot \ln\left(\frac{P_\theta}{P_{H2O}}\right)^2 \qquad \Delta rxnG_2 = 50.739 \cdot \frac{kJ}{mole}$$

$$\Delta rxnG_4 \quad \Delta rxnG_{\theta 4} \cdot R \cdot T \cdot \ln\left(\frac{P_\theta}{P_{H2O}}\right)^4 \qquad \Delta rxnG_4 = 90.608 \cdot \frac{kJ}{mole}$$

$MgCl_2 \cdot 4H_2O$ minimizes $\Delta rxnG$ and is therefore the most stable.

$$\Delta rxnG_6 \quad \Delta rxnG_{\theta 6} \cdot R \cdot T \cdot \ln\left(\frac{P_\theta}{P_{H2O}}\right)^6 \qquad \Delta rxnG_6 = 739.824 \cdot \frac{kJ}{mole}$$

CHAPTER 7   *Solutions*

SECTION 7.1

7.1

It is easiest to calculate the mole fractions of both components first, and then use eq. 7.3 to obtain the molality.

$$\text{mass}_{\text{acetone}} = 10.0 \cdot \text{gm} \qquad \text{mass}_{\text{ethanol}} = 450 \cdot \text{gm} \qquad \text{MW}_{\text{acetone}} = 58.08 \cdot \frac{\text{gm}}{\text{mole}}$$

$$\text{moles}_{\text{acetone}} = \frac{\text{mass}_{\text{acetone}}}{\text{MW}_{\text{acetone}}} \qquad \text{moles}_{\text{ethanol}} = \frac{\text{mass}_{\text{ethanol}}}{\text{MW}_{\text{ethanol}}} \qquad \text{MW}_{\text{ethanol}} = 46.07 \cdot \frac{\text{gm}}{\text{mole}}$$

$$\text{moles}_{\text{total}} = \text{moles}_{\text{acetone}} + \text{moles}_{\text{ethanol}}$$

$$X_{\text{acetone}} = \frac{\text{moles}_{\text{acetone}}}{\text{moles}_{\text{total}}} \qquad X_{\text{ethanol}} = \frac{\text{moles}_{\text{ethanol}}}{\text{moles}_{\text{total}}}$$

$$X_{\text{acetone}} = 0.017 \qquad X_{\text{ethanol}} = 0.983$$

$$m(X_2, M_1) = \frac{1000 \cdot \text{gm}}{M_1} \cdot \frac{X_2}{1 - X_2} \qquad \text{eq. 7.3}$$

$$m(X_{\text{acetone}}, \text{MW}_{\text{ethanol}}) = 0.383 \quad \text{Units are moles of acetone per kg ethanol}$$

7.3

We'll calculate the mole fraction of sucrose first, and then use eq. 7.3 to obtain the molality.

$$\text{mass}_{\text{sucrose}} = 25.5 \cdot \text{gm} \qquad \text{mass}_{\text{water}} = 250 \cdot \text{gm} \qquad \text{MW}_{\text{sucrose}} = 342.300 \cdot \frac{\text{gm}}{\text{mole}}$$

$$\text{moles}_{\text{sucrose}} = \frac{\text{mass}_{\text{sucrose}}}{\text{MW}_{\text{sucrose}}} \qquad \text{moles}_{\text{water}} = \frac{\text{mass}_{\text{water}}}{\text{MW}_{\text{water}}} \qquad \text{MW}_{\text{water}} = 18.015 \cdot \frac{\text{gm}}{\text{mole}}$$

$$X_{\text{sucrose}} = \frac{\text{moles}_{\text{sucrose}}}{\text{moles}_{\text{sucrose}} + \text{moles}_{\text{water}}}$$

$$X_{sucrose} = 0.00534$$

$$m(X_2, M_1) = \frac{1000 \cdot gm}{M_1} \cdot \frac{X_2}{1 - X_2} \qquad \text{eq. 7.3}$$

$$m(X_{sucrose}, MW_{water}) = 0.298 \qquad \text{Units are moles of sucrose per kg water.}$$

The molality could also be determined by multiplying the given amounts of sucrose and water by 4. The result is 102 g of sucrose in 1000 g of water. 102 g of sucrose is 0.298 moles.

7.5

Since we have 60.12 g of NaCl in 1 kg of water, all we have to do to find the molality is determine how many moles of NaCl there are.

$$MW_{NaCl} = 58.443 \cdot \frac{gm}{mole} \qquad mass_{NaCl} = 60.12 \cdot gm$$

$$moles_{NaCl} = \frac{mass_{NaCl}}{MW_{NaCl}}$$

$$m = moles_{NaCl}$$

$$m = 1.0287 \qquad \text{Molality units are moles of solute per kg of solvent.}$$

Concentration is moles of solute per liter of solution. We can use the density and total mass of the solution to determine the total volume of solution. Then, the concentration is easily determined.

$$\rho = 1.03887 \cdot \frac{gm}{cm^3} \qquad mass_{total} = 1 \cdot kg + 60.12 \cdot gm$$

$$V_{total} = \frac{mass_{total}}{\rho} \qquad V_{total} = 1.020 \cdot liter$$

$$c = \frac{moles_{NaCl}}{V_{total}} \qquad c = 1.00807 \cdot \frac{mole}{liter}$$

SECTION 7.2

7.7

The volume change on mixing can be estimated using eq. 7.10. Vpm denotes a partial molar volume. Vm denotes a pure component molar volume.

Water $\quad MW_{water} = 18.015 \cdot \dfrac{gm}{mole} \qquad moles_{water} = 2 \cdot mole \qquad \rho_{water} = 0.997 \cdot \dfrac{gm}{cm^3}$

$\qquad\qquad Vm_{water} = \dfrac{MW_{water}}{\rho_{water}} \qquad Vpm_{water} = 17 \cdot \dfrac{cm^3}{mole}$

Ethanol $\quad MW_{ethanol} \quad 46.069 \cdot \dfrac{gm}{mole} \qquad moles_{ethanol} \quad moles_{water} \cdot \begin{pmatrix} 0.4 \\ 0.6 \end{pmatrix} \quad \rho_{ethanol} \quad 0.7893 \cdot \dfrac{gm}{cm^3}$

$\qquad\qquad Vm_{ethanol} \quad \dfrac{MW_{ethanol}}{\rho_{ethanol}} \qquad Vpm_{ethanol} \quad 57 \cdot \dfrac{cm^3}{mole}$

$\Delta mixV \quad moles_{water} \cdot \left(Vpm_{water} \quad Vm_{water} \right) + moles_{ethanol} \cdot \left(Vpm_{ethanol} \quad Vm_{ethanol} \right) \qquad$ eq. 7.10

$\Delta mixV = 4.0 \cdot cm^3$

SECTION 7.3

7.9

Let's first list all of the distinct chemical species that we can identify: 1) water, 2) Ag + ion, 3) N(
ion, 4) AgCl, 5) AgBr, 6) Cl- ion, 7) Br- ion.

The number of components is the number of distinct species minus the number of constrai(
on the system. Constraints include things such as mass balance, electrical neutrality, and chemi(
equilibrium requirements.

Constraints here are 1) mass balance/electrical neutrality between Ag + ions and all negati(
ions, 2) equilibrium between solid AgCl and the dissociated salt, 3) equilibrium between solid AgBr a(
the dissociated salt. Thus, 7 - 3 = 4 components.

You may be wondering why the first constraint was not three constraints. After all, there (
three mass balance/electrical neutrality requirements that must be satisfied individually by each of t(
three salts ($AgNO_3$, AgCl, AgBr). This is perfectly valid. However, if this is how you choose t(
constraints, you have to distinguish between Ag + derived from each salt. Thus, Ag + becomes thr(
components. Then there would be 9 components, 5 constraints, and still 4 components.

The different phases present are 1) aqueous solution, 2) vapor, 3) solid AgCl, and 4) so(
AgBr.

Gibbs' Phase Rule gives the degrees of freedom as F = components + 2 - phases = 2. Thu(
if P is fixed, the temperature can still vary.

7.11

(a)   There is only one component because the two species present, the monomer and the dim(
are constrained by the reaction equilibrium between the two. The gas phase is the only phase prese(

(b)   There is only one component which can be independently varied because the three disti(
species present and bound by two constraints: 1) the reaction equilibrium and 2) mass balar(
between CO and Cl_2 (they must be formed in equal amounts if none is added). The gas phase is ag(
the only phase present.

(c)   There are two components because the second constraint discussed in part (b) is no lon(
present. There gas phase is the only phase.

(d)   There are two components because there are 3 distinct species and 1 constraint (the reaction equilibrium of $NH_4Cl = HCl + NH_3$). There is no mass balance linking HCl and NH_3 since there are arbitrary amounts of these gases. There are two phases, solid NH_4Cl and gas.

7.13

There are three distinct species: 1) ammonia (g), 2) carbon dioxide (g), and 3) ammonium carbamate (s). There is one constraint, that of reaction equilibrium. Thus there are two components. There are also two phases, solid and gas. The degree of freedom is $F = c + 2 - p = 2$.

SECTION 7.5

7.15
_____

(a) An ideal solution is one that obey's Raoult's Law, eq. 7.21. From the given data, we can calculate the partial pressures for each component using this equation. The total pressure is just the sum of the two partial pressures. Pure component pressures are denoted here as Pstar. All data and calculations are for 80 C.

$$\text{Pstar}_{benzene} = 760 \cdot torr \qquad\qquad \text{Pstar}_{toluene} = 350 \cdot torr$$

$$P_{benzene}(X_{benzene}) = X_{benzene} \cdot \text{Pstar}_{benzene} \qquad P_{toluene}(X_{benzene}) = (1 - X_{benzene}) \cdot \text{Pstar}_{toluene}$$

$$P_{benzene}(0.2) = 152 \cdot torr \qquad\qquad P_{toluene}(0.2) = 280 \cdot torr$$

$$P_{total}(X_{benzene}) = P_{benzene}(X_{benzene}) + P_{toluene}(X_{benzene})$$

$$P_{total}(0.2) = 432 \cdot torr$$

(b) Boiling occurs when the total vapor pressure equals the ambient pressure.

$$X_{guess} = 0.5$$

$$X_{benzene} = root(P_{total}(X_{guess}) - 500 \cdot torr, X_{guess})$$

$$X_{benzene} = 0.366$$

$$X_{toluene} = 1 - X_{benzene}$$

$$X_{toluene} = 0.634$$

7.17
_____

Raoult's Law, eq. 7.21, predicts partial pressures based on liquid mole fractions and pure component vapor pressures. Pure component vapor pressures are denoted here as Pstar.

$$Pstar(a,b,T) = 10^{\left[\frac{-0.05223 \cdot a}{\left(\frac{T}{K}\right)} + b\right]} \cdot torr \qquad T = 296.15 \cdot K$$

$$a_{bromide} = 32430 \quad b_{bromide} = 7.821 \qquad\qquad a_{chloride} = 28894 \quad b_{chloride} = 7.593$$

$$Pstar_{bromide} = Pstar(a_{bromide}, b_{bromide}, T) \qquad Pstar_{chloride} = Pstar(a_{chloride}, b_{chloride}, T)$$

$$Pstar_{bromide} = 126.339 \cdot torr \qquad\qquad Pstar_{chloride} = 314.165 \cdot torr$$

188

Calculation of partial pressures:

$X_{bromide} = 0.381$

$P_{bromide} = X_{bromide} \cdot P_{star\ bromide}$

$P_{bromide} = 48.135 \cdot torr$

$X_{chloride} = 1 - X_{bromide}$

$P_{chloride} = X_{chloride} \cdot P_{star\ chloride}$

$P_{chloride} = 194.468 \cdot torr$

$P_{total} = P_{bromide} + P_{chloride}$ $P_{total} = 243 \cdot torr$

7.19

(a) The activities may be estimated using eq. 7.25, assuming the fugacity may be estimated as the partial pressure. Dalton's Law can be used to estimate the partial pressure. The activity coefficient can be estimated using eq. 7.26.

$X_1 = 0.9006$ $P_{star\ 1} = 130.4 \cdot torr$ $P_{star\ 2} = 43.9 \cdot torr$

$Y_1 = 0.6667$ $P = 185.9 \cdot torr$

$a_1 = \dfrac{Y_1 \cdot P}{P_{star\ 1}}$ $a_1 = 0.950$ $a_2 = \dfrac{(1 - Y_1) \cdot P}{P_{star\ 2}}$ $a_2 = 1.411$

$\gamma_1 = \dfrac{a_1}{X_1}$ $\gamma_1 = 1.055$ $\gamma_2 = \dfrac{a_2}{(1 - X_1)}$ $\gamma_2 = 14.20$

(b) Raoult's Law is eq. 7.21 in the text

$P_{1RL} = X_1 \cdot P_{star\ 1}$ $P_{2RL} = (1 - X_1) \cdot P_{star\ 2}$

$P_{1RL} = 117.438 \cdot torr$ $P_{2RL} = 4.364 \cdot torr$

$P_{RL} = P_{1RL} + P_{2RL}$

$P_{RL} = 121.8 \cdot torr$

(a) Using eqs. 7.25 and 7.26, the activity coefficient may be estimated as $\gamma_i = P_i / (Pstar_i * X_i)$, where Pstar represents the pure component vapor pressure.

$i = 1..7$

$P_i =$	$X_{2_i} =$	$Y_{2_i} =$
942.6·torr	0.1312	0.0243
909.6·torr	0.2040	0.0300
883.3·torr	0.2714	0.0342
868.4·torr	0.3360	0.0362
830.2·torr	0.4425	0.0411
786.8·torr	0.5578	0.0451
758.7·torr	0.6036	0.0489

$X_{1_i} := 1 - X_{2_i}$ $\qquad$ $Y_{1_i} := 1 - Y_{2_i}$

$Pstar_1 = 1008 \cdot torr$ $\qquad$ $Pstar_2 = 48.3 \cdot torr$

$P_{1_i} = Y_{1_i} \cdot P_i$ $\qquad$ $P_{2_i} = Y_{2_i} \cdot P_i$

$$\gamma_{1_i} = \frac{P_{1_i}}{Pstar_1 \cdot X_{1_i}} \qquad \gamma_{2_i} = \frac{P_{2_i}}{Pstar_2 \cdot X_{2_i}}$$

γ_{1_i}	γ_{2_i}
1.050	3.615
1.100	2.769
1.162	2.305
1.250	1.937
1.417	1.596
1.686	1.317
1.806	1.273

(b) We can use eq. 7.27 to calculate $\Delta mixG^{ex}$ if we divide through by (n1+n2). Thus, the moles become mole fractions. But what is the mole fraction? Since we are not given data on the sizes of the liquid and vapor phases, we'll assume the majority is in the liquid phase and thus use the liquid mole fractions.

$$\Delta mixG_{ex_i} = R \cdot T \cdot \left(X_{1_i} \cdot \ln\left(\gamma_{1_i}\right) + X_{2_i} \cdot \ln\left(\gamma_{2_i}\right) \right)$$

$R = 8.31451 \cdot \dfrac{joule}{mole \cdot K}$ $\qquad$ $T = 363.15 \cdot K$

$\Delta mixG_{ex_i}$

$\left(\dfrac{joule}{mole}\right)$
637.472
855.725
$1.014 \cdot 10^3$
$1.119 \cdot 10^3$
$1.211 \cdot 10^3$
$1.161 \cdot 10^3$
$1.147 \cdot 10^3$

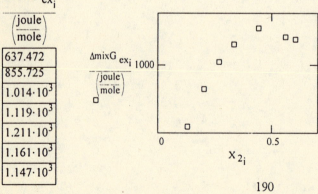

190

7.23

Table 7.2 gives values for w, the first derivative with respect to T, and the second derivative with respect to T. If we start with the second derivative, we can gradually integrate these derivatives back to give w(T).

$$\left(\frac{\partial^2 w}{\partial T^2}\right) = -0.046$$

Integrating with respect to T gives

$$\left(\frac{\partial w}{\partial T}\right) = -0.046T + c_1$$

We can determine the value of the integration constant, c_1, using the known value for the first derivative. Table 7.2 gives the first derivative as -7.03 at T = 293 K. This means c_1 = 6.448.

$$\left(\frac{\partial w}{\partial T}\right) = -0.046T + 6.448$$

Integrating again gives

$$w = -0.023T^2 + 6.448T + c_2$$

Since Table 7.2 gives w = 1275 at T = 293 K, c2 = 1360.263.

$$w(T) = 1360.263 + 6.448T - 0.023T^2$$

7.25

(a) If we assume an ideal solution, then Raoult's Law, eq. 7.21, applies. Pstar denotes the pure component vapor pressure.

$\text{Pstar}_1 = 118 \cdot \text{torr}$ $X_1 = 0.25$ $P_1 = X_1 \cdot \text{Pstar}_1$ $P_1 = 29.5 \cdot \text{torr}$

$\text{Pstar}_2 = 122 \cdot \text{torr}$ $X_2 = 0.75$ $P_2 = X_2 \cdot \text{Pstar}_2$ $P_2 = 91.5 \cdot \text{torr}$

$P_{total} = P_1 + P_2$ $P_{total} = 121 \cdot \text{torr}$

(b) Regular solution theory tries to correct for the differing interactions between like-like and like-unlike molecules. The parameter, w, was derived as a function of temperature in problem 7.23, and can be used here to calculate w at 303 K.

$$w(T) = (1360.263 + 6.448 \cdot T - 0.023 \cdot T^2) \cdot \frac{joule}{mole} \qquad w(303) = 1.202 \cdot 10^3 \cdot \frac{joule}{mole}$$

The predicted activity coefficients are given by eq. 7.28.

$$\gamma_1 = \exp\left(\frac{X_2^2 \cdot w(303)}{R \cdot T}\right) \qquad \gamma_2 = \exp\left(\frac{X_1^2 \cdot w(303)}{R \cdot T}\right) \qquad T = 303 \cdot K \qquad R = 8.31451 \cdot \frac{joule}{mole \cdot K}$$

$$\gamma_1 = 1.308 \qquad\qquad\qquad \gamma_2 = 1.03$$

The partial pressures are then given by eqs. 7.25 and 7.26.

$$P_1 = X_1 \cdot \gamma_1 \cdot Pstar_1 \qquad\qquad P_2 = X_2 \cdot \gamma_2 \cdot Pstar_2$$

$$P_1 = 38.585 \cdot torr \qquad\qquad P_2 = 94.271 \cdot torr$$

$$P_{total} = P_1 + P_2 \qquad P_{total} = 133 \cdot torr$$

SECTION 7.6

7.27

The Henry's Law constant is defined by eq. 7.31. If we plot P_2/X_2 as a function of X_2, k_x can be found by extrapolating back to $X_2 = 0$.

$i = 0..4$

$P_{2_i} =$ $X_{2_i} =$ $y_i := \dfrac{P_{2_i}}{X_{2_i}}$ $k_x = \text{intercept}(X_2, y)$ $m = \text{slope}(X_2, y)$

P_{2_i}	X_{2_i}
200·torr	0.040
400·torr	0.087
600·torr	0.151
700·torr	0.193
760·torr	0.226

$k_x = 7.05 \cdot \text{atm}$ $y_{fit}(x_2) = k_x + m \cdot x_2$

$x_2 = 0, 0.1 .. 1$

The Henry's Law constant is $k_x = 7.05$ atm.

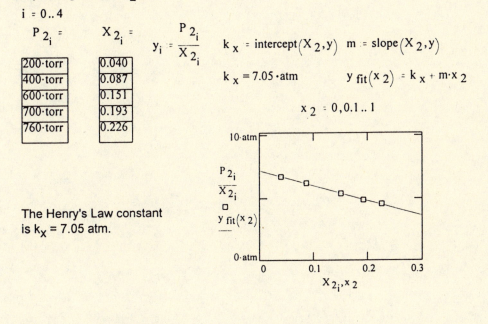

7.29

Henry's Law, eq. 7.30, gives us the Kr liquid phase mole fraction. Keep in mind that the 100 atm is the partial pressure of Kr.

$k_x = 2.00 \cdot 10^4 \cdot \text{atm}$ $P_2 = 100 \cdot \text{atm}$

$X_2 = \dfrac{P_2}{k_x}$ $X_2 = 0.00500$

Now we just have to find how many grams of Kr is equivalent to this mole fraction in 1000 g of water.

$$M_1 = 18.015 \cdot \frac{gm}{mole} \qquad M_2 = 83.80 \cdot \frac{gm}{mole} \qquad TOL = 10^{-5}$$

$$n_1 = \frac{1000 \cdot gm}{M_1} \qquad n_{2guess} = 1 \cdot mole$$

$$n_2 = root\left(\frac{n_{2guess}}{n_1 + n_{2guess}} - X_2, n_{2guess}\right)$$

$$n_2 = 0.279$$

$$mass_{Kr} = n_2 \cdot M_2$$

$$mass_{Kr} = 23.4 \cdot gm$$

SECTION 7.7

7.31

The osmotic coefficient data can be fit to a polynomial, and the activity coefficients of n-octane can be estimated with eq. 7.37. A curve fitting program gives

$$\phi(m) = 1 - 0.397281 \cdot m + 0.0359014 \cdot m^2 + 0.117675 \cdot m^3 - 0.00871332 \cdot m^4$$

$$\gamma_{2m}(m) = \exp\left(\phi(m) - 1 + \int_0^m \frac{\phi(mx) - 1}{mx} \, dmx\right)$$

$$\gamma_{2m}(0.01) = 0.992 \qquad \gamma_{2m}(0.1) = 0.925 \qquad \gamma_{2m}(0.5) = 0.695$$

7.33

We can use eq. 7.36 to estimate the osmotic coefficient. The solvent activity may be estimated as the partial pressure over the pure component pressure (Pstar). The solute molality is given in the problem.

$$P_{H2O} = 23.6798 \cdot torr \qquad Pstar_{H2O} = 23.7675 \cdot torr \qquad M_1 = 18.015$$

$$a_1 = \frac{P_{H2O}}{Pstar_{H2O}} \qquad a_1 = 0.996 \qquad m = 0.15$$

$$\phi = \frac{-1000 \cdot \ln(a_1)}{m \cdot M_1} \qquad \phi = 1.368$$

7.35

The osmotic coefficient can be calculated with eq. 7.36 (using $v = 1$ for nonelectrolytes). Then, the quantity $(\phi-1)/m$ can be fit to a linear function of m, which can be used in eq. 7.37 to calculate the desired activity coefficients.

$i = 0..9$ $\qquad m_i =$ $\qquad a_{1_i} =$ $\qquad \phi_i = \dfrac{-1000 \cdot \ln\left(a_{1_i}\right)}{m_i \cdot M_1}$ $\qquad M_1 = 18.015$

m_i	a_{1_i}
0.3	0.99448
0.4	0.99258
0.5	0.99067
0.6	0.98872
0.7	0.98672
0.8	0.98472
0.9	0.98267
1.0	0.98059
1.2	0.97634
1.4	0.97193

$$y_i = \frac{\phi_i - 1}{m_i}$$

$s = \text{slope}(m, y)$ $\qquad b = \text{intercept}(m, y)$

$s = 0.010$ $\qquad\qquad b = 0.078$

$y_{fit}(m) = b + s \cdot m$

$\phi_{fit}(m) = y_{fit}(m) \cdot m + 1$ $\qquad molal = 0, 0.1 .. 1.5$

$$\gamma_{2m}(m) = \exp\left(\phi_{fit}(m) - 1 + \int_0^m y_{fit}(mx)\, dmx\right)$$

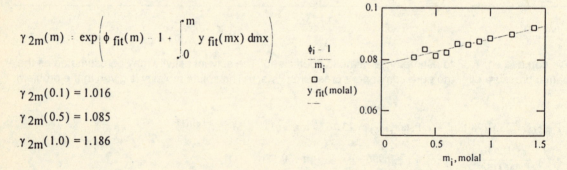

$\dfrac{\phi_i - 1}{m_i}$ □

$y_{fit}(molal)$ - - -

$\gamma_{2m}(0.1) = 1.016$

$\gamma_{2m}(0.5) = 1.085$

$\gamma_{2m}(1.0) = 1.186$

SECTION 7.8

7.37

The first thing we have to do is calculate the molality of ethanol in this solution, since the freezing point depression equations use molality. In 1000 g of water, we need 1000/9 g of ethanol to make a 10% ethanol solution by weight.

$$\text{mass}_{\text{ethanol}} = \frac{1000 \cdot \text{gm}}{9} \qquad \text{MW}_{\text{ethanol}} = 46.069 \cdot \frac{\text{gm}}{\text{mole}}$$

$$m = \frac{\text{mass}_{\text{ethanol}}}{\text{MW}_{\text{ethanol}}} \qquad m = 2.412$$

Now we can use eq. 7.4 to determine the freezing point depression.

$$K_f = 1.860 \qquad \text{Tstar}_f = 273.15 \cdot K$$

$$\theta = K_f m \qquad \theta = 4.5$$

$$T_f = \text{Tstar}_f - \theta \cdot K$$

$$T_f = 268.66 \cdot K$$

7.39

The boiling point elevation derivation directly parallels the freezing point depression derivation in the text. We can start with

$$\mu_1(vap) = \mu_1(solution)$$

With eqs. 7.16, 7.17, and the Raoult's Law reference state (pure liquid), the solution chemical potential becomes

$$\mu_1^v = \mu_1^* + RT\ln(a_1)$$

Solving for $\ln(a_1)$,

$$\ln(a_1) = \frac{1}{RT}(\mu_1^v - \mu_1^*)$$

Taking the partial derivative with respect to T gives the following (remember that T and μ are functions of temperature),

$$\left(\frac{\partial \ln a_1}{\partial T}\right) = -\frac{1}{RT^2}(\mu_1^v - \mu_1^*) + \frac{1}{RT}\left[\left(\frac{\partial \mu_1^v}{\partial T}\right) - \left(\frac{\partial \mu_1^*}{\partial T}\right)\right]$$

Using $\mu = \underline{G} = \underline{H} - T\underline{S}$ and $\left(\dfrac{\partial \underline{G}}{\partial T}\right)_P = -\underline{S}$ gives the following

$$\frac{\partial \ln a_1}{\partial T} = -\frac{1}{RT^2}(H_1^v - H_1^\bullet)$$

$$\frac{\partial \ln a_1}{\partial T} = -\frac{\Delta_{vap}H}{RT^2}$$

Integrating from $T_b^\bullet$ and $a_1 = 1$ to T_b and a_1 gives

$$\ln a_1 = \int_{T_b^\bullet}^{T_b} -\frac{\Delta_{vap}H}{RT^2} dT$$

If we assume the boiling point elevation is small, then the integrand may be assumed to be constant and T may be replaced with $T_b^\bullet$. Defining $\theta = T_b - T_b^\bullet$, we now have

$$\ln a_1 = -\frac{\Delta_{vap}H}{RT_b^{\bullet 2}}\theta \qquad (1)$$

The osmotic coefficient was given by eq. 7.36 as

$$\phi = \frac{-1000(g/kg)\ln a_1}{mM_1} \qquad (2)$$

Substituting eq. (1) into eq. (2) gives

$$\phi = \left[\frac{1000(g/kg)\Delta_{vap}H}{M_1 RT_b^{\bullet 2}}\right]\frac{\theta}{m} \qquad (3)$$

The grouping in the brackets can be used to define a boiling point elevation constant, K_b

$$K_b = \frac{M_1 RT_b^{\bullet 2}}{1000(g/kg)\Delta_{vap}H}$$

Thus, eq. (3) becomes

$$\phi = \frac{\theta}{K_b m}$$

Since ϕ approaches one in dilute solution, we have finally

$$\theta = K_b m$$

7.41

The necessary data for ethanol can be obtained from Table 4.2. Equations for boiling point elevation were derived in problem 7.39.

$$\text{Tstar}_b = 351.7 \cdot K \qquad \Delta vH = 38.58 \cdot \frac{kJ}{mole} \qquad M_1 = 46.069 \cdot \frac{gm}{mole} \qquad kJ \equiv 1000 \cdot joule$$

$$m = 0.825 \cdot \frac{mole}{kg} \qquad\qquad\qquad\qquad\qquad R \equiv 8.31451 \cdot \frac{joule}{mole \cdot K}$$

$$K_b = \frac{M_1 \cdot R \cdot \text{Tstar}_b{}^2}{\left(\frac{1000 \cdot gm}{kg}\right) \cdot \Delta vH} \qquad K_b = 1.228 \cdot \frac{K \cdot kg}{mole}$$

$$\Delta T_b = K_b \cdot m \qquad \Delta T_b = 1.0 \cdot K$$

$$T_b = \text{Tstar}_b + \Delta T_b$$

$$T_b = 352.7 \cdot K$$

7.43

The average molecular weight can be calculated using eq. 7.43, noting that the "c" is a molar concentration. We are given mass concentrations, and the two are related by the molecular weight. Thus, we may estimate the molecular weight as (mass concentration * R * T / osmotic pressure) at each data point, and then extrapolate back to infinite dilution.

$$i = 0..3$$

$$c_{mass_i} = \qquad Pi_i = \qquad MW_i = \frac{c_{mass_i} \cdot R \cdot T}{Pi_i} \qquad T = 298.15 \cdot K \qquad R \equiv 8.31451 \cdot \frac{joule}{mole \cdot K}$$

c_{mass_i}	Pi_i
$20.0 \cdot \frac{gm}{liter}$	$210.3 \cdot Pa$
	$150.4 \cdot Pa$
$15.0 \cdot \frac{gm}{liter}$	$100.5 \cdot Pa$
	$49.5 \cdot Pa$
$10.0 \cdot \frac{gm}{liter}$	
$5.0 \cdot \frac{gm}{liter}$	

$$m = \text{slope}\left(c_{mass}, MW\right) \qquad b = \text{intercept}\left(c_{mass}, MW\right)$$

$$m = -867.245 \cdot \frac{liter}{mole} \qquad b = 2.559 \cdot 10^5 \cdot \frac{gm}{mole}$$

$$MW_{fit}(c) = b + m \cdot c \qquad c := 0 \cdot \frac{gm}{liter}, 1 \cdot \frac{gm}{liter} .. 30 \cdot \frac{gm}{liter}$$

The molecular weight extrapolated to infinite dilution is just the y-intercept of the straight line fit.

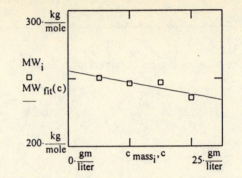

$MW := b$

$$MW = 256 \cdot \frac{kg}{mole}$$

7.45

The hydrostatic pressure is given by ρgh, while the osmotic pressure is given in the problem. All we have to do is find the depth at which the two become equal.

$$\rho := 1.03 \cdot \frac{gm}{cm^3} \qquad g = 9.81 \cdot \frac{m}{sec^2} \qquad P_{static}(h) := \rho \cdot g \cdot h \qquad P_{osmotic} := 23 \cdot atm$$

$$h_{guess} := 10 \cdot m$$

$$h := root\left(P_{static}\left(h_{guess}\right) - P_{osmotic}, h_{guess}\right)$$

$h = 231 \cdot m$ At a depth below 231 m, the hydrostatic pressure would exceed the osmotic pressure, and fresh water would flow into the pipe.

7.47

The osmotic pressure is approximated by eq. 7.43.

$$c := 0.2 \cdot \frac{mole}{liter} \qquad R := 8.31451 \cdot \frac{joule}{mole \cdot K} \qquad T := 293.15 \cdot K$$

$$Pi := c \cdot R \cdot T \qquad Pi = 4.81 \cdot atm$$

This is reasonably close to the observed value of 5.06 atm.

SECTION 7.9

7.49

The equilibrium constant is the ratio of the activities of boric acid in the various solvents. At infinite dilution, the Henry's law activity (concentration scale) is just the concentration. Thus, the equilibrium constant is just the ratio of the concentrations in the limit of infinite dilution. We can plot c (acid in H_2O) / c (acid in alcohol) as a function of either concentration and then extrapolate to zero.

$$i = 0..1$$

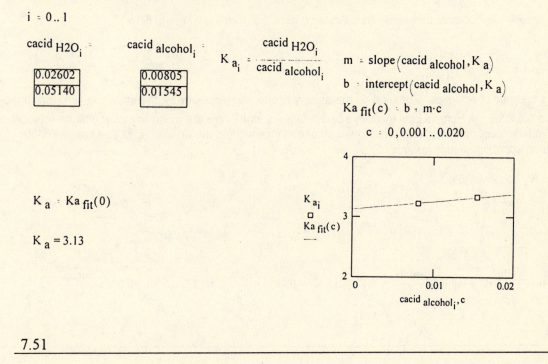

$$cacid\ H2O_i = \begin{array}{|c|} \hline 0.02602 \\ \hline 0.05140 \\ \hline \end{array}$$

$$cacid\ alcohol_i = \begin{array}{|c|} \hline 0.00805 \\ \hline 0.01545 \\ \hline \end{array}$$

$$K_{a_i} = \frac{cacid\ H2O_i}{cacid\ alcohol_i}$$

$$m = slope\left(cacid\ alcohol, K_a\right)$$

$$b = intercept\left(cacid\ alcohol, K_a\right)$$

$$Ka_{fit}(c) = b + m \cdot c$$

$$c = 0, 0.001 .. 0.020$$

$$K_a = Ka_{fit}(0)$$

$$K_a = 3.13$$

7.51

For the "reaction" I_2(solid) = I_2 (ao), we can calculate $\Delta rxnG_\theta(T)$ using data from Tables 6.1 and 8.3. Table 6.1 gives data for pure components, while Table 8.3 gives data for components in solution (HL molality scale). Since we are asked to calculate the solubility at several temperatures, we need to derive a temperature dependent expression for $\Delta rxnG_\theta(T)$. This can be done by assuming constant $\Delta rxnH_\theta$ and $\Delta rxnS_\theta$.

$$\Delta fH_{I2solid} = 0 \cdot \frac{kJ}{mole}$$

$$S_{I2solid} = 116.135 \cdot \frac{joule}{mole \cdot K}$$

$$kJ \equiv 1000 \cdot joule$$

$$R \equiv 8.31451 \cdot \frac{joule}{mole \cdot K}$$

$$\Delta fH_{I2ao} = 22.6 \cdot \frac{kJ}{mole}$$

$$S_{I2ao} = 137.2 \cdot \frac{joule}{mole \cdot K}$$

$$\Delta rxnH_{\theta} = \Delta fH_{I2ao} - \Delta fH_{I2solid}$$

$$\Delta rxnS_{\theta} = S_{I2ao} - S_{I2solid}$$

$$\Delta rxnH_{\theta} = 22.6 \cdot \frac{kJ}{mole}$$

$$\Delta rxnS_{\theta} = 21.065 \cdot \frac{joule}{mole \cdot K}$$

$$\Delta rxnG_{\theta}(T) = \Delta rxnH_{\theta} - T \cdot \Delta rxnS_{\theta}$$

The equilibrium constant can easily be calculated from the $\Delta rxnG_{\theta}$ using eq. 6.15

$$K_a(T) = \exp\left(\frac{-\Delta rxnG_{\theta}(T)}{R \cdot T}\right)$$

K_a is the ratio of activities. For solid I2, the activity is one. For aqueous I_2, the activity is $(\gamma m)/(m^{\theta})$, since Table 8.3 uses the Henry's Law molality scale reference state. We are going to assume dilute solution, so the activity coefficient will be approximated as one. Once we find the answer, we'll see that the dilute solution approximation is justified.

Thus K_a becomes just

$$K_a = \frac{m}{m_{\theta}}$$

$$m_{\theta} \equiv 1 \cdot \frac{mole}{kg}$$

$$m(T) = K_a(T) \cdot m_{\theta}$$

$$m(273.15 \cdot K) = 0.00060 \cdot \frac{mole}{kg} \quad m(298.15 \cdot K) = 0.0014 \cdot \frac{mole}{kg} \quad m(373.15 \cdot K) = 0.0086 \cdot \frac{mole}{kg}$$

7.53

For the "reaction" Xe (g) = Xe (ao), we can calculate $\Delta rxnG_{\theta}(T)$ by assuming constant $\Delta rxnH_{\theta}$ and $\Delta rxnS_{\theta}$.

$$\Delta rxnH_{\theta} = -17.6 \cdot \frac{kJ}{mole}$$

$$\Delta rxnS_{\theta} = -103.98 \cdot \frac{joule}{mole \cdot K}$$

$$kJ \equiv 1000 \cdot joule$$

$$R \equiv 8.31451 \cdot \frac{joule}{mole \cdot K}$$

$$\Delta rxnG_{\theta}(T) = \Delta rxnH_{\theta} - T \cdot \Delta rxnS_{\theta}$$

The equilibrium constant can easily be calculated from the $\Delta rxnG_{\theta}$ using eq. 6.15

$$K_a(T) = \exp\left(\frac{-\Delta rxnG_\theta(T)}{R \cdot T}\right)$$

K_a is the ratio of activities. For Xe (g), the activity is (P/P^θ). For Xe (ao), the activity is $(\gamma m)/(m^\theta)$. We are going to assume dilute solution, so the activity coefficient will be approximated as one. Once we find the answer, we'll see that the dilute solution approximation is justified.

Thus K_a becomes just

$$K_a = \frac{P_\theta}{P} \cdot \frac{m}{m_\theta} \qquad\qquad P = 1 \cdot bar \qquad P_\theta = 1 \cdot bar \qquad m_\theta = 1 \cdot \frac{mole}{kg}$$

$$bar = 10^5 \cdot Pa$$

$$m(T) = K_a(T) \cdot m_\theta \cdot \frac{P}{P_\theta}$$

$$m(273.15 \cdot K) = 0.0086 \cdot \frac{mole}{kg} \qquad m(298.15 \cdot K) = 0.0045 \cdot \frac{mole}{kg} \qquad m(373.15 \cdot K) = 0.0011 \cdot \frac{mole}{kg}$$

7.55

The first thing we need to do is calculate the equilibrium constant, K_a. For the reaction H_2S (g) = H_2S (aq), the equilibrium constant is

$$K_a = \frac{a(H2S, aq)}{a(H2S, g)} = \left(\frac{\gamma \cdot m}{m_\theta}\right) \cdot \left(\frac{P_\theta}{P_{H2S}}\right)$$

In the limit of infinite dilution, the activity coefficient is one. We will calculate K_a for each data point using the above expression (with $\gamma = 1$). Then, K_a at infinite dilution will be found by regressing the calculated K_a values vs molarity and finding the intercept at m=0.

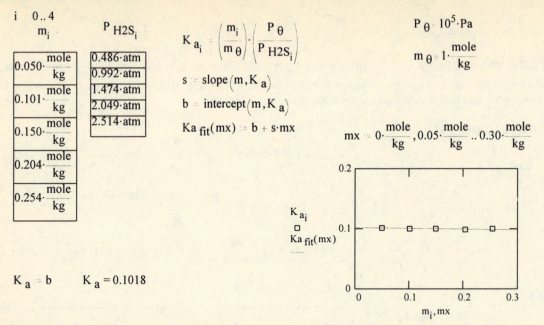

$$i := 0 .. 4$$

$$K_{a_i} := \left(\frac{m_i}{m_\theta}\right)\cdot\left(\frac{P_\theta}{P_{H2S_i}}\right) \qquad P_\theta := 10^5\cdot Pa$$

m_i	P_{H2S_i}
$0.050\cdot\dfrac{mole}{kg}$	$0.486\cdot atm$
	$0.992\cdot atm$
$0.101\cdot\dfrac{mole}{kg}$	$1.474\cdot atm$
	$2.049\cdot atm$
$0.150\cdot\dfrac{mole}{kg}$	$2.514\cdot atm$
$0.204\cdot\dfrac{mole}{kg}$	
$0.254\cdot\dfrac{mole}{kg}$	

$$m_\theta := 1\cdot\frac{mole}{kg}$$

$$s := slope\left(m, K_a\right)$$

$$b := intercept\left(m, K_a\right)$$

$$Ka_{fit}(mx) := b + s\cdot mx$$

$$mx := 0\cdot\frac{mole}{kg}, 0.05\cdot\frac{mole}{kg} .. 0.30\cdot\frac{mole}{kg}$$

$$K_a := b \qquad K_a = 0.1018$$

The equilibrium constant is related to $\Delta rxnG^\theta$ by eq. 6.15. However, we also know that $\Delta rxnG^\theta$ for this reaction is $\Delta fG^\theta(aq) - \Delta fG^\theta(g)$. $\Delta fG^\theta(g)$ can be found in Table 6.1, so we may readily solve for $\Delta fG^\theta(aq)$.

$$\Delta rxnG_\theta := -R\cdot T\cdot\ln\left(K_a\right) \qquad \Delta rxnG_\theta = 5.663\cdot\frac{kJ}{mole} \qquad T := 298.15\cdot K \qquad R = 8.31451\cdot\frac{joule}{mole\cdot K}$$

$$\Delta fG_{\theta gas} := -33.56\cdot\frac{kJ}{mole} \qquad\qquad\qquad kJ = 1000\cdot joule$$

$$\Delta fG_{\theta aq} := \Delta rxnG_\theta + \Delta fG_{\theta gas} \qquad \Delta fG_{\theta aq} = -27.90\cdot\frac{kJ}{mole}$$

This is in excellent agreement with the value given in Table 8.3, -27.83 kJ/mole.

7.57

The first thing we need to do is calculate the equilibrium constant, K_a. For the reaction glycine (s) = glycine (aq), the equilibrium constant is

$$K_a = \frac{\gamma_{2m} \cdot m}{m_\theta}$$

$$\gamma_{2m} = 0.872 \qquad m = 3.3 \cdot \frac{mole}{kg} \qquad m_\theta \equiv 1 \cdot \frac{mole}{kg}$$

$$K_a = 2.878$$

The equilibrium constant is related to $\Delta rxnG^\theta$ by eq. 6.15. However, we also know that $\Delta rxnG^\theta$ for this reaction is $\Delta fG^\theta(aq) - \Delta fG^\theta(s)$. $\Delta fG^\theta(s)$ can be found in Table 7.7, so we may readily solve for $\Delta fG^\theta(aq)$.

$$\Delta rxnG_\theta = -R \cdot T \cdot \ln(K_a) \qquad \Delta rxnG_\theta = -2.62 \cdot \frac{kJ}{mole} \qquad T = 298.15 \cdot K \qquad R = 8.31451 \cdot \frac{joule}{mole \cdot K}$$

$$\Delta fG_{\theta solid} = -370.7 \cdot \frac{kJ}{mole} \qquad\qquad kJ = 1000 \cdot joule$$

$$\Delta fG_{\theta aq} = \Delta rxnG_\theta + \Delta fG_{\theta solid} \qquad \Delta fG_{\theta aq} = -373.3 \cdot \frac{kJ}{mole}$$

This is in excellent agreement with the value given in Table 7.7, -373.0 kJ/mole.

7.59

Given

$$X_2(T) = \exp\left[-6.0579 - \frac{1152.5}{\left(\frac{T}{K}\right)} \right] \qquad P = 1 \cdot atm \qquad P_\theta = 10^5 \cdot Pa$$

For the reaction, He(g) = He(NMA), the equilibrium constant can be written as follows assuming an ideal solution (HL, X scale).

$$K_a(T) = \frac{X_2(T)}{\left(\frac{P}{P_\theta}\right)}$$

With an expression for the equilibrium constant, eq. 6.15 can be used to obtain $\Delta rxnG_\theta(T)$.

$$\Delta rxnG_{\theta}(T) = -R \cdot T \cdot \ln\left(K_a(T)\right) \qquad\qquad R = 8.31451 \cdot \frac{joule}{mole \cdot K}$$

An expression for $\Delta rxnH_{\theta}(T)$ can be found using eq. 6.23. Taking the derivative of $\ln(K_a(T))$ with respect to $1/T$ gives $-\Delta rxnH_{\theta}/R$.

$$\Delta rxnH_{\theta} = -R \cdot (-1152.5 \cdot K)$$

The "reaction" here is just the formation reaction for He in solution in NMA, since He(g) is already in its most stable form at standard conditions. Therefore, $\Delta fG(He, NMA:X) = \Delta rxnG_{\theta}$ and $\Delta fH(He, NMA:X) = \Delta rxnH_{\theta}$

$$\Delta fG(T) := \Delta rxnG_{\theta}(T) \qquad\qquad \Delta fH := \Delta rxnH_{\theta} \qquad\qquad kJ = 1000 \cdot joule$$

$$\Delta fG(298.15 \cdot K) = 24.63 \cdot \frac{kJ}{mole} \qquad \Delta fH = 9.58 \cdot \frac{kJ}{mole}$$

SECTION 7.10

7.61

This system is similar to that shown in Figure 7.19(c).

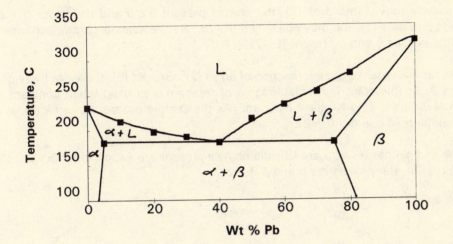

7.63

α represents the solution that is rich in component A. β represents the solutions that is rich in component B.

(1) α
(2) α + vapor
(3) vapor + β
(4) β
(5) α + β

Cooling histories:

Degrees of Freedom (F) = c + 2 - p.
Composition of single phase regions are given by extending dotted line down to composition ax
Compositions of two phase regions are given by the ends of horizontal tie lines spanning that area
that temperature.

All samples start as a homogenous vapor (F = 2 + 2-1 = 3).

(a) As the sample moves into area (2), the phases present are α and the vapor (F = 2). When t
sample passes into area (1), the only phase is α (F = 3). As the sample passes into area (5), there a
two immiscible phases α and β present (F = 2).

(b) When the sample reaches the intersection of areas (2), (3), and (5), there are three phases presen
vapor, α, and β. At this point F = 1; this degree of freedom was used when constant pressure w
specified in creating the T vs X phase diagram. As the sample passes into area (5), the two phas
present are immisicible α and β (F = 2).

(c) When the sample passes into are (3), the phases present are vapor and β (F = 2). As the samp
passes into area (5), the phases are α and β (F = 2).

CHAPTER 8   *Ionic Solutions*

SECTION 8.1

8.1

From eq. 8.1, we can calculate the "osmotic factor," i. We are given the molality of the acid, and the freezing point depression constant of water is found in Table 7.6.

$$K_f = 1.860 \cdot \frac{K}{\left(\frac{mole}{kg}\right)} \qquad \theta = 0.0208 \cdot K \qquad m = 0.01 \cdot \frac{mole}{kg}$$

$$i = \frac{\theta}{K_f m} \qquad i = 1.118$$

The degree of ionization (or dissociation) can be found using eq. 8.2. $\nu = 2$ since a weak acid should only dissociate into two ions at most.

$$\nu = 2 \qquad \alpha = \frac{i - 1}{\nu - 1} \qquad \alpha = 0.118$$

8.3

The activity coefficients can be evaluated by using eq. 7.37, but we first need an expression for j(m). We can use the given freezing point depression data to estimate the osmotic coefficient ϕ, using eq. 8.8 ignoring the b term. j is calculated as $1-\phi$. Finally, the desired fit for j(m) can be found by regressing (j/m) vs 1/sqrt(m).

$$i = 0..5$$

$$\nu = 2 \qquad K_f = 1.860 \qquad \phi_i = \frac{\theta_i}{\nu \cdot K_f m_i} \qquad j_i = 1 - \phi_i$$

$m_i =$	$\theta_i =$
0.003612	0.01316
0.006690	0.02421
0.009872	0.03509
0.016215	0.05712
0.030369	0.10541
0.048335	0.16359

$$x_i = \frac{1}{\sqrt{m_i}} \qquad y_i = \frac{j_i}{m_i}$$

$$s = slope(x, y) \qquad b = intercept(x, y)$$

$$y_{fit}(molal) = b + s \cdot \frac{1}{\sqrt{molal}}$$

$$molal = 0.003, 0.004 .. 0.050$$

$$j_{fit}(molal) = molal \cdot y_{fit}(molal)$$

$$\gamma_{mean}(m) = \exp\left(-j_{fit}(m) - \int_0^m \frac{j_{fit}(m)}{m} \, dm\right)$$

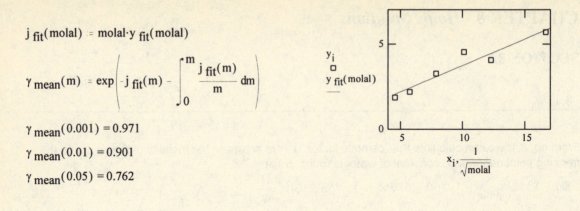

$$\gamma_{mean}(0.001) = 0.971$$

$$\gamma_{mean}(0.01) = 0.901$$

$$\gamma_{mean}(0.05) = 0.762$$

8.5

The distance of closest approach is the a_o parameter in the Debye-Huckel equation, eq. 8.10. For a 1:1 electrolyte like KCl, $z_+ = z_- = 1$, and $I = m$. With these simplifications and a little rearrangement, eq. 8.10 becomes

$$\frac{1}{\ln(\gamma_{mean})} = \frac{-B \cdot a_o}{\alpha} + \frac{1}{\alpha \cdot \sqrt{m}}$$

Thus, ao can be determined by regressing $1/\ln(\gamma_{mean})$ vs $1/sqrt(m)$ and finding the intercept. Only experimental activity coefficients up to m = 0.1 mol/kg will be used since the DH equation is only valid in this range. Values for α and B can be found in Table 8.2 for 25 C.

$i = 0 .. 5$

$\alpha = 1.177 \quad B = 0.329$

$m_i =$	$\gamma\,\text{mean}_i =$
0.001	0.9648
0.005	0.927
0.01	0.901
0.02	0.868
0.05	0.816
0.1	0.769

$$x_i := \frac{1}{\sqrt{m_i}} \qquad y_i := \frac{1}{\ln\left(\gamma\,\text{mean}_i\right)}$$

$$s := \text{slope}(x,y) \qquad b := \text{intercept}(x,y) \qquad y_{\text{fit}}(\text{molal}) := b + s \cdot \frac{1}{\sqrt{\text{molal}}}$$

$$\text{molal} := 0.001 , 0.002 .. 0.1$$

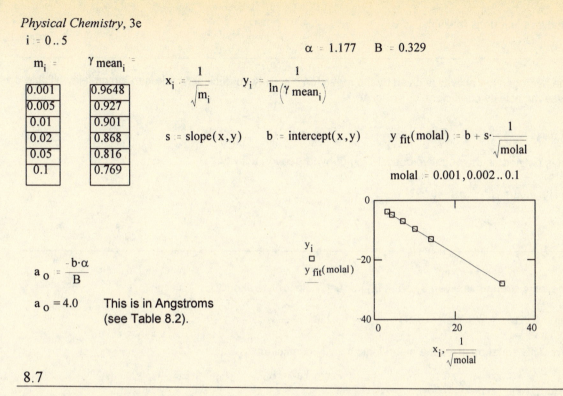

$$a_o := \frac{-b \cdot \alpha}{B}$$

$a_o = 4.0$ This is in Angstroms (see Table 8.2).

8.7

(a) The DHG equation is given by eq. 8.13. For a 1:1 electrolyte, the DHG equation can be solved for β to give

$$\beta := \frac{1}{2 \cdot m} \cdot \left(\ln(\gamma\,\text{mean}) + \frac{\alpha \cdot \sqrt{m}}{1 + \sqrt{m}} \right) \qquad m := 0.05 \quad \alpha := 1.177 \quad \gamma\,\text{mean} := 0.830$$

$\beta = 0.288$

(b) The DHG equation can be used to predict activity coefficients at other molalities with this β value.

$$\gamma\,\text{mean}(m) := \exp\left(\frac{-\alpha \cdot \sqrt{m}}{1 + \sqrt{m}} + 2 \cdot \beta \cdot m \right)$$

$i := 0 .. 5 \quad m_i :=$ $\gamma\,\text{actual}_i =$

m_i	$\gamma\,\text{mean}(m_i)$	$\gamma\,\text{actual}_i$
0.01	0.904	0.904
0.02	0.874	0.875
0.05	0.830	0.830
0.1	0.798	0.796
0.2	0.780	0.767
0.5	0.819	0.757

Remember, the DHG equation does not work well for ionic strengths greater than 0.1 mole/kg.

8.9
___

The hydrogen ion activity is given by the mean ionic activity coefficient * hydrogen ion molality. Table 8.1 gives the activity coefficient for this case.

$$\gamma_{mean} = 0.809 \qquad m_H = 1 \qquad a_H = \gamma_{mean} \cdot m_H \qquad a_H = 0.809$$

pH is defined as the negative base 10 log of hydrogen ion activity.

$$pH = -\log(a_H) \qquad pH = 0.0921$$

8.11
___

The ionic strength is given by eq. 8.9. For a 1:1 electrolyte like KCl,

$$m = 0.0250 \qquad I = \frac{1}{2} \cdot (m + m) \qquad I = 0.0250$$

The DHG equation gives the mean ionic activity coefficient

$$\gamma_{mean}(\alpha) = \exp\left(\frac{-\alpha \cdot \sqrt{I}}{1 + \sqrt{I}}\right)$$

From Table 8.2,

$$\alpha_0 = 1.133$$
$$\alpha_{25} = 1.177$$
$$\alpha_{100} = 1.372$$

$$\gamma_{mean}(\alpha_0) = 0.857$$

$$\gamma_{mean}(\alpha_{25}) = 0.852$$

$$\gamma_{mean}(\alpha_{100}) = 0.829$$

The mean ionic activity is given by

$$a_{mean} = \left(m_{plus} \cdot m_{minus} \cdot \gamma_{mean}^{\nu}\right)^{\frac{1}{\nu}}$$

Since

$$m_{plus} = m \qquad m_{minus} = m \qquad \nu = 2$$

$$a_{mean}(\alpha) = \left(m_{plus} \cdot m_{minus} \cdot \gamma_{mean}(\alpha)^{\nu}\right)^{\frac{1}{\nu}}$$

$$a_{mean}(\alpha_0) = 0.0214$$

$$a_{mean}(\alpha_{25}) = 0.0213$$

$$a_{mean}(\alpha_{100}) = 0.0207$$

8.13

The ionic strength is given by eq. 8.9. For a 3:2 electrolyte like $La_2(SO_4)_3$,

$$m = 0.00150 \qquad I = \frac{1}{2} \cdot \left[3^2 \cdot 2 \cdot m + 2^2 \cdot (3 \cdot m) \right] \qquad I = 0.0225$$

The DHG equation gives the mean ionic activity coefficient

$$\gamma_{mean} := \exp\left(\frac{-\alpha \cdot 3 \cdot 2 \cdot \sqrt{I}}{1 + \sqrt{I}} \right) \qquad\qquad \text{From Table 8.2,} \qquad \alpha := 1.177$$

$$\gamma_{mean} = 0.398$$

SECTION 8.2
8.15

For the reaction, DCAH = DCA + H, the equilibrium constant is

$$K_a = \frac{(\gamma_{mean} \cdot m_{DCA}) \cdot (\gamma_{mean} \cdot m_H)}{\gamma_{DCAH} \cdot m_{DCAH}} \qquad\qquad K_a = 3.32 \cdot 10^{-2}$$

(a) Assuming ideal solution, all activity coefficients are one. We can introduce ξ as the degree of dissociation.

$$m_o = 0.125 \qquad m_{DCA}(\xi) = \xi \cdot m_o \qquad m_H(\xi) = \xi \cdot m_o \qquad m_{DCAH}(\xi) = (1 - \xi) \cdot m_o$$

$$\xi_{guess} := 0.5$$

$$\xi_{ideal} := root\left(\frac{m_{DCA}(\xi_{guess}) \cdot m_H(\xi_{guess})}{m_{DCAH}(\xi_{guess})} - K_a, \xi_{guess} \right)$$

$$\xi_{ideal} = 0.40$$

Student's Solutions Manual

(b) Using DHG for the mean ionic activity coefficient, we'll still assume ideal solution for the neutral acid. The DHG equation requires the ionic strength, which depends on the degree of dissociation, so iterative solution is required.

$$I(\xi) := \frac{1}{2} \cdot \left(m_{DCA}(\xi) + m_H(\xi) \right) \qquad \gamma_{mean}(\xi) := \exp\left(\frac{-\alpha \cdot \sqrt{I(\xi)}}{1 + \sqrt{I(\xi)}} \right) \qquad \alpha := 1.177$$

from Table 8.2

First Approximation, Ideal dissociation

$$\gamma_{mean}\left(\xi_{ideal} \right) = 0.807$$

$$\xi_{1st} := root\left(\frac{m_{DCA}\left(\xi_{ideal} \right) \cdot m_H\left(\xi_{ideal} \right) \cdot \gamma_{mean}\left(\xi_{ideal} \right)^2}{m_{DCAH}\left(\xi_{ideal} \right)} - K_a, \xi_{ideal} \right)$$

$$\xi_{1st} = 0.469$$

Second Approximation

$$\gamma_{mean}\left(\xi_{1st} \right) = 0.795$$

$$\xi_{2nd} := root\left(\frac{m_{DCA}\left(\xi_{1st} \right) \cdot m_H\left(\xi_{1st} \right) \cdot \gamma_{mean}\left(\xi_{1st} \right)^2}{m_{DCAH}\left(\xi_{1st} \right)} - K_a, \xi_{1st} \right)$$

$$\xi_{2nd} = 0.472$$

Third Approximation

$$\gamma_{mean}\left(\xi_{2nd} \right) = 0.795$$

$$\xi_{3rd} := root\left(\frac{m_{DCA}\left(\xi_{2nd} \right) \cdot m_H\left(\xi_{2nd} \right) \cdot \gamma_{mean}\left(\xi_{2nd} \right)^2}{m_{DCAH}\left(\xi_{2nd} \right)} - K_a, \xi_{2nd} \right)$$

$$\xi_{3rd} = 0.472$$

The iteration has converged.

pH is defined as the $- \log_{10}(a_H)$

$$pH := -\log\left(\gamma_{mean}\left(\xi_{3rd} \right) \cdot m_H\left(\xi_{3rd} \right) \right) \qquad pH = 1.33$$

Recalculate the equilibrium constant to check:

$$Ka_{check} := \frac{m_{DCA}\left(\xi_{3rd} \right) \cdot m_H\left(\xi_{3rd} \right) \cdot \gamma_{mean}\left(\xi_{3rd} \right)^2}{m_{DCAH}\left(\xi_{3rd} \right)} \qquad Ka_{check} = 3.32 \cdot 10^{-2}$$

8.17

For the reaction given, the equilibrium constant can be calculated from data in Tables 6.1 and 8.3.

$$\Delta fG_{CH3NH2} = 20.77 \cdot \frac{kJ}{mole} \qquad \Delta fG_{H2Oliq} = -237.129 \cdot \frac{kJ}{mole} \qquad kJ = 1000 \cdot joule \qquad R = 8.31451 \cdot \frac{joule}{mole \cdot K}$$

$$\Delta fG_{CH3NH3} = -39.86 \cdot \frac{kJ}{mole} \qquad \Delta fG_{OH} = -157.244 \cdot \frac{kJ}{mole} \qquad T = 298.15 \cdot K$$

$$\Delta rxnG_{\theta} = \Delta fG_{CH3NH3} + \Delta fG_{OH} - \Delta fG_{CH3NH2} - \Delta fG_{H2Oliq} \qquad \Delta rxnG_{\theta} = 19.255 \cdot \frac{kJ}{mole}$$

$$K_a = \exp\left(\frac{-\Delta rxnG_{\theta}}{R \cdot T}\right) \qquad K_a = 4.233 \cdot 10^{-4}$$

The equilibrium constant is given by
$$K_a = \frac{(\gamma_{mean} \cdot m_{CH3NH3}) \cdot (\gamma_{mean} \cdot m_{OH})}{(\gamma_{CH3NH2} \cdot m_{CH3NH2}) \cdot a_{H2O}}$$

Water is referenced to the RL reference state. Since X_{H2O} is almost one, we will assume the activity of water is also one. We will also assume the activity coefficient for the neutral base is one. Defining ξ as the degree of dissociation, the molalities are expressed as

$$m_o = 0.178 \qquad m_{CH3NH3}(\xi) = \xi \cdot m_o \qquad m_{OH}(\xi) = \xi \cdot m_o \qquad m_{CH3NH2}(\xi) = (1 - \xi) \cdot m_o$$

We will use the DHG equation to estimate the mean ionic activity coefficient. However, this quantity depends on the ionic strength, which in turn depends on the degree of dissociation. Since we don't know what ξ is yet, we will assume an ideal solution first, and then iterate successively.

$$I(\xi) = \frac{1}{2} \cdot (m_{CH3NH3}(\xi) + m_{OH}(\xi)) \qquad \gamma_{mean}(\xi) = \exp\left(\frac{-\alpha \cdot \sqrt{I(\xi)}}{1 + \sqrt{I(\xi)}}\right) \qquad \alpha = 1.177$$

from Table 8.2

$$\xi_{guess} = 0.5 \qquad\qquad TOL = 10^{-6}$$

$$\xi_{ideal} = root\left(\frac{m_{CH3NH3}(\xi_{guess}) \cdot m_{OH}(\xi_{guess})}{m_{CH3NH2}(\xi_{guess})} - K_a, \xi_{guess}\right) \qquad \xi_{ideal} = 0.048$$

First Approximation

$$\gamma_{mean}(\xi_{ideal}) = 0.906$$

$$\xi_{1st} = root\left(\frac{m_{CH3NH3}(\xi_{ideal}) \cdot m_{OH}(\xi_{ideal}) \cdot \gamma_{mean}(\xi_{ideal})^2}{m_{CH3NH2}(\xi_{ideal})} - K_a, \xi_{ideal}\right) \qquad \xi_{1st} = 0.053$$

Second Approximation

$$\gamma_{mean}(\xi_{1st}) = 0.901$$

$$\xi_{2nd} = root\left(\frac{m_{CH3NH3}(\xi_{1st}) \cdot m_{OH}(\xi_{1st}) \cdot \gamma_{mean}(\xi_{1st})^2}{m_{CH3NH2}(\xi_{1st})} - K_{a}, \xi_{1st}\right) \qquad \xi_{2nd} = 0.053$$

The iteration has converged.

Recalculate the equilibrium constant to check:

$$Ka_{check} = \frac{m_{CH3NH3}(\xi_{2nd}) \cdot m_{OH}(\xi_{2nd}) \cdot \gamma_{mean}(\xi_{2nd})^2}{m_{CH3NH2}(\xi_{2nd})} \qquad Ka_{check} = 4.233 \cdot 10^{-4}$$

The concentration of hydroxide ion and the pH of the solution are given by:

$$m_{OH}(\xi_{2nd}) = 0.00937 \qquad pH = 14 + log(\gamma_{mean}(\xi_{2nd}) \cdot m_{OH}(\xi_{2nd})) \qquad pH = 11.927$$

Remember that pH + pOH = 14 is a useful rule of thumb. If you didn't know this, just do the calculations for $H_2O = H^+ + OH^-$ and you'll find that $-log[a(H^+)] - log[a(OH^-)] = 14.00$.

8.19

For the reaction, HCOOH = HCOO + H, the equilibrium constant can be calculated from data in Table 8.3.

$$\Delta fG_{HCOOH} = -372.3 \cdot \frac{kJ}{mole} \qquad \Delta fG_{HCOO} = -351.0 \cdot \frac{kJ}{mole} \qquad \Delta fG_{H} = 0 \cdot \frac{kJ}{mole} \qquad kJ = 1000 \cdot joule$$

$$\Delta rxnG_{\theta} = \Delta fG_{HCOO} + \Delta fG_{H} - \Delta fG_{HCOOH} \qquad\qquad R = 8.31451 \cdot \frac{joule}{mole \cdot K}$$

$$\Delta rxnG_{\theta} = 21.3 \cdot \frac{kJ}{mole} \qquad\qquad T = 298.15 \cdot K$$

$$K_{a} = exp\left(\frac{-\Delta rxnG_{\theta}}{R \cdot T}\right) \qquad K_{a} = 1.855 \cdot 10^{-4}$$

The equilibrium constant can also be expressed as

$$K_{a} = \frac{(\gamma_{mean} \cdot m_{HCOO}) \cdot (\gamma_{mean} \cdot m_{H})}{\gamma_{HCOOH} \cdot m_{HCOOH}}$$

We can introduce ξ as the degree of dissociation.

$$m_{o} = 0.022 \qquad m_{HCOOH}(\xi) = (1 - \xi) \cdot m_{o} \qquad m_{HCOO}(\xi) = \xi \cdot m_{o} \qquad m_{H}(\xi) = \xi \cdot m_{o}$$

Using DHG for the mean ionic activity coefficient, we'll assume ideal solution for the neutral acid. The DHG equation requires the ionic strength, which depends on the degree of dissociation, so iterative solution is required. As a starting point, we'll assume ideal solution to calculate the first ξ.

$$I(\xi) = \frac{1}{2} \cdot \left(m_{HCOO}(\xi) + m_H(\xi) \right) \qquad \gamma_{mean}(\xi) = \exp\left(\frac{-\alpha \cdot \sqrt{I(\xi)}}{1 + \sqrt{I(\xi)}} \right)$$

$\alpha = 1.177$

from Table 8.2

$$\xi_{guess} := 0.5$$

$$\text{TOL} = 10^{-8}$$

$$\xi_{ideal} = root\left(\frac{m_{HCOO}(\xi_{guess}) \cdot m_H(\xi_{guess})}{m_{HCOOH}(\xi_{guess})} - K_a, \xi_{guess} \right)$$

$$\xi_{ideal} = 0.0877$$

First Approximation (using ideal dissociation to calculate γ_{mean})

$$\gamma_{mean}(\xi_{ideal}) = 0.952$$

$$\xi_{1st} = root\left(\frac{m_{HCOO}(\xi_{ideal}) \cdot m_H(\xi_{ideal}) \cdot \gamma_{mean}(\xi_{ideal})^2}{m_{HCOOH}(\xi_{ideal})} - K_a, \xi_{ideal} \right)$$

$$\xi_{1st} = 0.0921$$

Second Approximation

$$\gamma_{mean}(\xi_{1st}) = 0.951$$

$$\xi_{2nd} = root\left(\frac{m_{HCOO}(\xi_{1st}) \cdot m_H(\xi_{1st}) \cdot \gamma_{mean}(\xi_{1st})^2}{m_{HCOOH}(\xi_{1st})} - K_a, \xi_{1st} \right)$$

$$\xi_{2nd} = 0.0921$$

The iteration has converged.

Recalculate the equilibrium constant to check:

$$Ka_{check} := \frac{m_{HCOO}(\xi_{2nd}) \cdot m_H(\xi_{2nd}) \cdot \gamma_{mean}(\xi_{2nd})^2}{m_{HCOOH}(\xi_{2nd})} \qquad Ka_{check} = 1.855 \cdot 10^{-4}$$

For the reaction, CH3COOH = CH3COO + H, the equilibrium constant can be calculated from data in Table 8.3

$$\Delta fG_{CH3COOH} = -396.46 \cdot \frac{kJ}{mole} \qquad \Delta fG_{CH3COO} = -369.31 \cdot \frac{kJ}{mole} \qquad \Delta fG_H = 0 \cdot \frac{kJ}{mole}$$

$$\Delta rxnG_\theta = \Delta fG_{CH3COO} + \Delta fG_H - \Delta fG_{CH3COOH} \qquad\qquad kJ = 1000 \cdot joule$$

$$\Delta rxnG_\theta = 27.15 \cdot \frac{kJ}{mole} \qquad\qquad R = 8.31451 \cdot \frac{joule}{mole \cdot K}$$

$$K_a = exp\left(\frac{-\Delta rxnG_\theta}{R \cdot T}\right) \qquad K_a = 1.752 \cdot 10^{-5} \qquad\qquad T = 298.15 \cdot K$$

The equilibrium constant can also be expressed as

$$K_a = \frac{(\gamma_{mean} \cdot m_{CH3COO}) \cdot (\gamma_{mean} \cdot m_H)}{\gamma_{CH3COOH} \cdot m_{CH3COOH}}$$

We can introduce ξ as the degree of dissociation and express the molalities as follows. Since HCl is a very strong acid, we can assume that it is totally dissociated.

$$m_0 = 0.012 \qquad m_{CH3COOH}(\xi) = (1 - \xi) \cdot m_0 \qquad m_{CH3COO}(\xi) = \xi \cdot m_0$$

$$m_H(\xi) = \xi \cdot m_0 + 0.05 \qquad m_{Cl} = 0.05$$

Using DHG for the mean ionic activity coefficient, we'll assume ideal solution for the neutral acetic acid. The DHG equation requires the ionic strength, which depends on the degree of dissociation, so iterative solution is required. As a starting point, we'll assume ideal solution to calculate the first ξ.

$$I(\xi) = \frac{1}{2} \cdot (m_{CH3COO}(\xi) + m_H(\xi) + m_{Cl}) \qquad \gamma_{mean}(\xi) = exp\left(\frac{-\alpha \cdot \sqrt{I(\xi)}}{1 + \sqrt{I(\xi)}}\right) \qquad \begin{array}{l} \alpha = 1.177 \\[4pt] \text{from Table 8.2} \end{array}$$

$$\xi_{guess} = 0.5$$

$$TOL = 10^{-8}$$

$$\xi_{ideal} = root\left(\frac{m_{CH3COO}(\xi_{guess}) \cdot m_H(\xi_{guess})}{m_{CH3COOH}(\xi_{guess})} - K_a, \xi_{guess}\right)$$

$$\xi_{ideal} = 3.5026 \cdot 10^{-4}$$

First Approximation (using ideal dissociation to calculate γ_{mean})

$$\gamma_{mean}(\xi_{ideal}) = 0.806$$

$$\xi_{1st} = root\left(\frac{m_{CH3COO}(\xi_{ideal}) \cdot m_H(\xi_{ideal}) \cdot \gamma_{mean}(\xi_{ideal})^2}{m_{CH3COOH}(\xi_{ideal})} - K_a, \xi_{ideal}\right) \qquad \xi_{1st} = 0.00054$$

Second Approximation

$$\gamma_{mean}(\xi_{1st}) = 0.806$$

$$\xi_{2nd} = root\left(\frac{m_{CH3COO}(\xi_{1st}) \cdot m_H(\xi_{1st}) \cdot \gamma_{mean}(\xi_{1st})^2}{m_{CH3COOH}(\xi_{1st})} - K_a, \xi_{1st}\right) \qquad \xi_{2nd} = 0.00054$$

The iteration has converged. Recalculate the equilibrium constant to check:

$$Ka_{check} = \frac{m_{CH3COO}(\xi_{2nd}) \cdot m_H(\xi_{2nd}) \cdot \gamma_{mean}(\xi_{2nd})^2}{m_{CH3COOH}(\xi_{2nd})} \qquad Ka_{check} = 1.752 \cdot 10^{-5}$$

The pH is defined as $-\log(a_H)$

$$pH = -\log(\gamma_{mean}(\xi_{2nd}) \cdot m_H(\xi_{2nd})) \qquad pH = 1.394$$

8.23

For the reaction, Ag_2SO_4 (s) = 2 Ag (ao) + SO_4 (ao), the equilibrium constant can be calculated from data in Tables 6.1 and 8.3.

$$\Delta fG_{Ag2SO4} = -618.41 \cdot \frac{kJ}{mole} \qquad \Delta fG_{Ag} = 77.107 \cdot \frac{kJ}{mole} \qquad \Delta fG_{SO4} = -744.53 \cdot \frac{kJ}{mole} \qquad kJ = 1000 \cdot joule$$

$$\Delta rxnG_\theta = 2 \cdot \Delta fG_{Ag} + \Delta fG_{SO4} - \Delta fG_{Ag2SO4} \qquad R = 8.31451 \cdot \frac{joule}{mole \cdot K}$$

$$\Delta rxnG_\theta = 28.094 \cdot \frac{kJ}{mole} \qquad T = 298.15 \cdot K$$

$$K_a = exp\left(\frac{-\Delta rxnG_\theta}{R \cdot T}\right) \qquad K_a = 1.197 \cdot 10^{-5}$$

The equilibrium constant can also be expressed as

$$K_a = \frac{(\gamma_{mean} \cdot m_{Ag})^2 \cdot (\gamma_{mean} \cdot m_{SO4})}{a_{Ag2SO4}}$$

We can introduce S as the moles of Ag_2SO_4 soluble in 1 kg of water. Since $CuSO_4$ is very soluble in water, we will assume that it is totally dissociated.

$$m_{Ag}(S) := 2 \cdot S \qquad m_{SO4}(S) := S + 0.012 \qquad m_{Cu} := 0.012$$

We will use DHG for the mean ionic activity coefficient. The activity of the solid salt is one. The DHG equation requires the ionic strength, which depends on the solubility, so an iterative solution is required. As a starting point, we'll assume ideal solution to calculate the first estimate for S.

$$I(S) := \frac{1}{2} \cdot \left(1^2 \cdot m_{Ag}(S) + 2^2 \cdot m_{SO4}(S) + 2^2 \cdot m_{Cu}\right) \qquad \gamma_{mean}(S) := \exp\left(\frac{-\alpha \cdot 2 \cdot \sqrt{I(S)}}{1 + \sqrt{I(S)}}\right) \qquad \begin{array}{l} \alpha := 1.177 \\ \text{from Table 8.2} \end{array}$$

$$S_{guess} := 0.1 \qquad\qquad\qquad\qquad TOL := 10^{-10}$$

$$S_{ideal} := root\left(m_{Ag}(S_{guess})^2 \cdot m_{SO4}(S_{guess}) - K_a, S_{guess}\right) \qquad\qquad S_{ideal} = 0.0113$$

First Approximation (using ideal solubility to calculate γ_{mean})

$$\gamma_{mean}(S_{ideal}) = 0.592$$

$$S_{1st} := root\left(m_{Ag}(S_{ideal})^2 \cdot m_{SO4}(S_{ideal}) \cdot \gamma_{mean}(S_{ideal})^3 - K_a, S_{ideal}\right) \qquad S_{1st} = 0.0228$$

Second Approximation

$$\gamma_{mean}(S_{1st}) = 0.550$$

$$S_{2nd} := root\left(m_{Ag}(S_{1st})^2 \cdot m_{SO4}(S_{1st}) \cdot \gamma_{mean}(S_{1st})^3 - K_a, S_{1st}\right) \qquad S_{2nd} = 0.0228$$

The iteration has converged.

Recalculate the equilibrium constant to check:

$$Ka_{check} := m_{Ag}(S_{2nd})^2 \cdot m_{SO4}(S_{2nd}) \cdot \gamma_{mean}(S_{2nd})^3 \qquad\qquad Ka_{check} = 1.197 \cdot 10^{-5}$$

8.25

Since we now have two independent processes by which the salt may solubilize, we have to calculate the individual solubilities and add them together.

For the reaction, $AgCl(s) = Ag^+ (ao) + Cl^- (ao)$, the equilibrium constant at 100 C can be calculated from data in Tables 6.1 and 8.3. We will assume $\Delta rxnH_\theta$ and $\Delta rxnS_\theta$ are independent of T.

$$\Delta fH_{AgCl} = -127.068 \cdot \frac{kJ}{mole} \qquad \Delta fH_{Ag} = 105.579 \cdot \frac{kJ}{mole} \qquad \Delta fH_{Cl} = -167.159 \cdot \frac{kJ}{mole} \qquad kJ \equiv 1000 \cdot joule$$

$$\Delta rxnH_\theta = \Delta fH_{Ag} + \Delta fH_{Cl} - \Delta fH_{AgCl} \qquad \Delta rxnH_\theta = 65.488 \cdot \frac{kJ}{mole} \qquad\qquad R = 8.31451 \cdot \frac{joule}{mole \cdot K}$$

$$S_{AgCl} = 96.2 \cdot \frac{joule}{mole \cdot K} \qquad S_{Ag} = 72.68 \cdot \frac{joule}{mole \cdot K} \qquad S_{Cl} = 56.5 \cdot \frac{joule}{mole \cdot K}$$

$$\Delta rxnS_\theta = S_{Ag} + S_{Cl} - S_{AgCl} \qquad\qquad \Delta rxnS_\theta = 32.98 \cdot \frac{joule}{mole \cdot K}$$

$$\Delta rxnG_\theta(T) = \Delta rxnH_\theta - T \cdot \Delta rxnS_\theta \qquad\qquad \Delta rxnG_\theta(373.15 \cdot K) = 53.182 \cdot \frac{kJ}{mole}$$

$$K_a(T) = exp\left(\frac{-\Delta rxnG_\theta(T)}{R \cdot T}\right) \qquad\qquad K_a(373.15 \cdot K) = 3.595 \cdot 10^{-8}$$

The equilibrium constant can also be expressed as

$$K_a = \frac{(\gamma_{mean} \cdot m_{Ag}) \cdot (\gamma_{mean} \cdot m_{Cl})}{a_{AgCl}}$$

We can introduce S_1 as the moles of AgCl soluble in 1 kg of water by ionization. Since $NaNO_3$ is very soluble in water, we will assume it is totally dissociated.

$$m_{Ag}(S_1) = S_1 \qquad m_{Cl}(S_1) = S_1 \qquad m_{Na} = 0.034 \qquad m_{NO3} = 0.034$$

We will use DHG for the mean ionic activity coefficient. The activity of the solid salt is one. The DHG equation requires the ionic strength, which depends on the solubility, so an iterative solution is required. As a starting point, we'll assume ideal solution to calculate the first estimate for S_1.

$$I(S_1) = \frac{1}{2} \cdot (m_{Ag}(S_1) + m_{Cl}(S_1) + m_{Na} + m_{NO3})$$

$$\gamma_{mean}(S_1) = exp\left(\frac{-\alpha \cdot \sqrt{I(S_1)}}{1 + \sqrt{I(S_1)}}\right) \qquad \alpha = 1.372 \text{ from Table 8.2 for 100 C}$$

$S_{guess} = 0.1$ $\qquad\qquad\qquad\qquad\qquad\qquad\qquad\qquad$ $TOL = 10^{-10}$

$S_{id} = root(m_{Ag}(S_{guess}) \cdot m_{Cl}(S_{guess}) - K_a(373.15 \cdot K), S_{guess})$ $\qquad$ $S_{id} = 1.896 \cdot 10^{-4}$

Successive Approximations

$S_{1st} = root(m_{Ag}(S_{id}) \cdot m_{Cl}(S_{id}) \cdot \gamma_{mean}(S_{id})^2 - K_a(373.15 \cdot K), S_{id})$ $\qquad$ $S_{1st} = 2.349 \cdot 10^{-4}$

$S_{2nd} = root(m_{Ag}(S_{1st}) \cdot m_{Cl}(S_{1st}) \cdot \gamma_{mean}(S_{1st})^2 - K_a(373.15 \cdot K), S_{1st})$ $\qquad$ $S_{2nd} = 2.349 \cdot 10^{-4}$

The iteration has converged. Recalculate the equilibrium constant to check:

$Ka_{check} = m_{Ag}(S_{2nd}) \cdot m_{Cl}(S_{2nd}) \cdot \gamma_{mean}(S_{2nd})^2$ $\qquad\qquad$ $Ka_{check} = 3.595 \cdot 10^{-8}$

Therefore, the solubility of AgCl through the dissociation is $\qquad S_1 = S_{2nd}$ $\qquad$ $S_1 = 2.349 \cdot 10^{-4}$

For the reaction AgCl (s) = AgCl (ao), we can find the equilibrium constant using data in Tables 6.1 and 8.3. We will again assume that $\Delta rxnH_\theta$ and $\Delta rxnS_\theta$ are independent of temperature.

$$\Delta fH_{AgCls} = -127.068 \cdot \frac{kJ}{mole} \qquad \Delta fH_{AgClao} = -72.8 \cdot \frac{kJ}{mole} \qquad \Delta rxnH_\theta = \Delta fH_{AgClao} - \Delta fH_{AgCls}$$

$$S_{AgCls} = 96.2 \cdot \frac{joule}{mole \cdot K} \qquad S_{AgClao} = 154.0 \cdot \frac{joule}{mole \cdot K} \qquad \Delta rxnS_\theta = S_{AgClao} - S_{AgCls}$$

$$\Delta rxnG_\theta(T) = \Delta rxnH_\theta - T \cdot \Delta rxnS_\theta$$

$$K_a(T) = exp\left(\frac{-\Delta rxnG_\theta(T)}{R \cdot T}\right) \qquad K_a(373.15 \cdot K) = 2.647 \cdot 10^{-5}$$

The equilibrium constant can also be expressed as

$$K_a = \frac{\gamma_{AgClao} \cdot m_{AgClao}}{a_{AgCls}}$$

We will assume that the activity of the solid salt and the activity coefficient of the neutral solute are equal to one. (This ideal assumption is not valid for ions.) Thus, we see that the solubility of AgCl through the dissolution of the neutral salt is just K_a.

$$S_2 = K_a(373.15 \cdot K) \qquad S_2 = 2.647 \cdot 10^{-5}$$

$$S_{total} = S_1 + S_2 \qquad S_{total} = 2.614 \cdot 10^{-4} \qquad \text{The units are moles of salt per kg water.}$$

8.27

Since there are now three processes by which TlCl can solubilize, we have to calculate the individual solubilities of each process.

Process 1: TlCl (s) = TlCl (ao)

$$K_1 = 7.167 \cdot 10^{-4}$$

The equilibrium constant can also be expressed as

$$K_1 = \frac{m_{TlClao} \cdot \gamma_{TlClao}}{a_{TlCls}}$$

We may assume that the activity of the solid salt is again one. Since there are no ions, it is reasonable to assume an ideal solution so K_1 is the solubility of TlCl through process 1.

$$S_1 = K_1 \qquad S_1 = 7.167 \cdot 10^{-4}$$

Process 2: $TlCl(s) = Tl+(ao) + Cl-(ao)$

$$K_2 = 1.861 \cdot 10^{-4}$$

The equilibrium constant may also be expressed as

$$K_2 = \left(m_{Tl} \cdot \gamma_{mean}\right) \cdot \left(m_{Cl} \cdot \gamma_{mean}\right)$$

In writing this expression, we have assumed the activity of the solid salt is one. We will use the DHG equation to calculate the mean ionic activity coefficient. However, since this quantity is dependent on ionic strength and we don't know the ionic strength until we calculate the solubility, an iterative solution is needed.

We can introduce S_2 as the moles of TlCl soluble in 1 kg of water through process 2. Since KCl is very soluble in water, we will assume it is totally ionized.

$$m_{Tl}(S_2) = S_2 \qquad m_{Cl}(S_2) = S_2 + 0.05 \qquad m_K = 0.05$$

$$I(S_2) = \frac{1}{2} \cdot \left(m_{Tl}(S_2) + m_{Cl}(S_2) + m_K\right) \qquad \gamma_{mean}(S_2) = \exp\left(\frac{-\alpha \cdot \sqrt{I(S_2)}}{1 + \sqrt{I(S_2)}}\right) \qquad \begin{array}{l} \alpha = 1.177 \\ \text{from Table 8.2} \end{array}$$

Successive Approximations $\qquad\qquad\qquad\qquad\qquad\qquad\qquad\qquad\qquad$ TOL $= 10^{-8}$

$$S_{guess} = 0.001$$

$$S_{1st} = root\left(m_{Tl}(S_{guess}) \cdot m_{Cl}(S_{guess}) \cdot \gamma_{mean}(S_{guess})^2 \quad K_2, S_{guess}\right) \qquad S_{1st} = 0.0053$$

$$S_{2nd} = root\left(m_{Tl}(S_{1st}) \cdot m_{Cl}(S_{1st}) \cdot \gamma_{mean}(S_{1st})^2 - K_2, S_{1st}\right) \qquad S_{2nd} = 0.0053$$

Therefore: $\quad S_2 = S_{2nd} \qquad S_2 = 0.00527$

The iteration has converged. Recalculate the equilibrium constant to check:

$$K2_{check} = m_{Tl}(S_{2nd}) \cdot m_{Cl}(S_{2nd}) \cdot \gamma_{mean}(S_{2nd})^2 \qquad K2_{check} = 1.861 \cdot 10^{-4}$$

Process 3: $TlCl(s) + Cl-(ao) = TlCl_2-(ao)$

$$K_3 = 2.724 \cdot 10^{-4}$$

The equilibrium constant may be expressed as

$$K_3 = \frac{m_{TlCl2} \cdot \gamma_{TlCl2}}{m_{Cl} \cdot \gamma_{Cl}}$$

$$S_3 = m_{TlCl2} \qquad S_3 = 1.362 \cdot 10^{-5}$$

Although we assumed the molality of the chloride ion was due only to the KCl, there is actually some interaction between processes 2 and 3. Process 2 produces some chloride ions, while process 3 uses them. Rigorously, each process should take into account the other. However, compared to the chloride ions produced by KCl, neither process changes the chloride molality significantly. Therefore, we can treat the processes as independent.

$$S_{total} = S_1 + S_2 + S_3 \qquad S_{total} = 0.0060 \qquad \text{Units are moles of TlCl per kg water.}$$

Although it is not reasonable to assume that ionic activity coefficients equal one, we have to assume that the activity coefficients here are equal, for lack of a better way to estimate them.

$$m_{Cl} = 0.05 \qquad m_{TlCl2} = K_3 \cdot m_{Cl}$$

8.29

The solubility of iodine will be calculated separately for each process, and then added together.

Process 1: For the reaction, I_2 (s) = I_2 (ao), the equilibrium constant can be calculated from data in Tables 6.1 and 8.3.

$$\Delta fG_{I2solid} = 0 \cdot \frac{kJ}{mole} \qquad \Delta fG_{I2ao} = 16.40 \cdot \frac{kJ}{mole} \qquad kJ = 1000 \cdot joule$$

$$\Delta rxnG_\theta = \Delta fG_{I2ao} - \Delta fG_{I2solid} \qquad \Delta rxnG_\theta = 16.4 \cdot \frac{kJ}{mole} \qquad R = 8.31451 \cdot \frac{joule}{mole \cdot K}$$

$$K_a = \exp\left(\frac{-\Delta rxnG_\theta}{R \cdot T}\right) \qquad K_a = 1.339 \cdot 10^{-3} \qquad T = 298.15 \cdot K$$

The equilibrium constant can also be expressed as

$$K_a = \frac{(\gamma_{I2ao} \cdot m_{I2ao})}{a_{I2solid}}$$ The activity of the solid and the activity coefficient of the neutral solute can be assumed to be one. (This assumption is not valid for ionic solutes.)

Thus, the solubility of iodine through process 1 is just K_a. (Actually, the standard molality of 1 mol/kg should rigorously be included to make the units correct. Units will added at the end.)

$$m_{I2ao} = K_a \qquad m_{I2ao} = 1.339 \cdot 10^{-3}$$

Process 2: For the reaction, I_2 (s) + I- (ao) = I_3- (ao), the equilibrium constant is calculated as

$$\Delta fG_{I2solid} = 0 \cdot \frac{kJ}{mole} \qquad \Delta fG_{Iao} = -51.57 \cdot \frac{kJ}{mole} \qquad \Delta fG_{I3ao} = -51.4 \cdot \frac{kJ}{mole}$$

$$\Delta rxnG_{\theta} = \Delta fG_{I3ao} - \Delta fG_{I2solid} - \Delta fG_{Iao} \qquad \Delta rxnG_{\theta} = 170 \cdot \frac{joule}{mole}$$

$$K_a = \exp\left(\frac{-\Delta rxnG_{\theta}}{R \cdot T}\right) \qquad K_a = 0.934$$

The equilibrium constant can be expressed as

$$K_a = \frac{m_{I3ao} \cdot \gamma_{I3ao}}{a_{I2solid} \cdot m_{Iao} \cdot \gamma_{Iao}}$$

The activity of the solid is one. It will also be assumed that the ionic activity coefficients of I- and I_3- are approximately equal. In calculating the molalities, NaI is presumably totally dissociated, since it is very soluble in water.

$$m_{Iao}(m_{I3ao}) = 0.015 - m_{I3ao} \qquad \text{since the formation of } I_3\text{- consumes I-}$$

$$m_{guess} = 0.1$$

$$m_{I3ao} = root(m_{Iao}(m_{guess}) \cdot K_a - m_{guess}, m_{guess}) \qquad \text{This solves the equilibrium expression for the molality of } I_3\text{-.}$$

$$m_{I3ao} = 0.00724$$

The total solubility is the sum of the solubilities for the two processes.

$$S = m_{I2ao} + m_{I3ao} \qquad S = 0.00858 \qquad \text{Units are mole of iodine per kg of water.}$$

8.31

(a) The equilibrium constant can be calculated using data in Table 8.3 and eq. 6.15.

$$\Delta fG_{Fe2} = -78.90 \cdot \frac{kJ}{mole} \quad \Delta fG_{Cu2} = 65.49 \cdot \frac{kJ}{mole} \quad \Delta fG_{Fe3} = -4.7 \cdot \frac{kJ}{mole} \quad \Delta fG_{Cu1} = 49.98 \cdot \frac{kJ}{mole}$$

$$\Delta rxnG_{\theta} = \Delta fG_{Fe3} + \Delta fG_{Cu1} - \Delta fG_{Fe2} - \Delta fG_{Cu2} \qquad \Delta rxnG_{\theta} = 58.69 \cdot \frac{kJ}{mole} \qquad T = 298.15 \cdot K$$

$$K_a = \exp\left(\frac{-\Delta rxnG_{\theta}}{R \cdot T}\right) \qquad K_a = 5.224 \cdot 10^{-11}$$

$$kJ \equiv 1000 \cdot joule$$

$$R \equiv 8.31451 \cdot \frac{joule}{mole \cdot K}$$

(b) The equilibrium constant can be expressed as follows for an ideal solution.

$$K_a = \frac{m_{Fe3} \cdot m_{Cu1}}{m_{Fe2} \cdot m_{Cu2}}$$

The molalities can be expressed in terms of the extent of reaction, ξ.

$$m_{Fe3}(\xi) = \xi \qquad m_{Cu1}(\xi) = \xi \qquad m_{Fe2}(\xi) = 0.001 - \xi \qquad m_{Cu2}(\xi) = 0.001 - \xi$$

$$\xi_{guess} = 0.0005 \qquad\qquad\qquad TOL = 10^{-15}$$

$$\xi = root\left(\frac{m_{Fe3}(\xi_{guess}) \cdot m_{Cu1}(\xi_{guess})}{m_{Fe2}(\xi_{guess}) \cdot m_{Cu2}(\xi_{guess})} - K_a, \xi_{guess}\right) \qquad \xi = 7.228 \cdot 10^{-9}$$

$$m_{Cu1}(\xi) = 7.2 \cdot 10^{-9} \qquad \text{Units are mole Cu+ per kg water.}$$

Recalculate the equilibrium constant to check:

$$Ka_{check} = \frac{m_{Fe3}(\xi) \cdot m_{Cu1}(\xi)}{m_{Fe2}(\xi) \cdot m_{Cu2}(\xi)} \qquad Ka_{check} = 5.224 \cdot 10^{-11}$$

8.33

The molality scale Henry's Law constant was defined in Chapter 7 as

$$k_m = \lim_{m \to 0} \frac{P_2}{m}$$

If you consider the equilibrium for the reaction, A (ao) = A (g), k_m is like an equilibrium constant und ideal conditions (activity/fugacity coefficients = 1) and without standard states (which are ju constants anyway). This is the reason why this proportionality is constant (at constant T) in the lim of dilute solution where conditions approach ideal.

The same reasoning can be used to derive the HL constant for HCl. The reaction we have consider is then H+ (ao) + Cl- (ao) = HCl (g). The equilibrium constant for this reaction is

$$K_a = \frac{a_{HCl,g}}{a_{H+,ao} a_{Cl-,ao}}$$

In the limit of ideal solution, the equilibrium constant (without standard states) is just

$$k_m = \lim_{m \to 0}\left(\frac{P}{m^2}\right)$$

since for HCl, $m = m_{H+,ao} = m_{Cl-,ao}$.

SECTION 8.3

8.35

The measured resistance can be used to calculate the conductivity (κ) using eq. 8.21. Then, if we can assume $\Lambda_o = \Lambda$, we can calculate this value using eq. 8.27 and Table 8.6. Eq. 8.25 will then give the equivalent concentration, which is easily converted to ordinary concentration.

$$A = 7.2 \cdot cm^2 \qquad l = 1.2 \cdot cm \qquad R = 13 \cdot \Omega \qquad\qquad\qquad equiv = 1$$

$$\kappa = \frac{l}{A \cdot R} \qquad \kappa = 0.013 \cdot \frac{siemens}{cm}$$

$$\text{From Table 8.6} \qquad \lambda o_{Ca} = 59.50 \cdot \frac{cm^2 \cdot siemens}{equiv} \qquad\qquad \lambda o_{NO3} = 71.44 \cdot \frac{cm^2 \cdot siemens}{equiv}$$

$$\Lambda = \lambda o_{Ca} + \lambda o_{NO3}$$

$$c_{equiv} = \frac{\kappa}{\Lambda} \qquad c_{equiv} = 0.098 \cdot \frac{mole}{liter}$$

Since $v_i z_i = 2$ for $Ca(NO_3)_2$,

$$c = \frac{c_{equiv}}{2} \qquad c = 0.049 \cdot \frac{mole}{liter}$$

8.37

We can use data in Table 8.6 and eqs 8.27 and 8.29 to calculate the equivalent conductivities and cation transference numbers, respectively.

(a) rubidium acetate

$$\lambda o_{Rb} = 77.8 \cdot \frac{cm^2 \cdot siemens}{equiv} \qquad \lambda o_{acetate} = 40.9 \cdot \frac{cm^2 \cdot siemens}{equiv} \qquad\qquad equiv = 1$$

$$\Lambda_o = \lambda o_{Rb} + \lambda o_{acetate} \qquad \Lambda_o = 118.7 \cdot \frac{cm^2 \cdot siemens}{equiv}$$

$$t_{Rb} = \frac{\lambda o_{Rb}}{\Lambda_o} \qquad t_{Rb} = 0.66$$

Student's Solutions Manual

(b) ammonium sulphate

$$\lambda o_{NH4} \quad 73.4 \cdot \frac{cm^2 \cdot siemens}{equiv} \qquad \lambda o_{SO4} \quad 80 \cdot \frac{cm^2 \cdot siemens}{equiv}$$

$$\Lambda_o \quad \lambda o_{NH4} \quad \lambda o_{SO4} \qquad \Lambda_o = 153.4 \cdot \frac{cm^2 \cdot siemens}{equiv}$$

$$t_{NH4} \quad \frac{\lambda o_{NH4}}{\Lambda_o} \qquad t_{NH4} = 0.48$$

(c) $K_3Fe(CN)_6$

$$\lambda o_K \quad 73.50 \cdot \frac{cm^2 \cdot siemens}{equiv} \qquad \lambda o_{FeCN6} \quad 99.1 \cdot \frac{cm^2 \cdot siemens}{equiv}$$

$$\Lambda_o \quad \lambda o_K \cdot \lambda o_{FeCN6} \qquad \Lambda_o = 172.6 \cdot \frac{cm^2 \cdot siemens}{equiv}$$

$$t_K \quad \frac{\lambda o_K}{\Lambda_o} \qquad t_K = 0.43$$

8.39

The inverse of the resistivity is the conductivity (κ) of the solution. Eq. 8.25 can be used to then calculate the equivalent conductivity. The ionic conductivity of the hydrogen ion can then be calculated using the cation transference number and eq. 8.29.

$$\rho = 25.554 \cdot \Omega \cdot cm \qquad \kappa \quad \frac{1}{\rho} \qquad c_{equiv} \quad 0.1 \cdot \frac{mole}{liter} \qquad t_H = 0.8314 \qquad equiv \quad 1$$

$$\Lambda \quad \frac{\kappa}{c_{equiv}} \qquad \Lambda = 391.328 \cdot \left(\frac{cm^2 \cdot siemens}{equiv} \right)$$

$$\lambda_H = t_H \cdot \Lambda \qquad \lambda_H = 325.4 \cdot \frac{cm^2 \cdot siemens}{equiv}$$

The value given in Table 8.6 is 349.8 (cm^2*S/equiv).

8.41

The data should be fit to a function of the form

$$\Lambda = \Lambda_0 + a_1 \cdot c^{0.5} + a_2 \cdot c + a_c \cdot c^{1.5} + \dots$$

For this problem, we used a second order regression in sqrt(c) to obtain the following fit

$$\Lambda(c) = 126.895 - 87.5336 \cdot c^{0.5} + 109.352 \cdot c$$

Thus, we see that $\Lambda_0 = 126.895$ (units are same as those in the problem). The standard deviation for this parameter is $\sigma = 0.025$.

It may be useful to see a graph of the data and the curve fit.

i   0..3

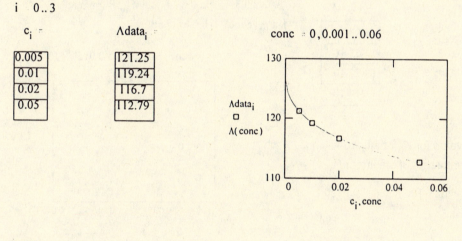

$c_i =$
0.005
0.01
0.02
0.05

$\Lambda data_i =$
121.25
119.24
116.7
112.79

conc = 0, 0.001 .. 0.06

8.43

For the reaction, PAH (ao) = PA- (ao) + H+ (ao), the thermodynamic dissociation constant can be evaluated as

$$K_a = \frac{C_{PA} \cdot C_H \cdot \gamma_{mean}^2}{C_{PAH} \cdot \gamma_{PAH}}$$

Student's Solutions Manual

DHG will be used to evaluate the mean ionic activity coefficient. The neutral acid activity coefficient will be assumed to be one.

$i := 0..3$

c_i	α_i
0.001563	0.3131
0.003125	0.2408
0.00625	0.182
0.0125	0.1379

$\alpha_{DH} := 4.58 \qquad I_i := \alpha_i \cdot c_i \qquad \gamma_{mean_i} := \exp\left(-\dfrac{\alpha_{DH} \cdot \sqrt{I_i}}{1 + \sqrt{I_i}}\right)$

The concentrations are expressed in terms of the degree dissociation, α.

$C_{PAH_i} := c_i \cdot (1 - \alpha_i) \qquad C_{PA_i} := c_i \cdot \alpha_i \qquad C_{H_i} := c_i \cdot \alpha_i$

$K_{a_i} := \dfrac{C_{PA_i} \cdot C_{H_i} \cdot (\gamma_{mean_i})^2}{C_{PAH_i}}$

$K_{a_i} =$

$1.83 \cdot 10^4$
$1.869 \cdot 10^4$
$1.877 \cdot 10^4$
$1.914 \cdot 10^4$

$m := slope(c, K_a) \qquad b := intercept(c, K_a)$

$Ka_{fit}(conc) := b + m \cdot conc$

$conc := 0, 0.001 .. 0.015$

The extrapolated value for K_a at infinite dilution is just the intercept of the linear fit.

$K_a := b \qquad K_a = 1.83 \cdot 10^4$

SECTION 8.4

8.45

Oxidation reaction (left): H_2 (g) = 2 H^+ (ao) + 2 e^-

Reduction reaction (right): 2 AgCl (s) + 2 e^- = 2 Ag (s) + 2 Cl^- (ao)

Cell reaction: 2 AgCl (s) + H_2 (g) = 2 H^+ (ao) + 2 Cl^- (ao) + 2 Ag (s)

Table 8.7 gives the standard EMF values for the half cell reactions (given as reduction potentials).

$\varepsilon_{left} = 0 \cdot volt$ $\qquad \varepsilon_{right} = 0.2225 \cdot volt$ $\qquad \varepsilon_\theta = \varepsilon_{right} - \varepsilon_{left}$ $\qquad \varepsilon_\theta = 0.2225 \cdot volt$

Since we are given the EMF of the cell, and we have just determined the standard EMF, we can use eq. 8.40 to calculate the activity quotient, Q.

$\varepsilon = \varepsilon_\theta - \dfrac{R \cdot T}{n \cdot F} \cdot \ln(Q)$ $\qquad \varepsilon = 0.20534 \cdot volt$ $\qquad T = 298.15 \cdot K$ $\qquad n = 2 \cdot mole$

$\qquad\qquad\qquad\qquad\qquad\qquad\qquad R = 8.31451 \cdot \dfrac{joule}{mole \cdot K}$ $\qquad F = 96485 \cdot \dfrac{coul}{mole}$

$Q = \exp\left[-\dfrac{(\varepsilon \cdot n \cdot F - \varepsilon_\theta \cdot n \cdot F)}{(R \cdot T)}\right]$ $\qquad Q = 3.803$

The activity quotient is also the ratio of component activities raised to their stoichiometric coefficients. We will assume that solid activities = 1. Thus, Q can be expressed as

$Q = \dfrac{m_H{}^2 \cdot m_{Cl}{}^2 \cdot \gamma_{mean}{}^4}{\left(\dfrac{P_{H2}}{P_\theta}\right)}$ $\qquad P_\theta = 1 \cdot atm$ (Table 8.7 uses this as the standard state)

$\qquad\qquad\qquad\qquad\qquad P_{H2} = 1 \cdot atm$ $\qquad m_H = 1.5346$ $\qquad m_{Cl} = 1.5346$

Now we can solve for the mean ionic activity coefficient.

$\gamma_{guess} = 1$

$\gamma_{mean} = root\left[\dfrac{m_H{}^2 \cdot m_{Cl}{}^2 \cdot \gamma_{guess}{}^4}{\left(\dfrac{P_{H2}}{P_\theta}\right)} - Q, \gamma_{guess}\right]$ $\qquad \gamma_{mean} = 0.910$

The pH is the - $\log_{10} (a_{H^+})$

$pH = -\log(\gamma_{mean} \cdot m_H)$ $\qquad pH = -0.145$

8.47

Oxidation reaction (left): $Cu (s) = Cu^{2+} (ao) + 2 e^-$

Reduction reaction (right): $PbSO_4 (s) + 2 e^- = Pb (s) + SO_4^{2-} (ao)$

Cell reaction: $Cu (s) + PbSO_4 (s) = Cu^{2+} (ao) + SO_4^{2-} (ao) + Pb (s)$

Table 8.7 gives the standard EMF values for the half cell reactions (given as reduction potentials).

$$\varepsilon_{left} = 0.337 \cdot volt \qquad \varepsilon_{right} = 0.3546 \cdot volt \qquad \varepsilon_\theta = \varepsilon_{right} - \varepsilon_{left} \qquad \varepsilon_\theta = -0.6916 \cdot volt$$

We still need to calculate the activity quotient, Q. The activity quotient is the ratio of component activities raised to their stoichiometric coefficients. We will assume that solid activities = 1. Thus, Q can be expressed as

$$Q = m_{Cu} \cdot m_{SO4} \cdot \gamma_{mean}^2 \qquad m_{Cu} = 0.2 \qquad m_{SO4} = 0.2 \qquad \gamma_{mean} = 0.11 \quad \text{From Table 8.1}$$

$$Q = 4.84 \cdot 10^{-4}$$

Eq. 8.40 can then be used to calculate the EMF.

$$\varepsilon = \varepsilon_\theta - \frac{R \cdot T}{n \cdot F} \cdot \ln(Q) \qquad\qquad T = 298.15 \cdot K \qquad n = 2 \cdot mole$$

$$R = 8.31451 \cdot \frac{joule}{mole \cdot K} \qquad F = 96485 \cdot \frac{coul}{mole}$$

$$\varepsilon = -0.594 \cdot volt$$

8.49

Oxidation reaction (left): $Pb (s) + 2 Br^- (ao) = PbBr_2 (s) + 2 e^-$

Reduction reaction (right): $Cu^{2+} (ao) + 2 e^- = Cu (s)$

Cell reaction: $Pb (s) + 2 Br^- (ao) + Cu^{2+} (ao) = PbBr_2 (s) + Cu (s)$

We are given the EMF of the cell. We can also calculate the activity quotient, Q, using DHG to estimate the mean ionic activity coefficient. If we assume the solid activities are one, Q may be expressed as follows

$$m_{Br} = 0.02 \quad m_{Cu} = 0.01 \quad I = \frac{1}{2} \cdot \left(2^2 \cdot m_{Cu} + 1^2 \cdot m_{Br}\right) \quad \alpha = 1.177 \quad \gamma_{mean} = \exp\left(\frac{-\alpha \cdot 2 \cdot \sqrt{I}}{1 + \sqrt{I}}\right)$$

$$Q = \frac{1}{m_{Br}^2 \cdot m_{Cu} \cdot \gamma_{mean}^3} \qquad Q = 7.091 \cdot 10^5$$

The standard EMF may now be estimated using eq. 8.40.

$$\varepsilon_\theta = \varepsilon + \frac{R \cdot T}{n \cdot F} \cdot \ln(Q) \qquad\qquad \varepsilon = 0.442 \cdot volt \quad T = 298.15 \cdot K \quad n = 2 \cdot mole$$

$$R = 8.31451 \cdot \frac{joule}{mole \cdot K} \qquad F = 96485 \cdot \frac{coul}{mole}$$

$$\varepsilon_\theta = 0.615 \cdot volt$$

8.51

Oxidation reaction (left): $Cu (s) = Cu^{2+} (ao) + 2 e^-$

Reduction reaction (right): $PbSO_4 (s) + 2 e^- = Pb (s) + SO_4^{2-} (ao)$

Cell reaction: $Cu (s) + PbSO_4 (s) = Cu^{2+} (ao) + SO_4^{2-} (ao) + Pb (s)$

Table 8.7 gives the standard EMF values for the half cell reactions (given as reduction potentials).

$$\varepsilon_{left} = 0.337 \cdot volt \qquad \varepsilon_{right} = -0.3546 \cdot volt \qquad \varepsilon_\theta = \varepsilon_{right} - \varepsilon_{left} \qquad \varepsilon_\theta = -0.6916 \cdot volt$$

We still need to calculate the activity quotient, Q. The activity quotient is the ratio of component activities raised to their stoichiometric coefficients. We will assume that solid activities = 1. Thus, Q can be expressed as

$$m_{Cu} = 0.03 \qquad m_{SO4} = 0.03 \qquad I = \frac{1}{2} \cdot \left(2^2 \cdot m_{Cu} + 2^2 \cdot m_{SO4}\right) \qquad \gamma_{mean} = \frac{-1.177 \cdot 4 \cdot \sqrt{I}}{1 + \sqrt{I}} \qquad \text{DHG equation}$$

$$Q = m_{Cu} \cdot m_{SO4} \cdot \gamma_{mean}^2$$

$$Q = 1.321 \cdot 10^{-3}$$

Eq. 8.40 can then be used to calculate the EMF.

$$\varepsilon = \varepsilon_\theta - \frac{R \cdot T}{n \cdot F} \cdot \ln(Q) \qquad\qquad T = 298.15 \cdot K \qquad n = 2 \cdot mole$$

$$\qquad\qquad\qquad R = 8.31451 \cdot \frac{joule}{mole \cdot K} \qquad F = 96485 \cdot \frac{coul}{mole}$$

$$\varepsilon = -0.606 \cdot volt$$

8.53

The following cell can be used

$$Zn(s) \,|\, ZnCl_2(ai,m) \,|\, Hg_2Cl_2(s) \,|\, Hg(liq)$$

The electrode reactions are

Oxidation: $\qquad\qquad Zn(s) = Zn^{2+}(ao) + 2e^- \qquad\qquad \mathscr{E}^\theta_{left} = -0.7628 \, volt$

Reduction: $\qquad\qquad Hg_2Cl_2(s) + 2e^- = 2Hg(liq) + 2Cl^-(ao) \qquad \mathscr{E}^\theta_{right} = 0.2680 \, volt$

Cell Reaction: $\qquad Zn(s) + Hg_2Cl_2(s) = Zn^{2+}(ao) + 2Cl^-(ao) + 2Hg(liq) \qquad \mathscr{E}^\theta = 1.0308 \, volt$

For this reaction, Q simplifies to

$$Q = (m)(2m)^2 \gamma_\pm^3 = 4m^3 \gamma_\pm^3$$

if we assume solid and liquid activities = 1.

The Nernst equation, eq. 8.40, can now be used to derive a formula relating measured EMFs to the mean ionic activity coefficient.

$$\mathscr{E} = 1.0308 \, volt - \frac{RT}{2\mathscr{F}} \ln(4m^3 \gamma_\pm^3)$$

Oxidation reaction (left): $1/2\ H_2\ (g) = H^+\ (ao) + e^-$

Reduction reaction (right): $AgCl\ (s) + e^- = Cl^-\ (ao) + Ag\ (s)$
Cell reaction: $1/2\ H_2\ (g) + AgCl\ (s) = HCl\ (ai) + Ag\ (s)$

The Nernst equation, eq. 8.40, for this cell can be expressed as follows. Solid activities were assumed to be one, as well as the H_2 gas activity (since $P/P_\theta = 1$)

$$\varepsilon = \varepsilon_\theta - \frac{R \cdot T}{n \cdot F} \cdot \ln\left(m^2 \cdot \gamma_{mean}^2\right) \qquad T = 298.15 \cdot K \quad n = 1 \cdot mole \quad F = 96485 \cdot \frac{coul}{mole} \quad R = 8.31451 \cdot \frac{joule}{mole \cdot K}$$

This can be rearranged to give

$$\varepsilon + \frac{2 \cdot R \cdot T}{n \cdot F} \cdot \ln(m) = \varepsilon_\theta - \frac{2 \cdot R \cdot T}{n \cdot F} \cdot \ln\left(\gamma_{mean}\right)$$

If the quantity on the left hand side is calculated from the data points and extrapolated to m = 0, the right hand side will become just ε_θ, since the mean ionic activity coefficient will approach 1. The quantity on the left and right sides of the above equation is called ε' in the text, and will be referred to here as ε_{prime}.

$i \quad 0..4$

m_i	ε_i	$\varepsilon\ prime_i = \varepsilon_i + \frac{2 \cdot R \cdot T}{n \cdot F} \cdot \ln(m_i)$	$\frac{\varepsilon\ prime_i}{volt}$
0.003215	0.52053·volt		
0.005619	0.49257·volt		0.225580
0.009138	0.46860·volt		0.226310
0.013407	0.44974·volt		0.227328
0.02563	0.41824·volt		0.228166
			0.229963

These ε_{prime} results were fit to a polynomial in sqrt(m) with the following result:

$$\varepsilon_{fit}(m) = 0.22283 + 0.0497872 \cdot m^{0.5} - 0.0324658 \cdot m$$

$$\varepsilon_\theta = \varepsilon_{fit}(0) \qquad \varepsilon_\theta = 0.2228$$

The standard deviation of the leading term in the polynomial fit is $\sigma = 0.0003$.

Here is a graph of the fit and the data.

molal $0, 0.0002 .. 0.03$

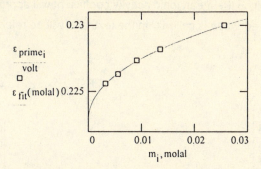

Another way to approach the problem is to calculate the quantity ε'' as described in the text. This uses the DHG equation to evaluate the mean ionic activity coefficient as it approaches one. The answer with this method is the same as given above, but with $\sigma = 0.00004$.

8.57

Oxidation reaction (left): $Zn\,(s) = Zn^{2+}\,(ao) + 2\,e^-$

Reduction reaction (right): $PbSO_4\,(s) + 2\,e^- = SO_4^{2-}\,(ao) + Pb\,(Hg)$

Cell reaction: $Zn\,(s) + PbSO_4\,(s) = ZnSO_4\,(ai) + Pb\,(Hg)$

The Nernst equation, eq. 8.40, for this cell can be expressed as follows, assuming the activities of condensed phases = 1.

$$\varepsilon = \varepsilon_\theta - \frac{R \cdot T}{n \cdot F} \cdot \ln\left(m^2 \cdot \gamma_{mean}^2\right) \qquad T = 298.15 \cdot K \quad n = 2 \cdot mole \quad F = 96485 \cdot \frac{coul}{mole} \quad R = 8.31451 \cdot \frac{joule}{mole \cdot K}$$

This can be rearranged to give

$$\varepsilon + \frac{2 \cdot R \cdot T}{n \cdot F} \cdot \ln(m) = \varepsilon_\theta - \frac{2 \cdot R \cdot T}{n \cdot F} \cdot \ln\left(\gamma_{mean}\right)$$

If the quantity on the left hand side is calculated from the data points and extrapolated to m = 0, the right hand side will become just ε_θ, since the mean ionic activity coefficient will approach 1. The quantity on the left and right sides of the above equation is called ε' in the text, and will be referred to here as ε_{prime}.

i   0..4

m_i	ε_i	$\varepsilon\ prime_i$ $\varepsilon_i + \dfrac{2 \cdot R \cdot T}{n \cdot F} \cdot \ln(m_i)$	$\dfrac{\varepsilon\ prime_i}{volt}$
0.0005	0.61144·volt		0.416151
0.001	0.59714·volt		0.419660
0.002	0.58319·volt		0.423519
0.005	0.56598·volt		0.429851
0.01	0.55353·volt		0.435210

These ε_{prime} results were fit to a polynomial in sqrt(m) with the following result:

$$\varepsilon\ _{fit}(m) = 0.407828 + 0.410941 \cdot m^{0.5} - 1.37626 \cdot m$$

$\varepsilon_\theta = \varepsilon_{fit}(0)$ $\varepsilon_\theta = 0.4078$ The standard deviation of the leading term in the polynomial fit is σ = 0.0006.

Here is a graph of the fit and the data.

molal   0, 0.0002 .. 0.012

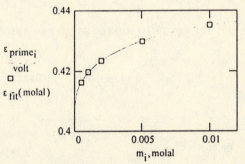

236

8.59

Cell reaction: $Zn\ (s) + 2\ MnO_2\ (s) + H_2O\ (liq) = Zn^{2+}\ (ao) + 2\ OH^-\ (ao) + Mn_2O_3\ (s)$

The maximum work obtainable from this cell is given by $(-w)_{max} = nF\varepsilon$. For a stoichiometric amount of reactancts,

$$n = 2 \cdot mole \qquad F = 96485 \cdot \frac{coul}{mole} \qquad \varepsilon = 1.5 \cdot volt$$

$$energy = n \cdot F \cdot \varepsilon \qquad energy = 289.455 \cdot kJ \qquad\qquad\qquad kJ = 1000 \cdot joule$$

Now we have to determine the mass of a stoichiometric amount of reactants.

$$mass_{Zn} = 65.39 \cdot gm \qquad mass_{MnO2} = 173.8736 \cdot gm \qquad mass_{H2O} = 18.01528 \cdot gm$$

$$energy_{specific} = \frac{energy}{mass_{Zn} + mass_{MnO2} + mass_{H2O}} \qquad energy_{specific} = 1.13 \cdot \frac{kJ}{gm}$$

The time this cell will last is the energy divided by the power.

$$time = \frac{(10 \cdot gm) \cdot energy_{specific}}{5 \cdot watt} \qquad time = 38 \cdot min$$

SECTION 8.5

8.61

Oxidation reaction (left): H_2 (g) = 2 H^+ (ao) + 2 e^-

Reduction reaction (right): 2 AgCl (s) + 2 e^- = 2 Ag (s) + 2 Cl^- (ao)

Cell reaction: H_2 (g) + 2 AgCl (s) = 2 Ag (s) + 2 HCl (ai)

With data from Tables 6.1 and 8.3, we can calculate $\Delta_{rxn}G_\theta$ at 60 C.

$$\Delta fH_{H2} = 0 \cdot \frac{kJ}{mole} \qquad \Delta fH_{AgCl} = -127.068 \cdot \frac{kJ}{mole} \qquad \Delta fH_{Ag} = 0 \cdot \frac{kJ}{mole} \qquad kJ = 1000 \cdot joule$$

$$R = 8.31451 \cdot \frac{joule}{mole \cdot K}$$

$$\Delta fH_H = 0 \cdot \frac{kJ}{mole} \qquad \Delta fH_{Cl} = -167.159 \cdot \frac{kJ}{mole}$$

$$\Delta rxnH_\theta = 2 \cdot \Delta fH_{Ag} + 2 \cdot \Delta fH_H + 2 \cdot \Delta fH_{Cl} - \Delta fH_{H2} - 2 \cdot \Delta fH_{AgCl} \qquad \Delta rxnH_\theta = -80.182 \cdot \frac{kJ}{mole}$$

$$S_{H2} = 130.684 \cdot \frac{joule}{mole \cdot K} \qquad S_{AgCl} = 96.2 \cdot \frac{joule}{mole \cdot K} \qquad S_{Ag} = 42.55 \cdot \frac{joule}{mole \cdot K}$$

$$S_H = 0 \cdot \frac{joule}{mole \cdot K} \qquad S_{Cl} = 56.5 \cdot \frac{joule}{mole \cdot K}$$

$$\Delta rxnS_\theta = 2 \cdot S_{Ag} + 2 \cdot S_H + 2 \cdot S_{Cl} - S_{H2} - 2 \cdot S_{AgCl} \qquad \Delta rxnS_\theta = 124.984 \cdot \frac{joule}{mole \cdot K}$$

$$\Delta rxnG_\theta(T) = \Delta rxnH_\theta - T \cdot \Delta rxnS_\theta \qquad \Delta rxnG_\theta(333.15 \cdot K) = -38.544 \cdot \frac{kJ}{mole}$$

Now we can use eq. 8.41 to calculate the standard EMF.

$$n = 2 \cdot mole \qquad F = 96485 \cdot \frac{coul}{mole}$$

$$\varepsilon_\theta = \frac{-\Delta rxnG_\theta(333.15 \cdot K)}{n \cdot F} \qquad \varepsilon_\theta = 0.1997 \cdot volt$$

8.63

Oxidation reaction (left): $1/2\ H_2\ (g) = H^+\ (ao) + e^-$

Reduction reaction (right): $AgBr\ (s) + e^- = Ag\ (s) + Br^-\ (ao)$

Cell reaction: $1/2\ H_2\ (g) + AgBr\ (s) = Ag\ (s) + HBr\ (ai)$

Eq. 8.43 can be used to calculate $\Delta rxnS_\theta$ with the given data

$$\varepsilon_{15} := 0.07595 \cdot volt \qquad \varepsilon_{25} := 0.07131 \cdot volt \qquad \varepsilon_{35} := 0.06597 \cdot volt$$

$$F \equiv 96485 \cdot \frac{coul}{mole} \qquad kJ \equiv 1000 \cdot joule$$

$$R \equiv 8.31451 \cdot \frac{joule}{mole \cdot K}$$

$$\Delta rxnS_\theta := n \cdot F \cdot \left(\frac{\varepsilon_{35} - \varepsilon_{15}}{20 \cdot K} \right) \qquad \Delta rxnS_\theta = -48.146 \cdot \frac{joule}{mole \cdot K} \qquad n := 1 \cdot mole$$

From Tables 6.1 and 8.3 we have the following data

$$S_{Ag} := 42.55 \cdot \frac{joule}{mole \cdot K} \qquad S_H := 0 \cdot \frac{joule}{mole \cdot K} \qquad S_{H2} := 130.684 \cdot \frac{joule}{mole \cdot K} \qquad S_{AgBr} := 107.1 \cdot \frac{joule}{mole \cdot K}$$

$$S_{Br} := \Delta rxnS_\theta - S_{Ag} - S_H + \frac{1}{2} \cdot S_{H2} + S_{AgBr} \qquad S_{Br} = 81.7 \cdot \frac{joule}{mole \cdot K}$$

The value given in Table 8.3 is 82.4 J/(mole*K).

8.65

Oxidation reaction (left): $1/2\ H_2$ (g) = H^+ (ao) + e^-

Reduction reaction (right): H_2O (liq) + e^- = $1/2\ H_2$ (g) + $(OH)^-$ (ao)

Cell reaction: H_2O (liq) = H^+ (ao) + $(OH)^-$ (ao)

Table 8.7 gives

$$\varepsilon_{left} = 0 \cdot volt \qquad \varepsilon_{right} = -0.82806 \cdot volt \qquad \varepsilon_\theta = \varepsilon_{right} - \varepsilon_{left}$$

Eq. 8.45 gives the equilibrium constant (also the ion product) as

$$K_a = exp\left(\frac{n \cdot F \cdot \varepsilon_\theta}{R \cdot T}\right) \qquad\qquad T = 298.15 \cdot K \quad n = 1 \cdot mole \quad F = 96485 \cdot \frac{coul}{mole}$$

$$K_a = 1.007 \cdot 10^{-14} \qquad\qquad\qquad\qquad R = 8.31451 \cdot \frac{joule}{mole \cdot K}$$

8.67

(a) Oxidation reaction (left): Pb (s) + 2 Cl^- (ao) = $PbCl_2$ (s) + 2 e^-

Reduction reaction (right): Hg_2Cl_2 (s) + 2 e^- = 2 Hg (liq) + 2 Cl^- (ao)

Cell reaction: Pb (s) + Hg_2Cl_2 (s) = $PbCl_2$ (s) + 2 Hg (liq)

(b) Free energies of formation from Table 6.1 can be used to calculate $\Delta rxnG_\theta$. (Remember, the free energy of formation of elements in their most stable form at standard pressure and a given temperature is zero). Then, eq. 8.41 can be used to calculate the standard EMF.

$$\Delta fG_{Pb} = 0 \cdot \frac{kJ}{mole} \quad \Delta fG_{Hg2Cl2} = -210.745 \cdot \frac{kJ}{mole} \quad \Delta fG_{PbCl2} = -314.10 \cdot \frac{kJ}{mole} \quad \Delta fG_{Hg} = 0 \cdot \frac{kJ}{mole}$$

$$\Delta rxnG_\theta = \Delta fG_{PbCl2} + 2 \cdot \Delta fG_{Hg} - \Delta fG_{Hg} - \Delta fG_{Hg2Cl2} \qquad \Delta rxnG_\theta = -103.355 \cdot \frac{kJ}{mole}$$

$$\varepsilon_{\theta25} = \frac{-\Delta rxnG_\theta}{n \cdot F} \qquad \varepsilon_{\theta25} = 0.5356 \cdot volt \qquad n = 2 \cdot mole \quad kJ = 1000 \cdot joule \quad F = 96485 \cdot \frac{coul}{mole}$$

Equation 8.40 (Nernst equation) relates the standard EMF to the actual EMF. Since we are dealing only with condensed phases, all activities are approximately one, and $\varepsilon = \varepsilon_\theta$.

(c) The temperature dependence of the standard EMF is given by eq. 8.43. Standard entropies can be found in Tables 3.2 and 6.1.

$$S_{Pb} = 64.81 \cdot \frac{joule}{mole \cdot K} \qquad S_{Hg2Cl2} = 192.5 \cdot \frac{joule}{mole \cdot K} \qquad S_{PbCl2} = 136.0 \cdot \frac{joule}{mole \cdot K} \qquad S_{Hg} = 76.02 \cdot \frac{joule}{mole \cdot K}$$

$$\Delta rxnS_\theta = S_{PbCl2} + 2 \cdot S_{Hg} - S_{Pb} - S_{Hg2Cl2} \qquad \Delta rxnS_\theta = 30.73 \cdot \frac{joule}{mole \cdot K}$$

$$d\varepsilon dT = \frac{\Delta rxnS_\theta}{n \cdot F}$$

$$\varepsilon_{\theta 18} = \varepsilon_{\theta 25} + d\varepsilon dT \cdot (18 - 25) \cdot K \qquad \varepsilon_{\theta 18} = 0.5345 \cdot volt$$

Again, for this cell the standard EMF is the actual EMF since all activities are one.

(d) When a cell is charged reversibly, the minimum amount of heat is released. This amount is given by $(\Delta rxnH_\theta - \Delta rxnG_\theta) = T \Delta rxnS_\theta$.

$$T = 298.15 \cdot K$$

$$heat_{reversible} = T \cdot \Delta rxnS_\theta \qquad heat_{reversible} = 9.16 \cdot \frac{kJ}{mole}$$

(e) If the cell were discharged totally irreversibly (as in a short circuit), the maximum amount of heat is released. This amount is given by $\Delta rxnH_\theta = \Delta rxnG_\theta + T \Delta rxnS_\theta$.

$$heat_{irreversible} = \Delta rxnG_\theta + T \cdot \Delta rxnS_\theta \qquad heat_{irreversible} = 94.19 \cdot \frac{kJ}{mole}$$

SECTION 8.6

8.69
_____

Oxidation reaction (left): Cl^- (ao) = 1/2 Cl_2 (g, 1 atm) + e^-

Reduction reaction (right): 1/2 Cl_2 (g, 0.3 atm) + e^- = Cl^- (ao)

Cell reaction: 1/2 Cl_2 (g, 0.3 atm) = 1/2 Cl_2 (g, 1 atm)

Since the two half cell reactions have the same standard EMF, the overall standard EMF is zero. Thus, the Nernst equation (eq. 8.40) simplifies to

$$\varepsilon(Q) = -\frac{R \cdot T}{n \cdot F} \cdot \ln(Q) \qquad T = 298.15 \cdot K \qquad n = 1 \cdot mole \qquad F = 96485 \cdot \frac{coul}{mole} \qquad R = 8.31451 \cdot \frac{joule}{mole \cdot K}$$

All that needs to be evaluated is the activity quotient, Q. With the ideal gas approximation, the chlorine activity is P/P_θ. Since the standard pressure cancels out, Q becomes just

$$Q = \left(\frac{1 \cdot atm}{0.3 \cdot atm}\right)^2 \qquad \varepsilon(Q) = -0.0155 \cdot volt$$

8.71
_____

The subscript 1 refers to the NaCl compartment with m = 0.1. The subscript 2 refers to the NaCl compartment with m = 0.02.

Oxidation reaction (left): Cl^- (m_1) = 1/2 Cl_2 (g) + e^-

Liquid Junction (cation): $t_{Na} Na^+$ (m_1) = $t_{Na} Na^+$ (m_2)

Liquid Junction (anion): $t_{Cl} Cl^-$ (m_2) = $t_{Cl} Cl^-$ (m_1)

Reduction reaction (right): 1/2 Cl_2 (g) + e^- = Cl^- (m_2)

Cell reaction: Cl^- (m_1) + $t_{Na} Na^+$ (m_1) + $t_{Cl} Cl^-$ (m_2) = Cl^- (m_2) + $t_{Na} Na^+$ (m_2) + $t_{Cl} Cl^-$ (m_1)

This cell reaction can be simplified to:

$$t_{Na} [Na^+ (m_1) + Cl^- (m_1)] = t_{Na} [Na^+ (m_2) + Cl^- (m_2)]$$

$$T = 298.15 \cdot K \qquad n = 1 \cdot mole$$

The Nernst equation for this cell is expressed below (since the standard EMF is zero here).

$$m_{Na1} = 0.1 \quad m_{Na2} = 0.02 \quad m_{Cl1} = 0.1 \quad m_{Cl2} = 0.02$$

$$\gamma_{mean1} = 0.778 \quad \gamma_{mean2} = 0.875 \quad \text{From Table 8.1}$$

$$\varepsilon = \frac{R \cdot T}{n \cdot F} \cdot \ln\left[\left(\frac{m_{Na2} \cdot m_{Cl2} \cdot \gamma_{mean2}^2}{m_{Na1} \cdot m_{Cl1} \cdot \gamma_{mean1}^2}\right)^{t_{Na}}\right]$$

$$t_{Na} = \frac{0.3902 + 0.3854}{2} \qquad \text{Average value from Table 8.4}$$

$$F = 96485 \cdot \frac{coul}{mole} \qquad R = 8.31451 \cdot \frac{joule}{mole \cdot K}$$

$$\varepsilon = 0.0297 \cdot volt$$

SECTION 8.7

8.73

We can use eq. 8.54 to estimate the liquid junction potential. Data can be found in Table 8.6.

$$c_1 = 0.1 \qquad c_2 = 0.2 \qquad T = 298.15 \cdot K \qquad R = 8.31451 \cdot \frac{joule}{mole \cdot K} \qquad F = 96485 \cdot coul \qquad equiv = 1$$

$$\lambda_K = 73.50 \cdot \frac{cm^2 \cdot siemens}{equiv} \qquad \lambda_{OH} = 197.6 \cdot \frac{cm^2 \cdot siemens}{equiv} \qquad \lambda_{Cl} = 76.34 \cdot \frac{cm^2 \cdot siemens}{equiv}$$

$$\Lambda_{KOH} = \lambda_K + \lambda_{OH} \qquad \Lambda_{KCl} = \lambda_K + \lambda_{Cl}$$

$$\varepsilon_J = \frac{R \cdot T}{F} \cdot \left[\frac{c_1 \cdot (\lambda_K - \lambda_{OH}) - c_2 \cdot (\lambda_K - \lambda_{Cl})}{c_1 \cdot \Lambda_{KOH} - c_2 \cdot \Lambda_{KCl}} \right] \cdot \ln\left(\frac{c_1 \cdot \Lambda_{KOH}}{c_2 \cdot \Lambda_{KCl}} \right)$$

$$\varepsilon_J = -0.0107 \cdot volt$$

8.75

We can use the bridge to separate the cell into 2 cells in series, with neither of the resulting ce having a liquid junction. The following can be used to give the desired overall cell reaction.

$$Zn \mid ZnSO_4(ai) \mid PbSO_4(s) \mid Pb-Pb(s) \mid PbSO_4(s) \mid CuSO_4(ai) \mid Cu$$

8.77

The liquid junction potential of the following bridge,

$$MX(c_1) \mid M''X''(c) \mid M'X'(c_2)$$

can be summed from the two individual liquid junctions present. The starting point for the evaluat of either potential is eq. 8.54.

$$\mathscr{E}_J = \frac{RT}{\mathscr{F}} \left[\frac{c_a(\lambda_+ - \lambda_-) - c_b(\lambda'_+ - \lambda'_-)}{c_a\Lambda - c_b\Lambda'} \right] \ln\left[\frac{c_a\Lambda}{c_b\Lambda'} \right]$$

where we have replaced the subscripts to avoid confusion with the salt bridge subscripts.

For the first junction, $c_a = c_1$ and $c_b = c$. Since $c >> c_1$, the terms in the left brackets containing c_a are neglible compared to those containing c_b. Thus, the first junction potential is given as

$$\mathscr{E}_{J1} = \frac{RT}{\mathscr{F}} \left[\frac{\lambda_+'' - \lambda_-''}{\Lambda''} \right] \ln \left[\frac{c_1 \Lambda}{c \Lambda''} \right]$$

The group in the left brackets is now easily seen to be the difference between the transference numbers for the concentrated salt species.

$$\mathscr{E}_{J1} = \frac{RT}{\mathscr{F}} (t_+'' - t_-'') \ln \left[\frac{c_1 \Lambda}{c \Lambda''} \right]$$

For the second junction, $c_a = c$ and $c_b = c_2$. Since $c >> c_2$, the terms in the left brackets of eq. 8.54 containing c_b are negligible to those containing c_a. Thus, the second junction potential is given as

$$\mathscr{E}_{J2} = \frac{RT}{\mathscr{F}} \left[\frac{\lambda_+'' - \lambda_-''}{\Lambda''} \right] \ln \left[\frac{c \Lambda''}{c_2 \Lambda'} \right]$$

$$\mathscr{E}_{J2} = \frac{RT}{\mathscr{F}} (t_+'' - t_-'') \ln \left[\frac{c \Lambda''}{c_2 \Lambda'} \right]$$

Combining the two junction potentials gives

$$\mathscr{E}_J = \mathscr{E}_{J1} + \mathscr{E}_{J2} = \frac{RT}{\mathscr{F}} (t_+'' - t_-'') \ln \left[\frac{c_1 \Lambda}{c_2 \Lambda'} \right]$$

Since $t_-'' = 1 - t_+''$, we have

$$\mathscr{E}_J = \frac{RT}{\mathscr{F}} (2t_+'' - 1) \ln \left[\frac{c_1 \Lambda}{c_2 \Lambda'} \right]$$

CHAPTER 9   *Transport Properties*
SECTION 9.1

9.1

We use equations (9.2), (9.9c), (9.8), and (9.3). The diameter, sigma, can be found on Table 1.7.

$$R \quad 8.31451 \cdot \frac{joule}{mole \cdot K} \qquad T \quad 250 \cdot K \qquad P \quad 3.55 \cdot torr \qquad M \quad 16.0428 \cdot 10^{-3} \cdot \frac{kg}{mole} \qquad nm \quad 10^{-9} \cdot m$$

$$L \quad 6.02 \cdot 10^{23} \cdot \frac{1}{mole} \qquad \sigma \quad 0.3817 \cdot nm \qquad \qquad \mu m \quad 10^{-6} \cdot m$$

$$v_{av} \quad \sqrt{\frac{8 \cdot R \cdot T}{\pi \cdot M}} \qquad v_{av} = 574.4 \cdot \frac{m}{sec}$$

$$\lambda \quad \frac{R \cdot T}{\sqrt{2} \cdot P \cdot L \cdot \pi \cdot \sigma^2} \qquad \lambda = 11.27 \cdot \mu m$$

$$nstar \quad \frac{P \cdot L}{R \cdot T}$$

$$z \quad \sqrt{2} \cdot v_{av} \cdot \pi \cdot \sigma^2 \cdot nstar \qquad z = 5.097 \cdot 10^7 \cdot \frac{1}{sec}$$

9.3

We use equations (9.2), (9.9c), (9.8), and (9.3). The diameters for N2 and O2 can be found on Table 1.7, and we weight these with their mole fractions in air to obtain an average diameter.

$$R \quad 8.31451 \cdot \frac{joule}{mole \cdot K} \qquad T \quad 570 \cdot K \qquad P \quad 2 \cdot 10^{-6} \cdot torr \qquad M \quad 29 \cdot 10^{-3} \cdot \frac{kg}{mole} \qquad nm \quad 10^{-9} \cdot m$$

$$L \quad 6.02 \cdot 10^{23} \cdot \frac{1}{mole} \qquad \sigma \quad (0.8) \cdot (0.3698 \cdot nm) \; + \; (0.2) \cdot (0.358 \cdot nm) \qquad \mu m \quad 10^{-6} \cdot m$$

$$v_{av} \quad \sqrt{\frac{8 \cdot R \cdot T}{\pi \cdot M}} \qquad v_{av} = 645.1 \cdot \frac{m}{sec}$$

$$\lambda \quad \frac{R \cdot T}{\sqrt{2} \cdot P \cdot L \cdot \pi \cdot \sigma^2} \qquad \lambda = 49.22 \cdot m$$

$$nstar \quad \frac{P \cdot L}{R \cdot T}$$

$$z \quad \sqrt{2} \cdot v_{av} \cdot \pi \cdot \sigma^2 \cdot nstar \qquad z = 13.11 \cdot \frac{1}{sec}$$

9.5

We use equations (9.2), (9.9c), (9.8), and (9.3). The diameter, sigma, can be found on Table 1.7.

$$R \quad 8.31451 \cdot \frac{joule}{mole \cdot K} \qquad T \quad 990 \cdot K \qquad M \quad 83.80 \cdot 10^{-3} \cdot \frac{kg}{mole} \qquad nm \quad 10^{-9} \cdot m$$

$$L \quad 6.02 \cdot 10^{23} \cdot \frac{1}{mole} \qquad \sigma \quad 0.360 \cdot nm \qquad i \quad 0..2 \qquad bar \quad 10^5 \cdot Pa \qquad \mu m \quad 10^{-6} \cdot m$$

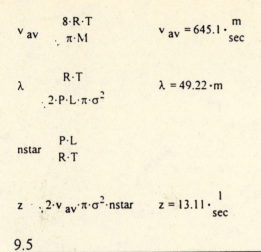

$$P_i$$

0.01·bar
1.00·bar
15.0·bar

$$v_{av} \quad \sqrt{\frac{8 \cdot R \cdot T}{\pi \cdot M}} \qquad v_{av} = 500.1 \cdot \frac{m}{sec}$$

$$\lambda_i \quad \frac{R \cdot T}{\sqrt{2} \cdot P_i \cdot L \cdot \pi \cdot \sigma^2} \qquad \lambda = \begin{matrix} 23.747 \\ 0.237 \\ 0.016 \end{matrix} \cdot \mu m$$

$$\text{ratio}_i \quad \frac{\lambda_i}{\sigma} \qquad\qquad \text{ratio} = \begin{array}{l} 6.596 \cdot 10^4 \\ 659.634 \\ 43.976 \end{array}$$

$$\text{nstar}_i \quad \frac{P_i \cdot L}{R \cdot T}$$

$$z_i \quad \cdot 2 \cdot \overline{v}_{av} \cdot \pi \cdot \sigma^2 \cdot \text{nstar}_i \qquad z = \begin{array}{l} 2.1 \cdot 10^7 \\ 2.1 \cdot 10^9 \\ 3.2 \cdot 10^{10} \end{array} \cdot \frac{1}{\text{sec}}$$

9.7

The relative velocity is given by eq. (9.11). First, we calcualte the reduced mass using eq. (9.10b).

$$R \quad 8.31451 \cdot \frac{\text{joule}}{\text{mole} \cdot K} \qquad M_A \quad 16.0428 \cdot 10^{-3} \cdot \frac{kg}{\text{mole}} \qquad M_B \quad 83.80 \cdot 10^{-3} \cdot \frac{kg}{\text{mole}} \qquad T \quad 350 \cdot K$$

$$L\mu \quad \frac{M_A \cdot M_B}{M_A \cdot M_B}$$

$$v_{AB} \quad \sqrt{\frac{8 \cdot R \cdot T}{\pi \cdot L\mu}} \qquad v_{AB} = 741.9 \cdot \frac{m}{\text{sec}}$$

9.9

The collision frequency is given by eq. (9.13a). First, we must calculate the reduced mass and the relative velocity. "A" denotes methane and "B" denotes krypton.

$$R \quad 8.31451 \cdot \frac{\text{joule}}{\text{mole} \cdot K} \qquad M_A \quad 16.0428 \cdot 10^{-3} \cdot \frac{kg}{\text{mole}} \qquad M_B \quad 83.80 \cdot 10^{-3} \cdot \frac{kg}{\text{mole}} \qquad T \quad 350 \cdot K$$

$$P_A \quad 55 \cdot \text{torr} \qquad P_B \quad 25 \cdot \text{torr} \qquad nm \quad 10^{-9} \cdot m \qquad \sigma_A \quad 0.3817 \cdot nm \qquad \sigma_B \quad 0.360 \cdot nm$$

248

$$L \quad 6.02 \cdot 10^{23} \cdot \frac{1}{mole}$$

$$\sigma_{AB} \quad \frac{1}{2} \cdot \sigma_A \cdot \sigma_B \qquad \text{eq. (9.12)}$$

$$L\mu \quad \frac{M_A \cdot M_B}{M_A \cdot M_B} \qquad \text{eq. (9.10b)}$$

$$v_{AB} \quad \sqrt{\frac{8 \cdot R \cdot T}{\pi \cdot L\mu}} \qquad v_{AB} = 742 \cdot \frac{m}{sec} \qquad \text{eq. (9.11)}$$

$$nstar_A \quad \frac{P_A \cdot L}{R \cdot T} \qquad nstar_B \quad \frac{P_B \cdot L}{R \cdot T}$$

In eq. (9.13a), "zAB" denotes the number of collisions of A with B, or in this case, the number of times methane collides with krypton.

$$z_{AB} \quad \pi \cdot \sigma_{AB}^2 \cdot v_{AB} \cdot nstar_B \qquad z_{AB} = 2.21 \cdot 10^8 \cdot \frac{1}{sec}$$

$$z_{BA} \quad \pi \cdot \sigma_{AB}^2 \cdot v_{AB} \cdot nstar_A \qquad z_{BA} = 4.862 \cdot 10^8 \cdot \frac{1}{sec}$$

9.11

The relative velocity is given by eq. (9.11), and the number of collisions is given by eq. (9.13a). In the following calculations, the first entry in the answer vecot denotes He, the second Ne, the third Ar, and the fourth Kr.

$$R \quad 8.31451 \cdot \frac{joule}{mole \cdot K} \qquad M_{CO2} \quad 44.0098 \cdot 10^{-3} \cdot \frac{kg}{mole} \qquad T \quad 298.15 \cdot K \qquad nm \quad 10^{-9} \cdot m$$

$$P_{CO2} \quad 5 \cdot torr \qquad \sigma_{CO2} \quad 0.4486 \cdot nm \qquad L \quad 6.02 \cdot 10^{23} \cdot \frac{1}{mole} \qquad i \quad 0..3$$

Data for other gases:

P_B $10 \cdot torr$

M_i	σ_i

M_i	σ_i
$4.002602 \cdot 10^{-3} \cdot \dfrac{kg}{mole}$	$0.263 \cdot nm$
	$0.2749 \cdot nm$
$20.179 \cdot 10^{-3} \cdot \dfrac{kg}{mole}$	$0.3405 \cdot nm$
	$0.360 \cdot nm$
$39.948 \cdot 10^{-3} \cdot \dfrac{kg}{mole}$	
$83.80 \cdot 10^{-3} \cdot \dfrac{kg}{mole}$	

$$nstar_B \quad \frac{P_B \cdot L}{R \cdot T} \qquad L\mu_i \quad \frac{M_{CO2} \cdot M_i}{M_{CO2} \cdot M_i}$$

$$\sigma mix_i \quad \frac{1}{2} \cdot \sigma_{CO2} \cdot \sigma_i \qquad v_{AB_i} \quad \frac{8 \cdot R \cdot T}{\pi \cdot L\mu_i} \qquad z_{AB_i} \quad \pi \cdot \sigma mix_i^2 \cdot v_{AB_i} \cdot nstar_B$$

$$v_{AB} = \begin{matrix} 1312 \\ 675 \\ 549 \\ 468 \end{matrix} \cdot \frac{m}{sec} \qquad z_{AB} = \begin{matrix} 1.689 \cdot 10^8 \\ 8.991 \cdot 10^7 \\ 8.694 \cdot 10^7 \\ 7.776 \cdot 10^7 \end{matrix} \cdot \frac{1}{sec}$$

9.13

The collision frequency of unlike molecules is given by eq. (9.13a). The collision frequency of like molecules is given by eq. (9.8). Below, "A" denotes helium and "B" denotes argon.

$$R \quad 8.31451 \cdot \frac{joule}{mole \cdot K} \qquad M_A \quad 4.002602 \cdot 10^{-3} \cdot \frac{kg}{mole} \qquad M_B \quad 39.948 \cdot 10^{-3} \cdot \frac{kg}{mole} \qquad L \quad 6.02 \cdot 10^{23} \cdot \frac{1}{mole}$$

$$nm \quad 10^{-9} \cdot m \qquad \sigma_A \quad 0.263 \cdot nm \qquad \sigma_B \quad 0.3405 \cdot nm \qquad T \quad 350 \cdot K$$

$$bar \quad 10^5 \cdot Pa \qquad P_A \quad 0.124 \cdot bar \qquad P_B \quad 0.345 \cdot bar$$

He-He collisions (zAA) and Ar-Ar collisions (zBB):

$$nstar_A \quad \frac{P_A \cdot L}{R \cdot T} \qquad v_A \quad \sqrt{\frac{8 \cdot R \cdot T}{\pi \cdot M_A}} \qquad z_{AA} \quad \sqrt{2} \cdot v_A \cdot \pi \cdot \sigma_A^2 \cdot nstar_A \qquad z_{AA} = 1.073 \cdot 10^9 \cdot \frac{1}{sec}$$

$$nstar_B \quad \frac{P_B \cdot L}{R \cdot T} \qquad v_B \quad \sqrt{\frac{8 \cdot R \cdot T}{\pi \cdot M_B}} \qquad z_{BB} \quad \sqrt{2} \cdot v_B \cdot \pi \cdot \sigma_B^2 \cdot nstar_B \qquad z_{BB} = 1.583 \cdot 10^9 \cdot \frac{1}{sec}$$

He-Ar collisions (zAB):

$$L\mu \quad \frac{M_A \cdot M_B}{M_A \cdot M_B} \qquad v_{AB} \quad \sqrt{\frac{8 \cdot R \cdot T}{\pi \cdot L\mu}}$$

$$\sigma_{AB} \quad \frac{1}{2} \sigma_A \cdot \sigma_B \qquad z_{AB} \quad \pi \cdot \sigma_{AB}^2 \cdot v_{AB} \cdot nstar_B \qquad z_{AB} = 2.914 \cdot 10^9 \cdot \frac{1}{sec}$$

To determine the total number of collisions, we use eqs. (9.14) and (9.15):

$$Z_{AA} \quad \frac{z_{AA} \cdot nstar_A}{2} \qquad Z_{AA} = 1.376 \cdot 10^{33} \cdot \frac{1}{m^3 \cdot sec}$$

$$Z_{BB} \quad \frac{z_{BB} \cdot nstar_B}{2} \qquad Z_{BB} = 5.65 \cdot 10^{33} \cdot \frac{1}{m^3 \cdot sec}$$

$$Z_{AB} \quad z_{AB} \cdot nstar_A \qquad Z_{AB} = 7.474 \cdot 10^{33} \cdot \frac{1}{m^3 \cdot sec}$$

SECTION 9.2

9.15

(a) The probablity is given by eq. (9.16). Since both numbers are odd, the parity factor is 1.

$$C(n,m) \quad \frac{n!}{m! \cdot (n \quad m)!} \qquad \rho_{nm} \quad 1$$

$$W(n,m) \quad \frac{1}{2^n} \cdot C\left(n, \frac{n \cdot m}{2}\right) \cdot \rho_{nm} \qquad W(11,5) = 0.0806$$

(b) We integrate the probability distribution function, eq. (9.19), between -6 and 6:

$$n \quad 11 \qquad \lambda \quad 1$$

$$P \quad \int_{.6}^{.6} \frac{1}{\sqrt{2 \cdot \pi \cdot n \cdot \lambda^2}} \cdot e^{\frac{x^2}{2 \cdot n \cdot \lambda^2}} \, dx \qquad P = 93 \cdot \%$$

9.17

This is similar to example 9.5. First, we use eq. (9.9c) to calculate the mean free path and eq. (9.8) to calculate the number of collisions.

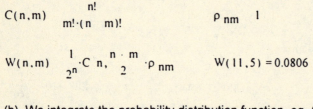

$$R \quad 8.31451 \cdot \frac{joule}{mole \cdot K} \qquad T \quad 300 \cdot K \qquad M \quad 39.948 \cdot 10^{3} \cdot \frac{kg}{mole} \qquad i \quad 0..2 \qquad nm \quad 10^{9} \cdot m$$

$$L \quad 6.02 \cdot 10^{23} \cdot \frac{1}{mole} \qquad \sigma \quad 0.3405 \cdot nm \qquad bar \quad 10^{5} \cdot Pa \qquad t \quad 5 \cdot min \qquad P_i$$

10·bar
1·bar
0.1·bar

$$v_{av} \quad \sqrt{\frac{8 \cdot R \cdot T}{\pi \cdot M}} \qquad v_{av} = 398.8 \cdot \frac{m}{sec} \qquad nstar_i \quad \frac{P_i \cdot L}{R \cdot T}$$

$$\lambda_i \quad \frac{R \cdot T}{2 \cdot P_i \cdot L \cdot \pi \cdot \sigma^2} \qquad \lambda = \begin{matrix} 8.044 \\ 80.438 \\ 804.382 \end{matrix} \cdot nm$$

$$z_i \quad 2 \cdot v_{av} \cdot \pi \cdot \sigma^2 \cdot nstar_i \qquad z = \begin{matrix} 4.957 \cdot 10^{10} \\ 4.957 \cdot 10^9 \\ 4.957 \cdot 10^8 \end{matrix} \cdot \frac{1}{sec}$$

$$\delta x_{rms_i} \quad z_i \cdot t \cdot \lambda_i \qquad \delta x_{rms} = \begin{matrix} 31 \\ 98.1 \\ 310.2 \end{matrix} \cdot mm$$

9.19

This is similar to example 9.5. First, we use eq. (9.9c) to calculate the mean free path and eq. (9.8) to calculate the number of collisions.

$$R \quad 8.31451 \cdot \frac{joule}{mole \cdot K} \qquad T \quad 250 \cdot K \qquad M \quad 28.0134 \cdot 10^{-3} \cdot \frac{kg}{mole} \qquad \sigma \quad 0.3698 \cdot 10^{-9} \cdot m$$

$$L \quad 6.02 \cdot 10^{23} \cdot \frac{1}{mole} \qquad \delta x_{rms} \quad 1 \cdot cm \qquad P \quad 2.5 \cdot atm$$

$$v_{av} \quad \frac{8 \cdot R \cdot T}{\pi \cdot M} \qquad v_{av} = 434.7 \cdot \frac{m}{sec} \qquad nstar \quad \frac{P \cdot L}{R \cdot T}$$

$$\lambda \quad \frac{R \cdot T}{2 \cdot P \cdot L \cdot \pi \cdot \sigma^2} \qquad \lambda = 2.243 \cdot 10^{-8} \cdot m$$

$$z \quad 2 \cdot v_{av} \cdot \pi \cdot \sigma^2 \cdot nstar \qquad z = 1.938 \cdot 10^{10} \cdot \frac{1}{sec}$$

$$t \quad \frac{\delta x_{rms}^2}{\lambda^2 \cdot z} \qquad t = 10.25 \cdot sec$$

9.21

We integrate the probability distribution function, eq. (9.23), between r1 and r2.

$$W(r,t)\cdot dr = \frac{1}{2\cdot D\cdot t}\cdot e^{\frac{r^2}{4\cdot D\cdot t}}\cdot r\cdot dr$$

Make the following substitution:

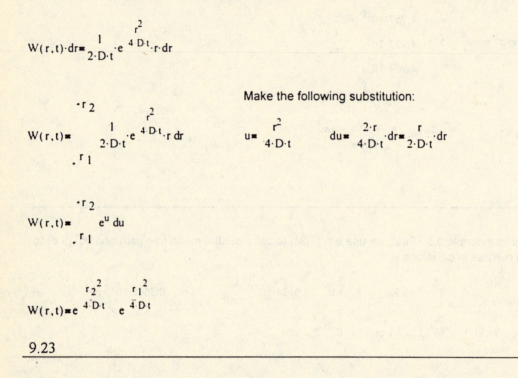

$$W(r,t) = \int_{r_1}^{r_2}\frac{1}{2\cdot D\cdot t}\cdot e^{\frac{r^2}{4\cdot D\cdot t}}\cdot r\,dr \qquad u = \frac{r^2}{4\cdot D\cdot t} \qquad du = \frac{2\cdot r}{4\cdot D\cdot t}\cdot dr = \frac{r}{2\cdot D\cdot t}\cdot dr$$

$$W(r,t) = \int_{r_1}^{r_2} e^u\,du$$

$$W(r,t) = e^{\frac{r_2^2}{4\cdot D\cdot t}}\quad e^{\frac{r_1^2}{4\cdot D\cdot t}}$$

9.23

The rms distance for a 1-dimensional case is given by eq. (9.22). We can derive the result for a 2-dimensional system from the probability distribution function. However, we can also use the fact that the 1-D result is not unique to the x-direction, but also applies to the y-direction (see example 9.9)

$$r^2 = x^2 \cdot y^2 = 4\cdot D\cdot t$$

$$r_{rms} = \sqrt{4\cdot D\cdot t}$$

For this problem,

$$D \quad 2.2\cdot 10^9\cdot\frac{m^2}{sec} \qquad r_{rms} \quad 15\cdot 10^9\cdot m \qquad ns \quad 10^9\cdot sec \qquad t \quad \frac{r_{rms}^2}{4\cdot D} \qquad t = 25.6\cdot ns$$

9.25

We can integrate the 2-dimensional probability distribution function, en (9.23), from 24 to infinity. However, since the probabilities must sum to one, we can also compute the probability that the sailor will not be in the water (perform the integration from 0 to 24) and subtract this probability from one. If you are working the problem on a computer, it does not make any difference, but if you are numerically evaluating the integral by hand, it is much easier to do it the second way. We substitue in eq. (9.21) for the diffusion coefficient to get the probability distribution function in terms of the number of steps and step size.

$$n \quad 159 \qquad \lambda \quad 1$$

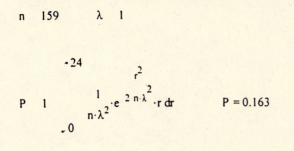

$$P \quad 1 \quad \int_{0}^{24} \frac{1}{n \cdot \lambda^2} \cdot e^{\frac{r^2}{2 \cdot n \cdot \lambda^2}} \cdot r \, dr \qquad P = 0.163$$

9.27

Using the result of the previous problem, we have::

$$n \quad 1000 \qquad \lambda \quad 1 \cdot cm$$

$$r_{av} \quad \sqrt{\frac{8 \cdot n}{\pi}} \cdot \lambda \qquad r_{av} = 50.46 \cdot cm$$

9.29

We integrate the 3D probability distribution function, eq. (9.24), from 0 to 1 cm.

$$D \quad 4.5 \cdot 10^{-5} \cdot \frac{cm^2}{sec} \qquad t \quad 1 \cdot hr$$

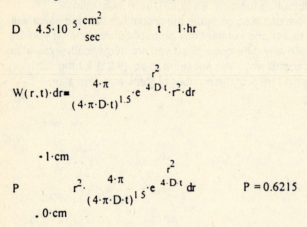

$$W(r,t) \cdot dr = \frac{4 \cdot \pi}{(4 \cdot \pi \cdot D \cdot t)^{1.5}} \cdot e^{\frac{r^2}{4 \cdot D \cdot t}} \cdot r^2 \cdot dr$$

$$P \quad \int_{0 \cdot cm}^{1 \cdot cm} r^2 \cdot \frac{4 \cdot \pi}{(4 \cdot \pi \cdot D \cdot t)^{1.5}} \cdot e^{\frac{r^2}{4 \cdot D \cdot t}} \, dr \qquad P = 0.6215$$

9.31

When there is no wall, we have:

$$C(n,m) \quad \frac{n!}{m! \cdot (n \quad m)!} \qquad \rho_{nm} \quad 1$$

$$W(n,m) \quad \frac{1}{2^n} \cdot C\left(n, \frac{n \quad m}{2}\right) \cdot \rho_{nm} \qquad W(10,0) = 0.2461$$

When the reflecting wall is in place, we use eq. (9.25):

$$W_r \, n,m,m_1 \quad W(n,m) \quad W\left(n, 2 \cdot m_1 \quad m\right)$$

$$W_r(10,0,3) = 0.29$$

We use eq. (9.26) to compute the probabilty, since there is a large number of steps:

$$n = 100 \qquad l = \sqrt{2} \qquad x_1 = 11$$

$$W(x,t) = \frac{1}{\sqrt{2\pi n\,l^2}}\, exp\left(-\frac{x^2}{n\cdot l^2}\right)$$

$$P_{abs} = 2\cdot \int_{x_1}^{100} W(x,n)\,dx \qquad P_{abs} = 19.2\cdot\%$$

SECTION 9.3

9.35

We use eq. (9.22). The diffusion constant can be found on Table 9.1.

$$D \quad 0.0059 \cdot 10^{-5} \cdot \frac{cm^2}{sec} \qquad week \quad 7 \cdot day$$

$$t \quad 1 \cdot week \qquad \delta x_{rms} \quad \sqrt{2 \cdot D \cdot t} \qquad \delta x_{rms} = 0.267 \cdot cm$$

$$t \quad 4 \cdot week \qquad \delta x_{rms} \quad \sqrt{2 \cdot D \cdot t} \qquad \delta x_{rms} = 0.534 \cdot cm$$

9.37

We rearrange eq. (9.22) and solve for time. The diffusion coefficient can be found on Table 9.1.

$$D \quad 0.67 \cdot 10^{-5} \cdot \frac{cm^2}{sec}$$

$$\delta x_{rms} \quad 1 \cdot mm \qquad t \quad \frac{\delta x_{rms}^2}{2 \cdot D} \qquad t = 12.4 \cdot min$$

9.39

Equation (9.33) gives the diffusion coefficient in terms of the average velocity and the mean free path. We use equations (9.2) and (9.9c) to calculate these. The diameter, sigma, can be found on Table 1.7.

$$R \quad 8.31451 \cdot \frac{joule}{mole \cdot K} \qquad nm \quad 10^{-9} \cdot m \qquad P \quad 1 \cdot atm \qquad M \quad 39.948 \cdot 10^{-3} \cdot \frac{kg}{mole}$$

$$L \quad 6.02 \cdot 10^{23} \cdot \frac{1}{mole} \qquad \sigma \quad 0.3405 \cdot nm \qquad i \quad 0..2$$

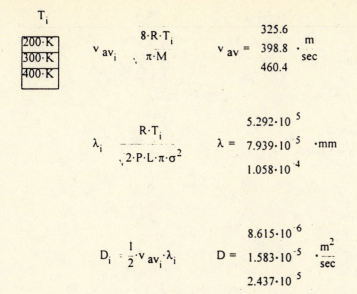

$$T_i$$

200·K
300·K
400·K

$$v\,av_i \quad \sqrt{\frac{8 \cdot R \cdot T_i}{\pi \cdot M}} \qquad v\,av = \begin{matrix} 325.6 \\ 398.8 \\ 460.4 \end{matrix} \cdot \frac{m}{sec}$$

$$\lambda_i \quad \frac{R \cdot T_i}{\sqrt{2} \cdot P \cdot L \cdot \pi \cdot \sigma^2} \qquad \lambda = \begin{matrix} 5.292 \cdot 10^{-5} \\ 7.939 \cdot 10^{-5} \\ 1.058 \cdot 10^{-4} \end{matrix} \cdot mm$$

$$D_i \quad \frac{1}{2} \cdot v\,av_i \cdot \lambda_i \qquad D = \begin{matrix} 8.615 \cdot 10^{-6} \\ 1.583 \cdot 10^{-5} \\ 2.437 \cdot 10^{-5} \end{matrix} \cdot \frac{m^2}{sec}$$

9.41

We use eq. (9.34). Here, the upper limit of integration is denoted "curlyQ."

$$D \quad 2.95 \cdot 10^{-5} \cdot \frac{cm^2}{sec} \qquad x \quad 1 \cdot mm \qquad i \quad 0..3 \qquad t_i$$

1·min
10·min
60·min
24·hr

$$curlyQ_i \quad \frac{x}{\sqrt{4 \cdot D \cdot t_i}} \qquad curlyQ = \begin{matrix} 1.188 \\ 0.376 \\ 0.153 \\ 0.031 \end{matrix}$$

$$c\_co_i \quad \frac{1}{2} \quad 1 \quad \frac{2}{\sqrt{\pi}} \cdot \int_0^{curlyQ_i} e^{-y^2} dy \qquad c\_co = \begin{matrix} 0.046 \\ 0.298 \\ 0.414 \\ 0.482 \end{matrix}$$

259

9.43

We rearrange eq. (9.22) and solve for time.

$$D \quad 10^{12} \cdot \frac{cm^2}{sec}$$

$$\delta x_{rms} \quad 0.1 \cdot 10^6 \cdot m \qquad t \quad \frac{\delta x_{rms}^2}{2 \cdot D} \qquad t = 50 \cdot sec$$

SECTION 9.4

9.45

We use eq. (9.37).

η $0.01 \cdot poise$ ΔP $3 \cdot torr$ r $0.1 \cdot cm$ l $5 \cdot cm$

flowrate $\dfrac{\pi \cdot r^4 \cdot \Delta P}{8 \cdot \eta \cdot l}$ flowrate $= 188.48 \cdot \dfrac{cm^3}{min}$

9.47

We use eq. (9.42). This is similar to example 9.18. The diffusion coefficient can be found on Table 9.1, and the viscosity of water can be found on Table 9.2.

T $293.15 \cdot K$ η $10.05 \cdot 10^3 \cdot poise$ nm $10^9 \cdot m$

D $0.0059 \cdot 10^5 \cdot \dfrac{cm^2}{sec}$ k $1.38066 \cdot 10^{16} \cdot \dfrac{erg}{K}$

r $\dfrac{k \cdot T}{6 \cdot \pi \cdot \eta \cdot D}$ $r = 36.212 \cdot nm$

9.49

This is similar to example 9.17. We use eq. (9.43). The diffusion coefficient given in Table 9.1 is at 20 C.

T_1 $293.15 \cdot K$ T_2 $323.15 \cdot K$ η_1 $10.05 \cdot 10^3 \cdot poise$ η_2 $5.494 \cdot 10^3 \cdot poise$

D_1 $0.4586 \cdot 10^5 \cdot \dfrac{cm^2}{sec}$

$\dfrac{D_1 \cdot \eta_1}{T_1} = \dfrac{D_2 \cdot \eta_2}{T_2}$ D_2 $D_1 \cdot \dfrac{\eta_1}{T_1 \cdot \eta_2} \cdot T_2$ $D_2 = 9.248 \cdot 10^6 \cdot \dfrac{cm^2}{sec}$

9.51

We start with eq. (9.37). This tells us that, if the length of the tube, radius, and pressure drop remain constant, the time required is proportional to the viscosity. For an ideal gas, the viscosity is proportional to the square root of the temperature:

$$\frac{\Delta V}{\Delta t} = \frac{\pi \cdot r^4 \cdot \Delta P}{8 \cdot \eta \cdot l} \qquad \eta = \frac{M \cdot \sqrt{\frac{8 \cdot R \cdot T}{\pi \cdot M}}}{2 \cdot \sqrt{2} \cdot \pi \cdot \sigma^2 \cdot L}$$

Thus, we have:

$$T_1 \quad 773.15 \cdot K \qquad T_2 \quad 293.15 \cdot K$$

$$\text{timeratio} \quad \sqrt{\frac{T_1}{T_2}} \qquad \text{timeratio} = 1.624 \qquad \text{It will take 1.62 times as long at 500 C.}$$

9.53

Equation (9.51) relates the molecular diameter to the viscosity.

$$T \quad 292.05 \cdot K \qquad \eta \quad 73.5 \cdot 10^{6} \cdot poise \qquad M \quad 74.1174 \cdot 10^{3} \cdot \frac{kg}{mole} \qquad R \quad 8.31451 \cdot \frac{joule}{mole \cdot K}$$

$$L \quad 6.02 \cdot 10^{23} \cdot \frac{1}{mole} \qquad nm \quad 10^{9} \cdot m$$

$$\eta = \frac{M \cdot \sqrt{\frac{8 \cdot R \cdot T}{\pi \cdot M}}}{2 \cdot \sqrt{2} \cdot \pi \cdot \sigma^2 \cdot L} \qquad \sigma \quad \sqrt{\frac{M \cdot \sqrt{\frac{8 \cdot R \cdot T}{\pi \cdot M}}}{2 \cdot \sqrt{2} \cdot \pi \cdot \eta \cdot L}} \qquad \sigma = 0.738 \cdot nm$$

9.55

We use eq. (9.48) to calculate the viscosity:

$$R \quad 8.31451 \cdot \frac{joule}{mole \cdot K} \qquad M \quad 17.0305 \cdot 10^{-3} \cdot \frac{kg}{mole} \qquad nm \quad 10^{-9} \cdot m \qquad \mu p \quad 10^{-6} \cdot poise$$

$$L \quad 6.02 \cdot 10^{23} \cdot \frac{1}{mole} \qquad k_s \quad 15.69 \cdot \frac{\mu p}{\cdot K} \qquad S \quad 510 \cdot K$$

$$T \quad 373.15 \cdot K \qquad \eta \quad \frac{k_s \cdot T}{1 \cdot \frac{S}{T}} \qquad \eta = 128.1 \cdot \mu p$$

$$T \quad 473.15 \cdot K \qquad \eta \quad \frac{k_s \cdot T}{1 \cdot \frac{S}{T}} \qquad \eta = 164.2 \cdot \mu p$$

We can compute the diameter by solving eq. (9.49) for the diameter:

$$\sigma \quad \frac{R \cdot M}{\pi^{1.5} \cdot k_s \cdot L} \qquad \sigma = 0.2675 \cdot nm$$

9.57

We fit the data to eq. (9.48) using a nonlinear regression program.

$$\eta = \frac{k_s \cdot T}{1 \cdot \frac{S}{T}}$$

Alternatively, we can linearize the equation as follows, and use a linear regression package to find the parameters:

$$\frac{T}{\eta} = \frac{S}{k_s} \cdot \frac{1}{T} + \frac{1}{k_s}$$

The parameters obtained using nonlinear regression are give below. We can compute the diameter by solving eq. (9.49) for the diameter:

$$R \quad 8.31451 \cdot \frac{joule}{mole \cdot K} \qquad M \quad 39.948 \cdot 10^{-3} \cdot \frac{kg}{mole} \qquad nm \quad 10^{-9} \cdot m \qquad \mu p \quad 10^{-6} \cdot poise$$

$$L \quad 6.02 \cdot 10^{23} \cdot \frac{1}{mole} \qquad k_s = 19.09 \cdot \frac{\mu p}{\sqrt{K}} \qquad S \quad 137.7 \cdot K$$

$$\sigma = \frac{\sqrt{R \cdot M}}{\pi^{1.5} \cdot k_s \cdot L} \qquad\qquad \sigma = 0.3001 \cdot nm$$

9.59

Equation (9.47) gives the Enskog relation, which can be used to predict viscosities at higher pressures. First, we need the molar volume from the van der Waals equation of state:

$$P \quad 100 \cdot atm \qquad T \quad 323.15 \cdot K \qquad a \quad 0.3649 \cdot \frac{Pa \cdot m^6}{mole^2} \qquad b \quad 42.74 \cdot 10^{-6} \cdot \frac{m^3}{mole}$$

$$R \quad 8.31451 \cdot \frac{joule}{mole \cdot K}$$

As a first guess, use ifeal gas law: $\qquad V \quad \frac{R \cdot T}{P}$

$$V_m \quad root \; P \cdot V^3 - (R \cdot T + b \cdot P) \cdot V^2 + a \cdot V + a \cdot b, V \qquad V_m = 149.351 \cdot \frac{cm^3}{mole}$$

The viscosity is then determined as follows. The value of bo can be found on Table 1.7.

$$\mu p \cdot 10^{6} \cdot \text{poise} \qquad b_o \cdot 113.9 \cdot \frac{cm^3}{mole} \qquad \eta_o \cdot 163 \cdot \mu p$$

$$\eta \cdot \eta_o \cdot 1 \cdot 0.175 \cdot \frac{b_o}{V_m} \cdot 0.865 \cdot \frac{b_o}{V_m}^2 \qquad \eta = 266.8 \cdot \mu p$$

9.61

The Mark-Houwink equation, eq. (9.54), can be used to obtain the intrinsic viscosity of polymeric solutions. The parameters can be found on Table 9.3:

$$K \cdot 8.1 \cdot 10^{2} \cdot \frac{cm^3}{gm} \qquad \alpha \cdot 0.5 \qquad M \cdot 1 \cdot 10^6 \qquad \text{int}\eta \cdot K \cdot M^\alpha \qquad \text{int}\eta = 81 \cdot \frac{cm^3}{gm}$$

For a dilute solution, we can take the limit of eq. (9.51) to obtain the specific viscosity:

$$c_m \cdot 20 \cdot \frac{mg}{cm^3} \qquad \eta_{sp} \cdot c_m \cdot \text{int}\eta \qquad \eta_{sp} = 1.62$$

Finally, we use eq. (9.50) to obtain the viscosity:

$$\eta_o \cdot 7.5 \cdot 10^{3} \cdot \text{poise} \qquad mp \cdot 10^{3} \cdot \text{poise}$$

$$\eta \cdot \eta_o \cdot (1 \cdot \eta_{sp}) \qquad \eta = 19.65 \cdot mp$$

9.63

We can rearrange the Mark-Houwink equation as follows:

$$\text{int}\eta = K \cdot M_v^\alpha \qquad M_v = \frac{\text{int}\eta}{K}^{\frac{1}{\alpha}}$$

The intrinsic viscosity can be written as follows, using eqs. (9.50) and (9.51)

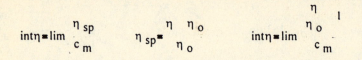

$$\mathrm{int}\eta = \lim \frac{\eta_{sp}}{c_m} \qquad \eta_{sp} = \frac{\eta - \eta_0}{\eta_0} \qquad \mathrm{int}\eta = \lim \frac{\frac{\eta}{\eta_0} - 1}{c_m}$$

So, to find the intrinsic viscosity, we regress (n/no - 1)/cm versus cm, and find the limit as cm goes to zero.

$$i \quad 0..5$$

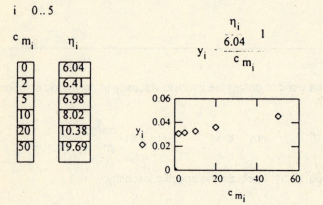

c_{m_i}	η_i
0	6.04
2	6.41
5	6.98
10	8.02
20	10.38
50	19.69

$$y_i = \frac{\frac{\eta_i}{6.04} - 1}{c_{m_i}}$$

The intrinsic viscosity seems to be about 0.003. We plug this result into the Mark-Houwink equation to find Mv.

$$\mathrm{int}\eta \quad 0.03 \cdot \frac{cm^3}{mg} \qquad K \quad 0.95 \cdot 10^{-2} \cdot \frac{cm^3}{gm} \qquad \alpha \quad 0.74$$

$$M_v \quad \frac{\mathrm{int}\eta}{K}^{\frac{1}{\alpha}} \qquad M_v = 5.357 \cdot 10^4$$

We can use eq. (9.56) to find Mn:

$$\frac{M_v}{M_n} = ((1-\alpha) \cdot \Gamma(1-\alpha))^{\frac{1}{\alpha}}$$

$$M_n = \frac{M_v}{((1-\alpha) \cdot \Gamma(1-\alpha))^{\frac{1}{\alpha}}} \qquad M_n = 2.85 \cdot 10^4$$

SECTION 9.5

9.65

We use eq. (9.57):

$$rev \quad 2 \cdot \pi \cdot rad \qquad \omega \quad 60000 \cdot \frac{rev}{min} \qquad x \quad 6 \cdot cm \qquad \Delta x \quad 5.1 \cdot mm \qquad \Delta t \quad 3 \cdot hr$$

$$s \quad \frac{\Delta x}{\Delta t} \cdot \frac{1}{\omega^2 \cdot x} \qquad s = 1.994 \cdot 10^{13} \cdot sec$$

9.67

(a) We use eq. (9.58):

$$R \quad 8.31451 \cdot \frac{joule}{mole \cdot K} \qquad D \quad 0.4586 \cdot 10^{5} \cdot \frac{cm^2}{sec} \qquad \rho_o \quad 0.9982 \cdot \frac{gm}{cm^3} \qquad \upsilon_s \quad 0.630 \cdot \frac{cm^3}{gm}$$

$$T \quad 293.15 \cdot K \qquad M \quad 342.2407 \cdot \frac{gm}{mole}$$

$$s \quad \frac{M \cdot D \cdot 1}{R \cdot T} \quad \rho_o \cdot \upsilon_s \qquad s = 2.39 \cdot 10^{14} \cdot sec$$

(b) We use eq. (9.57):

$$rev \quad 2 \cdot \pi \cdot rad \qquad x \quad 5 \cdot cm \qquad \Delta x \quad 1 \cdot mm \qquad \Delta t \quad 1 \cdot hr$$

$$\omega \quad \frac{\Delta x}{\Delta t} \cdot \frac{1}{s \cdot x} \qquad \omega = 1.456 \cdot 10^5 \cdot \frac{rev}{min}$$

9.69

We use eq. (9.59) and solve for the molecular weight.

$$R = 8.31451 \cdot \frac{\text{joule}}{\text{mole} \cdot \text{K}} \qquad \text{rev} = 2 \cdot \pi \cdot \text{rad} \qquad \omega = 80000 \cdot \frac{\text{rev}}{\text{min}} \qquad \rho_o = 0.9951 \cdot \frac{\text{gm}}{\text{cm}^3} \qquad \upsilon_s = 0.75 \cdot \frac{\text{cm}^3}{\text{gm}}$$

$$T = 296.15 \cdot \text{K} \qquad x_1 = 0.7 \cdot \text{cm} \qquad x_2 = 1 \cdot \text{cm} \qquad c2\_c1 = 4.2$$

$$M = \frac{\ln(c2\_c1)}{\left(x_2^2 - x_1^2\right)} \cdot \frac{2 \cdot R \cdot T}{1 - \rho_o \cdot \upsilon_s \cdot \omega^2}$$

$$M = 7.783 \cdot 10^3 \cdot \frac{\text{gm}}{\text{mole}}$$

CHAPTER 10   *Chemical Kinetics*

SECTION 10.2

10.1
___

For a 3/2 order reaction, the rate law will be something like:

$$\frac{dc_a}{dt} = k \cdot c_a^{1.5} \quad \text{with the units:} \quad \frac{\text{mole}}{\text{liter} \cdot \text{sec}} = k \cdot \frac{\text{mole}^{1.5}}{\text{liter}^{1.5}}$$

Thus, k must have units of:

$$\frac{1}{\sqrt{\text{mole}}} \cdot \frac{\sqrt{\text{liter}}}{\text{sec}}$$

10.3
___

This is similar to example 10.1. We use eq. (10.4):

$$\ln(v) = \ln(k) - n \cdot \ln(C) \quad \text{The slope of this plot will give the order of the reaction.}$$

$i \quad 0..14$

$n \quad 0..13$

t_i	C_i
$0 \cdot min$	1
$2 \cdot min$	0.9602
$3 \cdot min$	0.9427
$5 \cdot min$	0.9084
$6 \cdot min$	0.8911
$8 \cdot min$	0.8576
$9 \cdot min$	0.8412
$11 \cdot min$	0.8089
$12 \cdot min$	0.7957
$14 \cdot min$	0.7666
$15 \cdot min$	0.7530
$17 \cdot min$	0.7256
$18 \cdot min$	0.7107
$20 \cdot min$	0.6873
$21 \cdot min$	0.6735

$$v_n \quad \frac{C_n - C_{n-1}}{t_{n-1} - t_n} \cdot min \qquad C_{av_n} \quad \frac{C_n - C_{n-1}}{2}$$

$$y_n \quad \ln v_n \qquad\qquad x_n \quad \ln C_{av_n}$$

order   slope(x, y)

order = 1.073

10.5

We use the method of initial velocities to analyze this problem. First, we try to determine the rate law. From the first two data points, it appears that when we double [OCl-], we double the rate. From the first and third data points, the conclusion is that when we double [I-], we double the rate. Finally, from the first and fourth data points, we can see that when we decrease [OH-] by a factor of two, we double the rate. This suggests that the rate law is first order in [OCl-] and [I-], and is inversely proportional to [OH-]:

v = k [OCl-] [I-] / [OH-]

Now, we can use any of the data points to estimate k.

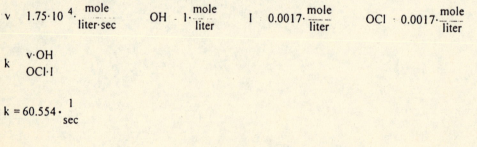

$$v \quad 1.75 \cdot 10^{-4} \cdot \frac{mole}{liter \cdot sec} \qquad OH \quad 1 \cdot \frac{mole}{liter} \qquad I \quad 0.0017 \cdot \frac{mole}{liter} \qquad OCl \quad 0.0017 \cdot \frac{mole}{liter}$$

$$k \quad \frac{v \cdot OH}{OCl \cdot I}$$

$$k = 60.554 \cdot \frac{1}{sec}$$

10.7

Since we do not have the partial pressures of the individual species, we must first derive them from the data given. Let us write the reaction as $A = 2B + C$, and define an extent of reaction, x. Then, the partial pressures of the species at any time will be:

$$P_A = P_o - x$$

$$P_B = 2 \cdot x$$

$$P_C = x$$

and the total pressure will be:

$$P = P_A + P_B + P_C + P_{N2} = P_o - x + 2 \cdot x + x + 4.2 \cdot \text{torr}$$

$$P = P_o + 2 \cdot x + 4.2 \cdot \text{torr}$$

$$x = \frac{P - 4.2 \cdot \text{torr} - P_o}{2}$$

Now, for a first order reaction:

$$\frac{dP_A}{dt} = -k \cdot P_A \qquad P_A = P_o - x$$

$$\frac{dx}{dt} = k \cdot (P_o - x)$$

$$\ln\left(\frac{P_o - x}{P_o}\right) = -k \cdot t$$

So, a plot of the left-hand side vs. t should be linear with a slope of $-k$.

$i := 0 .. 14$

$t_i :=$	$P_i :=$
0·min	173.5·torr
2·min	187.3·torr
3·min	193.4·torr
5·min	205.3·torr
6·min	211.3·torr
8·min	222.9·torr
9·min	228.6·torr
11·min	239.8·torr
12·min	244.4·torr
14·min	254.4·torr
15·min	259.2·torr
17·min	268.7·torr
18·min	273.9·torr
20·min	282.0·torr
21·min	286.8·torr

$$x_i \quad \dfrac{P_i \quad 4.2 \cdot torr \quad P_0}{2} \qquad y_i \quad \ln \dfrac{P_0 \quad x_i}{P_0}$$

$$k \quad slope(t, y)$$

$$k = 0.0185 \cdot \dfrac{1}{min}$$

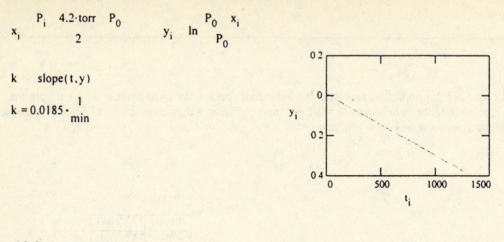

10.9

We use eq. (10.6). This shows that a graph of ln (l - linf) vs. t will be linear with a slope of - k.

$$\lambda \quad \lambda_{inf} = \lambda_o \quad \lambda_{inf} \cdot exp(\ k \cdot t)$$

$$i \quad 0 .. 14 \qquad l_{inf} \quad 4.95$$

t_i	l_i
1·sec	3.65
2·sec	2.87
3·sec	2.03
4·sec	0.992
5·sec	0.517
6·sec	0.274
8·sec	1.22
10·sec	1.99
12·sec	2.77
14·sec	3.17
16·sec	3.46
20·sec	3.91
24·sec	4.41
30·sec	4.76
35·sec	4.79

$$y_i \quad \ln \dfrac{l_i \quad l_{inf}}{l_0 \quad l_{inf}}$$

$$k \quad slope(t, y)$$

$$k = 0.122 \cdot \dfrac{1}{sec}$$

The relaxation time is defined as 1/k.

$$\tau_R \quad \dfrac{1}{k}$$

$$\tau_R = 8.201 \cdot sec$$

10.11

We can use the information given to calculate the rate constant. Since the reaction is first-order, we know that it follows an exponential decay. When the reaction is 20% complete, 80% of the reactant remains.

$conc = e^{k \cdot t}$ $\qquad$ $t_{0.8} = 20 \cdot min$ $\qquad$ $conc = 0.80$

$k = \dfrac{\ln(conc)}{t_{0.8}}$ $\qquad$ $k = 1.86 \cdot 10^{-4} \cdot \dfrac{1}{sec}$

When the reaction is 90% complete, the concentration of the reactant will be 10% of the original concentration.

$conc = 0.1$ $\qquad$ $t = \dfrac{\ln(conc)}{k}$ $\qquad$ $t = 206.4 \cdot min$

10.13

Since we are only given total pressure, we must relate this to the partial pressure of ethylene oxide and the extent of reaction. Let x be the extent of reaction. Then the partial pressures of the species will be:

$P_{EO} = P_o - x$

$P_{CH4} = x$

$P_{CO} = x$

The total pressure is given by:

$P = P_{EO} + P_{CH4} + P_{CO} = P_o - x + 2 \cdot x$

$P = P_o + x$

$x = P - P_o$ $\qquad$ $P_{EO} = P_o - x = P_o - (P - P_o) = 2 \cdot P_o - P$

$i = 0..9$	
t_i	P_i
$0 \cdot min$	$115.30 \cdot torr$
$6 \cdot min$	$122.91 \cdot torr$
$7 \cdot min$	$124.51 \cdot torr$
$8 \cdot min$	$126.18 \cdot torr$
$9 \cdot min$	$127.53 \cdot torr$
$10 \cdot min$	$129.10 \cdot torr$
$11 \cdot min$	$130.57 \cdot torr$
$12 \cdot min$	$132.02 \cdot torr$
$13 \cdot min$	$133.49 \cdot torr$
$18 \cdot min$	$140.16 \cdot torr$

Since the reaction is first-order. we have:

$$\frac{dP_{EO}}{dt} = k \cdot P_{EO}$$

$$\frac{d\ 2 \cdot P_0 \quad P}{dt} = k\ 2 \cdot P_0 \quad P$$

$$\ln \frac{2 \cdot P_0 \quad P}{2 \cdot P_0 \quad P_0} = k \cdot t$$ A plot of the left-hand side vs. t should yield a linear plot with a slope of - k.

$$y_i \quad \ln \frac{2 \cdot P_0 \quad P_i}{2 \cdot P_0 \quad P_0} \qquad k \quad slope(t,y) \qquad k = 1.371 \cdot 10^2 \cdot \frac{1}{min}$$

We can repeat this calculation without the zero point:

$$i \quad 0..8$$

t_i	P_{r_i}
6·min	122.91·torr
7·min	124.51·torr
8·min	126.18·torr
9·min	127.53·torr
10·min	129.10·torr
11·min	130.57·torr
12·min	132.02·torr
13·min	133.49·torr
18·min	140.16·torr

$$y_i \quad \ln \frac{2 \cdot P_0 \quad P_{r_i}}{2 \cdot P_0 \quad P_0} \qquad k \quad slope(t,y) \qquad k = 1.451 \cdot 10^2 \cdot \frac{1}{min}$$

10.15

The concentration will be proportional to the volume of titer. Since the reaction is pseudo-first order, we can plot ln (titer) vs. time to obtain the rate constant.

$$ml \quad 10^3 \cdot liter$$

$$i \quad 0..5$$

t_i	$Titer_i$
0·min	24.5·ml
15·min	18.1·ml
30·min	13.2·ml
45·min	9.7·ml
60·min	7.1·ml
75·min	5.2·ml

$$y_i \quad \ln \frac{Titer_i}{mL}$$

$$k \quad slope(t,y)$$

$$k = 2.07 \cdot 10^2 \cdot \frac{1}{min}$$

10.17

Since a and b are equal, we can use eq. (10.8), and solve for the time. When the reaction is 90% complete, the concentrations of the reactants will be 10% of the original value.

$$C_o \quad 0.365 \cdot \frac{mole}{liter} \qquad k \quad 1.23 \cdot \frac{liter}{mole \cdot sec} \qquad C_{final} \quad 0.1 \cdot C_o$$

$$\frac{1}{C_{final}} - \frac{1}{C_o} \quad k \cdot t \qquad t \quad \frac{\frac{1}{C_{final}} \quad \frac{1}{C_o}}{k} \qquad t = 20.05 \cdot sec$$

10.19

Since the reaction is second-order, we can use eq. (10.10):

$$\frac{1}{L \quad L_{inf}} = \frac{1}{L_o \quad L_{inf}} \cdot \frac{C_o \cdot k \cdot t}{L_o \quad L_{inf}} \qquad t_{inf} \quad 0.560$$

$$i \quad 0..8$$

t_i	L_i
0·min	1.560
5·min	1.315
7·min	1.247
9·min	1.193
15·min	1.064
18·min	1.020
20·min	0.994
25·min	0.945
27·min	0.923

$$L_{inf} \quad 0.560 \qquad C_o \quad 0.01 \cdot \frac{mole}{liter}$$

$$y_i \quad \frac{1}{L_i \quad L_{inf}}$$

$$intercept(t, y) = 1.003$$

$$k \quad \frac{slope(t, y)}{C_o \cdot intercept(t, y)}$$

$$k = 6.448 \cdot \frac{liter}{mole \cdot min}$$

10.21

We use eq. (10.11b):

$$k \cdot t = \frac{1}{(b - a)} \cdot \ln \frac{a \cdot (b - x)}{b \cdot (a - x)}$$

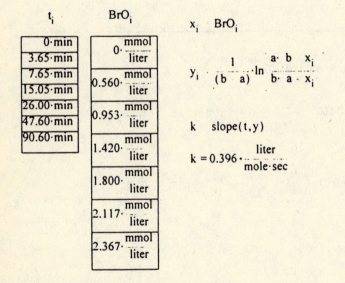

$a := 2.508 \cdot 10^{-3} \cdot \frac{mole}{liter}$ $b := 3.230 \cdot 10^{-3} \cdot \frac{mole}{liter}$ $mmol := 10^{-3} \cdot mole$

$i := 0 .. 6$

t_i	BrO_i
$0 \cdot min$	$0 \cdot \frac{mmol}{liter}$
$3.65 \cdot min$	
$7.65 \cdot min$	$0.560 \cdot \frac{mmol}{liter}$
$15.05 \cdot min$	
$26.00 \cdot min$	$0.953 \cdot \frac{mmol}{liter}$
$47.60 \cdot min$	
$90.60 \cdot min$	$1.420 \cdot \frac{mmol}{liter}$
	$1.800 \cdot \frac{mmol}{liter}$
	$2.117 \cdot \frac{mmol}{liter}$
	$2.367 \cdot \frac{mmol}{liter}$

$x_i := BrO_i$

$$y_i := \frac{1}{(b - a)} \cdot \ln \frac{a \cdot b - x_i}{b \cdot a - x_i}$$

$k := slope(t, y)$

$$k = 0.396 \cdot \frac{liter}{mole \cdot sec}$$

10.23

A second-order reaction will have a rate law of the form:

$$\frac{dC}{dt} = k \cdot C^2$$

$$C := p \cdot \lambda - \lambda_{inf}$$

$$\frac{d \, p \cdot \lambda - \lambda_{inf}}{dt} = k \cdot p^2 \cdot \lambda - \lambda_{inf}^2$$

$$p \cdot \frac{d(\lambda - \lambda_{inf})}{dt} = k \cdot p^2 \cdot (\lambda - \lambda_{inf})^2$$

$$\frac{d(\lambda - \lambda_{inf})}{dt} = k \cdot p \cdot (\lambda - \lambda_{inf})^2$$

$$\int_{\lambda_o}^{\lambda} \frac{1}{(\lambda - \lambda_{inf})^2} d(\lambda - \lambda_{inf}) = \int_0^t k \cdot p \, dt$$

$$\frac{1}{\lambda - \lambda_{inf}} - \frac{1}{\lambda_o - \lambda_{inf}} = k \cdot p \cdot t$$

$$\frac{1}{\lambda - \lambda_{inf}} - \frac{1}{\lambda_o - \lambda_{inf}} = k \cdot p \cdot t$$

We can find the value of the proportionality constant by using the inital concentration:

$$C_o = p \cdot (\lambda_o - \lambda_{inf}) \qquad p = \frac{C_o}{\lambda_o - \lambda_{inf}}$$

$$\frac{1}{\lambda - \lambda_{inf}} - \frac{1}{\lambda_o - \lambda_{inf}} = \frac{k \cdot C_o \cdot t}{\lambda_o - \lambda_{inf}}$$

10.25

Let x be the extent of reaction. Then, the concentrations of A and P at any time will be:

$$A = a - x \qquad P = p - x$$

We can substitute these into the rate law to obtain:

$$\frac{d(a - x)}{dt} = -k(a - x) \cdot (p - x)$$

Since a is a constant:

$$\frac{d(a - x)}{dt} = \frac{dx}{dt}$$

$$\frac{dx}{dt} = k \cdot (a - x) \cdot (p - x)$$

$$\int_{0}^{x_t} \frac{1}{k \cdot (a - x) \cdot (p - x)} dx = \int_{0}^{t} 1 \, dt$$

$$t = \frac{\ln|a - x_t| - \ln|p - x_t| - \ln(a) + \ln(p)}{((p - a) \cdot k)}$$

$$t = \frac{\ln\left(\frac{p - x_t}{a - x_t} \cdot \frac{a}{p}\right)}{k \cdot (p - a)} \qquad t = \frac{\ln\left(\frac{a \cdot p - x_t}{p \cdot a - x_t} - \ln\frac{a}{p}\right)}{k \cdot (p - a)}$$

SECTION 10.3

10.27

The dependence of the rate constant on temperature is given by eq. (10.18):

$$\frac{d \cdot \ln(k)}{dT} = \frac{E_a}{R \cdot T^2}$$

The rate will increase faster for the reaction with the larger activation energy. Even if you don't remember eq. (10.18), you should be able to tell this from the definition of an activation energy.

10.29

If both rate constants follow an Arrhenius dependence on temperature, and the pre-expoential factors are independent of temperature, we can compute the activation energy:

$$\frac{k_1}{k_2} = \frac{A \cdot e^{\frac{E_a}{R \, T_1}}}{A \cdot e^{\frac{E_a}{R \cdot T_2}}} = \exp\left[\frac{E_a}{R}\left(\frac{1}{T_1} - \frac{1}{T_2}\right)\right] \qquad E_a = \frac{R \cdot \ln\frac{k_1}{k_2}}{\frac{1}{T_2} - \frac{1}{T_1}}$$

$$R \quad 8.31451 \cdot \frac{joule}{mole \cdot K} \qquad kJ \quad 1000 \cdot joule \qquad k_{ratio} \quad \frac{1}{2} \qquad T_1 \quad 298.15 \cdot K \qquad T_2 = 308.15 \cdot K$$

$$E_a \quad \frac{R \cdot \ln k_{ratio}}{\frac{1}{T_2} - \frac{1}{T_1}} \qquad E_a = 53 \cdot kJ$$

10.31

We can use eq. (10.18):

$$\ln(k) = \ln(A) - \frac{E_a}{R \cdot T}$$

$$i = 0..9$$

$T_i =$	$k_i =$
273.1·K	$7.97 \cdot 10^{-7} \cdot \frac{1}{sec}$
288.1·K	
293.1·K	$1.04 \cdot 10^{-5} \cdot \frac{1}{sec}$
298.1·K	
308.1·K	$1.76 \cdot 10^{-5} \cdot \frac{1}{sec}$
313.1·K	
318.1·K	$3.38 \cdot 10^{-5} \cdot \frac{1}{sec}$
323.1·K	
328.1·K	$1.35 \cdot 10^{-4} \cdot \frac{1}{sec}$
338.1·K	
	$2.47 \cdot 10^{-4} \cdot \frac{1}{sec}$
	$4.98 \cdot 10^{-4} \cdot \frac{1}{sec}$
	$7.59 \cdot 10^{-4} \cdot \frac{1}{sec}$
	$1.50 \cdot 10^{-3} \cdot \frac{1}{sec}$
	$4.87 \cdot 10^{-3} \cdot \frac{1}{sec}$

$kJ = 1000 \cdot joule$ $\qquad R = 8.31451 \cdot \frac{joule}{mole \cdot K}$

$$y_i = \ln k_i \cdot sec$$

$$x_i = \frac{1}{T_i}$$

$$E_a = -R \cdot slope(x,y)$$

$$E_a = 101.401 \cdot \frac{kJ}{mole}$$

$$A = exp(intercept(x,y))$$

$$A = 2.105 \cdot 10^{13}$$

10.33

$$k = \frac{k_1 \sqrt{k_2}}{k_3} = \frac{A_1 e^{-Ea_1/RT} \sqrt{A_2 e^{-Ea_2/RT}}}{A_3 e^{-Ea_3/RT}}$$

$$\ln(k) = -\frac{Ea_{overall}}{RT} + \ln(A) = -\frac{Ea_1}{RT} - \frac{1}{2}\frac{Ea_2}{RT} + \frac{Ea_3}{RT} + \ln\left(\frac{A_1 \sqrt{A_2}}{A_3}\right)$$

We combine all the terms which are divided by RT to obtain the activation energy:

$$Ea_{overall} = Ea_1 + \frac{1}{2}Ea_2 - Ea_3$$

SECTION 10.4

10.35

We use eq. (10.23):

$$R \quad 8.31451 \cdot \frac{joule}{mole \cdot K} \qquad M_A \quad 15.0348 \cdot \frac{gm}{mole} \qquad M_B \cdot M_A \qquad L\mu : \frac{M_A \cdot M_B}{M_A - M_B}$$

$$p : 1 \qquad \sigma : 4 \cdot 10^{-8} \cdot cm \qquad S_{AB} \quad \pi \cdot \sigma^2 \qquad T : 500 \cdot K$$

$$L \cdot 6.022137 \cdot 10^{23} \cdot \frac{1}{mole} \qquad E_{min} \quad 0 \cdot \frac{joule}{mole}$$

$$k \quad p \cdot L \cdot S_{AB} \cdot \sqrt{\frac{8 \cdot R \cdot T}{\pi \cdot L\mu}} \cdot \exp \frac{E_{min}}{R \cdot T} \qquad k = 3.592 \cdot 10^{14} \cdot \frac{cm^3}{mole \cdot sec}$$

10.37

We use eq. (10.23):

$$R : 8.31451 \cdot \frac{joule}{mole \cdot K} \qquad M_A : 31.9988 \cdot \frac{gm}{mole} \qquad M_B : 2.0159 \cdot \frac{gm}{mole} \qquad L\mu : \frac{M_A \cdot M_B}{M_A - M_B}$$

$$p. : 1 \qquad \sigma_A \quad 0.358 \cdot 10^{-9} \cdot m \qquad \sigma_B \quad 0.287 \cdot 10^{-9} \cdot m \qquad \sigma_{AB} \quad \frac{1}{2} \cdot \sigma_A - \sigma_B$$

$$L \quad 6.022137 \cdot 10^{23} \cdot \frac{1}{mole} \qquad E_{min} : 0 \cdot \frac{joule}{mole} \qquad T : 1500 \cdot K \qquad S_{AB} \quad \pi \cdot \sigma_{AB}^2$$

$$k \quad p \cdot L \cdot S_{AB} \cdot \sqrt{\frac{8 \cdot R \cdot T}{\pi \cdot L\mu}} \cdot \exp \frac{E_{min}}{R \cdot T} \qquad k = 8.052 \cdot 10^{14} \cdot \frac{cm^3}{mole \cdot sec}$$

10.39

We use the formulas for a bimolecular reaction, eq. (10.28).

$R = 8.31451 \cdot \dfrac{joule}{mole \cdot K}$ $h = 6.62607 \cdot 10^{34} \cdot joule \cdot sec$ $k_b = 1.38066 \cdot 10^{23} \cdot \dfrac{joule}{K}$ $kJ = 1000 \cdot joule$

$E_a = 42 \cdot \dfrac{kJ}{mole}$ $A = 8 \cdot 10^{13} \cdot \dfrac{cm^3}{mole \cdot sec}$ $T = 300 \cdot K$ $C = 1 \cdot \dfrac{mole}{cm^3}$

$\Delta S = R \cdot \ln \dfrac{A \cdot h \cdot C}{e^2 \cdot k_b \cdot T}$ $\Delta S = 4.6 \cdot \dfrac{joule}{mole \cdot K}$

$\Delta H = E_a - 2 \cdot R \cdot T$ $\Delta H = 37 \cdot \dfrac{kJ}{mole}$

10.41

This is similar to example 10.9. We can write:

$$\ln \dfrac{k \cdot h}{k_b \cdot T} = \dfrac{\Delta S}{R} - \dfrac{\Delta H}{R \cdot T}$$

$i = 0..8$

T_i	k_i
556·K	$3.25 \cdot 10^7$
575·K	$1.22 \cdot 10^6$
625·K	$3.02 \cdot 10^5$
647·K	$8.59 \cdot 10^5$
666·K	$2.19 \cdot 10^4$
683·K	$5.12 \cdot 10^4$
700·K	$1.16 \cdot 10^3$
716·K	$2.50 \cdot 10^3$
781·K	$3.95 \cdot 10^2$

$R = 8.31451 \cdot \dfrac{joule}{mole \cdot K}$ $h = 6.62607 \cdot 10^{34} \cdot joule \cdot sec$ $kJ = 1000 \cdot joule$

$k_b = 1.38066 \cdot 10^{23} \cdot \dfrac{joule}{K}$

$y_i = \ln \dfrac{k_i \cdot h}{k_b \cdot T_i \cdot sec}$ $x_i = \dfrac{1}{T_i}$

$\Delta S = R \cdot intercept(x, y)$ $\Delta S = 49.749 \cdot \dfrac{joule}{mole \cdot K}$

$\Delta H = R \cdot slope(x, y)$ $\Delta H = 180.561 \cdot \dfrac{kJ}{mole}$

SECTION 10.5

10.43

We use L'Hopital's rule, taking derivatives with respect to k2, and taking the limit as k2 -> k1.

$$[B] = \frac{k_1 a \left(e^{-k_1 t} - e^{-k_2 t}\right)}{k_2 - k_1}$$

$$[B] = \text{limit} (k_2 \to k_1) \left[\frac{k_1 a \left(t \, e^{-k_2 t}\right)}{1}\right]$$

$$[B] = k_1 a \, t \, e^{-k_1 t}$$

10.45

We use L'Hopital's rule, taking derivatives with respect to k2 of eq. (10.34b), and taking the limit as k2 -> k1.

$$[B] = \frac{k_1 a \left(e^{-k_1 t} - e^{-k_2 t}\right)}{k_2 - k_1}$$

$$[B] = \text{limit} (k_2 \to k_1) \left[\frac{k_1 a \left(t \, e^{-k_2 t}\right)}{1}\right]$$

$$[B] = k_1 a \, t \, e^{-k_1 t}$$

SECTION 10.6

10.47

We must rearrange eq. (10.41):

$$k_{uni} = \frac{k_1 \cdot k_3 \cdot M}{k_3 - k_2 \cdot M}$$

$$\frac{1}{k_{uni}} = \frac{1}{k_1 \cdot M} + \frac{k_2}{k_1 \cdot k_3} \qquad M = \frac{N}{V} = \frac{P}{RT}$$

$$\frac{1}{k_{uni}} = \frac{R \cdot T}{k_1 \cdot P} + \frac{k_2}{k_1 \cdot k_3}$$

So, a plot of $1/k_{uni}$ vs. $1/P$ shoulw be linear if the Lindeman mechanism holds.

$$y_i = \frac{1}{k_i} \qquad x_i = \frac{1}{P_i}$$

$i = 0 .. 9$

$P_i =$	$k_i =$
84.1·torr	$2.98 \cdot 10^{-4} \cdot \frac{1}{sec}$
34.0·torr	
11.0·torr	$2.82 \cdot 10^{-4} \cdot \frac{1}{sec}$
6.07·torr	
2.89·torr	$2.23 \cdot 10^{-4} \cdot \frac{1}{sec}$
1.37·torr	
0.569·torr	$2.00 \cdot 10^{-4} \cdot \frac{1}{sec}$
0.170·torr	
0.120·torr	$1.54 \cdot 10^{-4} \cdot \frac{1}{sec}$
0.067·torr	
	$1.30 \cdot 10^{-4} \cdot \frac{1}{sec}$
	$0.875 \cdot 10^{-4} \cdot \frac{1}{sec}$
	$0.486 \cdot 10^{-4} \cdot \frac{1}{sec}$
	$0.392 \cdot 10^{-4} \cdot \frac{1}{sec}$
	$0.303 \cdot 10^{-4} \cdot \frac{1}{sec}$

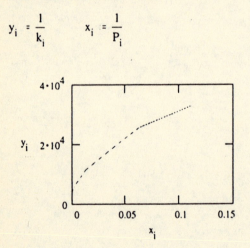

The graph deviates from linearity, indicating that the Lindeman mechanism does not explain this data very well.

SECTION 10.7

10.49

Example (10.16) shows the velocity for a photo-initiated reaction. The elementary steps and velocities are:

(1)	$Br_2 + h\nu \rightarrow 2Br$	$v_1 = \phi Ia$
(2)	$Br + H_2 \rightarrow HBr + H$	$v_2 = k_2 [Br] [H]$
(3)	$H + Br_2 \rightarrow HBr + Br$	$v_3 = k_3 [H] [Br_2]$
(4)	$H + HBr \rightarrow H_2 + Br$	$v_4 = k_4 [H] [HBr]$
(5)	$Br + Br \rightarrow Br_2$	$v_5 = k_5 [Br]^2$

The overall rate can be written in terms of the disappearance of Br_2:

$$- d[Br_2]/dt = v_1 + v_3 - v_5$$

Now, we use the steady-state approximation on the radical species:

$$0 = d[H]/dt = v_2 - v_3 - v_4$$
$$v_2 - v_3 = v_4$$

$$0 = d[Br]/dt = 2v_1 - v_2 + v_3 + v_4 - 2v_5$$
$$2v_1 = 2v_5$$
$$v_1 = v_5$$
$$\phi Ia = k_5[Br]^2$$
$$[Br] = [\phi Ia/k_5]^{1/2}$$

Now that we have an expression for the concentration of Br, we can use the equation for d[H]/dt to find [H]:

$$v_2 - v_3 = v_4$$

$$k_2 [Br] [H_2] - k_3 [H] [Br_2] = k_4 [H] [HBr]$$

$$k_2 [\phi Ia/k_5]^{1/2} [H_2] - k_3 [H] [Br_2] = k_4 [H] [HBr]$$

$$k_2 [H_2] [\phi Ia/k_5]^{1/2} = [H] (k_4 [HBr] + k_3 [Br_2])$$

$$[H] = \frac{k_2[H_2] \sqrt{\phi I_a}}{\sqrt{k_5} (k_4 [HBr] + k_3 [Br_2])}$$

Now, since v1 = v5, the overall rate is given by v3:

$$v = v_3 = k_3 [H] [Br_2] = \frac{k_2 k_3 [H_2] [Br_2] \sqrt{\phi I_a}}{\sqrt{k_5} (k_4 [HBr] + k_3 [Br_2])}$$

10.51

The elementary steps and velocities are:

(1) $\quad 2\,ROOH \rightarrow ROO + RO + H_2O \qquad\quad v_1 = k_1\,[ROOH]^2$

(2) $\quad ROO + RH \rightarrow ROOH + R \qquad\qquad\; v_2 = k_2\,[ROO]\,[RH]$

(3) $\quad R + O_2 \rightarrow ROO \qquad\qquad\qquad\qquad v_3 = k_3\,[R]\,[O_2]$

(4) $\quad 2\,ROO \rightarrow ROOR + O_2 \qquad\qquad\quad\; v_4 = k_4\,[ROO]^2$

The overall rate can be written in terms of the disappearance of O_2:

$$- d[O_2]/dt = v_3 . v_4$$

Now, we use the steady-state approximation on the radical species:

$0 = d[R]/dt = v_2 - v_3$ $\qquad\qquad\qquad$ $0 = d[ROO]/dt = v_1 - v_2 + v_3 - 2v_4$

$k_2\,[ROO]\,[RH] = k_3\,[R]\,[O_2]$ $\qquad\qquad$ $v_1 = 2v_4$

$[R] = k2\,[ROO]\,[RH]/k3\,[O2]$ $\qquad\qquad$ $k_1\,[ROOH]^2 = 2k_4\,[ROO]^2$

$\qquad\qquad\qquad\qquad\qquad\qquad\qquad\qquad$ $[ROO] = [ROOH]\,[k_1/2k_4]^{1/2}$

We can solve for the concentration of [R] now:

$$[R] = k_2\,[ROO]\,[RH]/k_3\,[O_2]$$

$$[R] = \left[\frac{k_1\,k_2^2}{2k_3^2\,k_4}\right]^{1/2}\frac{[[ROOH]\,[RH]]}{[O_2]}$$

Now, we know that the overall rate is given by v3 - v4. If we make the long-chain approximation, then the rate of proppagation is much larger than the rate of termination, and so we assume that the overall rate is given by v3.

$$v = v_3 = k_3\,[O_2]\,[R] = \left[\frac{k_1\,k_2^2}{2k_4}\right]^{1/2}[[ROOH]\,[RH]]$$

10.53

Example (10.16) shows the velocity for a photo-initiated reaction. The elementary steps and velocities are:

(1) $H_2 + h\nu \rightarrow 2H$ $\quad\quad\quad v_1 = \phi Ia$
(2) $SiH_4 + H \rightarrow H_2 + SiH_3$ $\quad v_2 = k_2 [SiH_4] [H]$
(3) $H_2 + SiH_3 \rightarrow SiH_4 + H$ $\quad v_3 = k_3 [H_2] [SiH_3]$
(4) $2H \rightarrow H_2$ $\quad\quad\quad\quad\quad v_4 = k_4 [H]^2$

We use the steady-state approximation on the radical species:

$0 = d[SiH_3]/dt = v_2 - v_3$
$k_2 [SiH_4] [H] = k_3 [H_2] [SiH_3]$

$0 = d[H]/dt = 2v_1 - v_2 + v_3 - 2v_4$
$2v_1 = 2v_4$
$v_1 = v_4$
$\phi Ia = k_4[H]^2$
$[H] = [\phi Ia/k_4]1/2$

Now we can solve for [SiH3]:

$k_2 [SiH_4] [H] = k_3 [H_2] [SiH_3]$

$$k_2 [SiH_4] \sqrt{\frac{I_a\phi}{k_4}} = k_3 [H_2] [SiH_3]$$

$$[SiH_3] = \frac{k_2 [SiH_4]}{k_3 [H_2]} \sqrt{\frac{I_a\phi}{k_4}}$$

[SiH3] is inversely proportional to [H2]. If the goal is to maximize [SiH3], then the concentration of hydrogen should be small.

10.55

The elementary steps and velocities are:

(1) $C_2H_6 \rightarrow 2CH_3$ $\quad\quad\quad\quad v_1 = k_1 [C_2H_6]$
(2) $CH_3 + C_2H_6 \rightarrow CH_4 + C_2H_5$ $\quad v_2 = k_2 [CH_3] [C_2H_6]$
(3) $C_2H_5 \rightarrow C_2H_4 + H$ $\quad\quad\quad v_3 = k_3 [C_2H_5]$
(4) $C_2H_6 + H \rightarrow H_2 + C_2H_5$ $\quad v_4 = k_4 [H] [C_2H_6]$
(5) $C_2H_5 + H \rightarrow C_2H_6$ $\quad\quad\quad v_5 = k_5 [H] [C_2H_5]$

The overall velocity can be written in terms of the appearance of ethylene:

$v = d[C_2H_4]/dt = v_3 = k_3 [C_2H_5]$

We use the steady-state approximation on the radical species:

$0 = d[CH_3]/dt = 2v_1 - v_2$
$0 = d[H]/dt = v_3 - v_4 - v_5$
$0 = d[C_2H_5]/dt = v_2 - v_3 + v_4 - v_5$

We can add the last two equations and substitute $2v1 = v2$. This gives us an expression for the concentration of H:

$2v_1 = 2v_5$
$k_1 [C_2H_6] = k_5 [H] [C_2H_5]$
$[H] = k_1 [C_2H_6] / k_5 [C_2H_5]$

Now, we use the equation from the steady-state assumption on [H] to solve for the concentration of the ethyl radical.

$0 = v_3 - v_4 - v_5$

We use the long chain approximation (the rate of propagation is much faster than termination), so v4 >> v5, and we have:

$v_3 = v_4$ $\qquad\qquad$ $k_3 [C_2H_5] = k_4 [H] [C_2H_6]$

$$k_3 [C_2H_5] = \frac{k_1 k_4 [C_2H_6]^2}{k_5 [C_2H_5]}$$

$$k_3 k_5 [C_2H_5]^2 = k_1 k_4 [C_2H_6]^2$$

$$[C_2H_5] = \sqrt{\frac{k_1 k_4}{k_3 k_5}} \, [C_2H_6]$$

Thus, the overall rate is given by:

$$v = k_3 [C_2H_5] = k_3 \sqrt{\frac{k_1 k_4}{k_3 k_5}} \, [C_2H_6] = k_{eff} [C_2H_6] \qquad \text{where} \qquad k_{eff} = \sqrt{\frac{k_1 k_3 k_4}{k_5}}$$

With this effective rate constant, the activation energy of the overall reaction will be:

$$Ea_{eff} = \frac{1}{2}\left(Ea_1 + Ea_3 + Ea_4 - Ea_5\right)$$

10.57

The elementary steps and velocities are:

(1) $X_2 \rightarrow 2X$ $v_1 = k_1 [X_2]$
(2) $X + RH \rightarrow R + HX$ $v_2 = k_2 [X] [RH]$
(3) $R + X_2 \rightarrow RX + X$ $v_3 = k_3 [R] [X_2]$
(4) $X + R \rightarrow RX$ $v_4 = k_4 [X] [R]$

The overall velocity can be written in terms of the appearance of HX:

$v = d[HX]/dt = v_2 = k_2 [X] [RH]$

We use the steady-state approximation on the radical species:

$0 = d[R]/dt = v_2 - v_3 - v_4$
$0 = d[X]/dt = 2v_1 - v_2 + v_3 - v_4$

We can add the last two equations to obtain an expression for [R]:

$2v_1 = 2v_4$
$k_1 [X_2] = k_4 [X] [R]$
$[R] = k_1 [X_2] / k_4 [X]$

Now, we use the equation from the steady-state assumption on [R] to solve for [X]:

$0 = v_2 - v_3 - v_4$

We use the long chain approximation (the rate of propagation is much faster than termination), so v3 >> v4, and we have:

$v_2 = v_3$ $k_2 [X] [RH] = k_3 [R] [X_2]$

$$k_2 [X] [RH] = \frac{k_1 k_3 [X_2]^2}{k_4 [X]}$$

$$k_2 k_4 [X]^2 [RH] = k_1 k_3 [X_2]^2$$

$$[X] = \sqrt{\frac{k_1 k_3}{k_2 k_4}} \frac{[X_2]}{[RH]^{1/2}}$$

Thus, the overall rate is given by:

$$v = k_2 [X] [RH] = k_2 \sqrt{\frac{k_1 k_3}{k_2 k_4}} \frac{[X_2]}{[RH]^{1/2}} [RH] = k_{eff} [X_2] [RH]^{1/2} \qquad \text{where} \qquad k_{eff} = \sqrt{\frac{k_1 k_2 k_3}{k_4}}$$

SECTION 10.8

10.59

The rate of propagation for these types of reactions is given by eq. (10.49):

$$v_p = k_p \, [M] \, [R_x \cdot]$$

To obtain $[R_x \cdot]$, we set the rate of initiation equal to the rate of termination:

$$Ia\phi = k_t \, [R_x \cdot]^2$$

$$[R_x \cdot] = \sqrt{\frac{Ia\phi}{k_t}}$$

$$v_p = k_p \, [M] \, [R_x \cdot] = \frac{k_p \sqrt{Ia\phi}}{k_t^{1/2}} \, [M]$$

The kinetic chain length can now be computed using eq. (10.44):

$$v = \frac{v_p}{v_i} = \frac{k_p \sqrt{Ia\phi}}{k_t^{1/2} Ia\phi} \, [M] = \frac{k_p \, [M]}{\sqrt{k_t Ia\phi}}$$

10.61

We use eq. (10.51) to compute the rate of polymerization:

$$k_p = 2.3 \cdot 10^3 \cdot \frac{liter}{mole \cdot sec} \qquad k_t = 2.9 \cdot 10^2 \cdot \frac{liter}{mole \cdot sec} \qquad k_i = 1.07 \cdot 10^{-5} \cdot sec \qquad kJ = 1000 \cdot joule$$

$$f = 1 \qquad\qquad M = 1 \cdot \frac{mole}{liter} \qquad\qquad In = 0.001 \cdot \frac{mole}{liter}$$

$$v_p = k_p \cdot \sqrt{\frac{k_i \cdot f}{k_t}} \cdot M \cdot In^{0.5} \qquad v_p = 0.014 \cdot \frac{mole}{liter}$$

To calcualte the activation energy, we define an effective rate constant:

$$k_{eff} = k_p \cdot \sqrt{\frac{k_i}{k_t}} \qquad Ea_p = 26 \cdot \frac{kJ}{mole} \qquad Ea_t = 13 \cdot \frac{kJ}{mole} \qquad Ea_i = 130 \cdot \frac{kJ}{mole}$$

$$Ea_{eff} = Ea_p + \frac{1}{2} \cdot Ea_i - \frac{1}{2} \cdot Ea_t \qquad Ea_{eff} = 84.5 \cdot \frac{kJ}{mole}$$

Since the activation energy is positive, the rate will increase with temperature. To look at the behavior of the kinetic chain length, we use eq. (10.52) and define an effective rate constnat again:

$$v = \frac{k_p \cdot M}{\sqrt{f \cdot k_i \cdot k_t \cdot In}} \qquad k_v = \frac{k_p}{\sqrt{k_i \cdot k_t}} \qquad Ea_v = Ea_p - \frac{1}{2} \cdot Ea_i - \frac{1}{2} \cdot Ea_t$$

$$Ea_v = -45.5 \cdot \frac{kJ}{mole}$$

Since the effective activation energy is zero, the kinetic chain length will decrease with temperature.

SECTION 10.9

10.63

We use eq. (10.56):

$$\frac{P}{V_{ads}} = \frac{1}{b \cdot V_{max}} + \frac{1}{V_{max}} \cdot P \qquad \text{Plot P/Vads vs. P.}$$

$$i = 0 .. 7$$

$$y_i = \frac{P_i}{V_{ads_i}}$$

$$V_{max} = \frac{1}{slope(P, y)} \qquad V_{max} = 39.072 \cdot \frac{cm^3}{gm}$$

$$b = \frac{1}{V_{max} \cdot intercept(P, y)} \qquad b = 0.069 \cdot \frac{1}{torr}$$

V_{ads_i} =	P_i =
$5.89 \cdot \dfrac{cm^3}{gm}$	$2.45 \cdot torr$
	$3.5 \cdot torr$
	$5.2 \cdot torr$
$7.76 \cdot \dfrac{cm^3}{gm}$	$7.2 \cdot torr$
	$11.2 \cdot torr$
$10.10 \cdot \dfrac{cm^3}{gm}$	$12.8 \cdot torr$
	$14.6 \cdot torr$
$12.35 \cdot \dfrac{cm^3}{gm}$	$16.1 \cdot torr$
$16.45 \cdot \dfrac{cm^3}{gm}$	
$18.05 \cdot \dfrac{cm^3}{gm}$	
$19.72 \cdot \dfrac{cm^3}{gm}$	
$21.10 \cdot \dfrac{cm^3}{gm}$	

10.65

Use eq. (10.56):

$$\frac{P}{V_{ads}} = \frac{1}{b \cdot V_{max}} + \frac{1}{V_{max}} \cdot P \qquad \text{Plot P/Vads vs. P.}$$

$i = 0 .. 4$

$P_i =$	$V_i =$
35.9·torr	3.70·cm^3
64.5·torr	5.09·cm^3
120·torr	6.70·cm^3
232·torr	8.48·cm^3
357·torr	9.92·cm^3

$$y_i = \frac{P_i}{V_i}$$

$$V_{max} = \frac{1}{slope(P,y)} \qquad V_{max} = 12.2 \cdot cm^3$$

$$b = \frac{1}{V_{max} \cdot intercept(P,y)} \qquad b = 0.011 \cdot \frac{1}{torr}$$

10.67

The most general expression for a Langmuir-type adsorption would be:

$$v = \frac{kS_0 b_{NO} P_{NO}}{1 + b_{NO} P_{NO} + b_{N2} P_{N2} + b_{O2} P_{O2}}$$

If the follwoing are true:

$b_{O2} P_{O2} \gg 1$

$b_{O2} P_{O2} \gg b_{N2} P_{N2}$

$b_{O2} P_{O2} \gg b_{NO} P_{NO}$

then we will have a rate law of the form:

$$v = \frac{k_{eff} P_{NO}}{P_{O2}}$$

10.69

If dissociation occurs upon adsorption, we have:

$$H_2 + 2S \rightarrow 2H_s, \qquad k = k_a$$

$$2H_s \rightarrow H_2 + 2H_s \qquad k = k_d$$

where S is a site and Hs is hydrogen on the surface.

At equilibrium, the rates of adsorption and desorption are equal:

$$k_a P_{H2} (1 - \theta)^2 = k_d \theta^2$$

$$\sqrt{\frac{k_a}{k_d}} P_{H2}^{1/2} (1 - \theta) = \theta$$

$$\sqrt{\frac{k_a}{k_d}} P_{H2}^{1/2} - \theta \sqrt{\frac{k_a}{k_d}} P_{H2}^{1/2} = \theta$$

$$\sqrt{\frac{k_a}{k_d}} P_{H2}^{1/2} = \theta \left(1 + \sqrt{\frac{k_a}{k_d}} P_{H2}^{1/2} \right)$$

$$\theta = \frac{\sqrt{\frac{k_a}{k_d}} P_{H2}^{1/2}}{\left(1 + \sqrt{\frac{k_a}{k_d}} P_{H2}^{1/2} \right)}$$

$$\theta = \frac{b P_{H2}^{1/2}}{\left(1 + b P_{H2}^{1/2} \right)} \qquad \text{where} \quad b = \sqrt{\frac{k_a}{k_d}}$$

SECTION 10.10

10.71

A material balance on the enzyme gives:

$E_0 = [E] + [ES] + [ESI]$

At equilibrium, from eq. (10.61), we have:

$$K_i' = \frac{[ES]\,[I]}{[ESI]} \quad \text{and} \quad K_m = \frac{[E]\,[S]}{[ES]} \quad \Rightarrow \quad [ES] = \frac{[E]\,[S]}{K_m}$$

$$[ESI] = \frac{[ES]\,[I]}{K_i'} = \frac{[E]\,[S]\,[I]}{K_i' K_m}$$

We can plug these into the enzyme balance:

$$E_0 = [E] + [ES] + [ESI] = [E] + \frac{[E]\,[S]}{K_m} + \frac{[E]\,[S]\,[I]}{K_i' K_m} = [E]\left(1 + \frac{[S]}{K_m} + \frac{[S]\,[I]}{K_i' K_m}\right)$$

$$[E] = \frac{E_0}{\left(1 + \dfrac{[S]}{K_m} + \dfrac{[S]\,[I]}{K_i' K_m}\right)}$$

The rate is given by:

$$v = k_2\,[ES] = \frac{k_2\,[E]\,[S]}{K_m} = \frac{k_2\,[S]\,E_0}{K_m\left(1 + \dfrac{[S]}{K_m} + \dfrac{[S]\,[I]}{K_i' K_m}\right)}$$

$$v = \frac{v_{max}\,[S]}{K_m + [S] + \dfrac{[S]\,[I]}{K_i'}} = \frac{v_{max}\,[S]}{K_m + [S]\left(1 + \dfrac{[I]}{K_i'}\right)}$$

To find the slope and intercept of the graphical method described in the problem:

$$\frac{1}{v} = \frac{K_m + [S]\left(1 + \dfrac{[I]}{K_I'}\right)}{v_{max}\,[S]}$$

$$\frac{1}{v} = \frac{K_m}{v_{max}}\frac{1}{[S]} + \frac{\left(1 + \dfrac{[I]}{K_I'}\right)}{v_{max}}$$

$$\text{slope} = \frac{K_m}{v_{max}} \qquad\qquad \text{intercept} = \frac{1}{v_{max}}\left(1 + \frac{[I]}{K_I'}\right)$$

10.73

This is similar to example 10.18. We use a Lineweaver-Burk plot, which is a plot of 1/v vs. 1/[S].

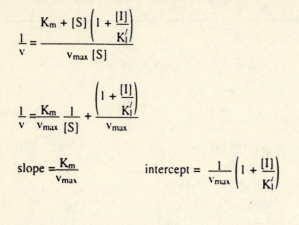

$$i = 0.. 8$$

$S_i =$	$v_i =$
0.03	0.14
0.04	0.165
0.05	0.18
0.0863	0.26
0.129	0.305
0.216	0.345
0.431	0.40
0.647	0.435
1.078	0.445

$$\frac{1}{v} = \frac{1}{v_{max}} + \frac{K_m}{v_{max}}\cdot\frac{1}{S}$$

$$y_i := \frac{1}{v_i} \qquad x_i = \frac{1}{S_i}$$

$$v_{max} = \frac{1}{\text{intercept}(x,y)} \qquad v_{max} = 0.47$$

$$K_m = v_{max}\cdot\text{slope}(x,y) \qquad K_m = 0.073$$

10.75

This is similar to example 10.18. We use a Lineweaver-Burk plot, which is a plot of $1/v$ vs. $1/[S]$.

$$i = 0..3$$

$$S_i = \qquad v_i =$$

$2.5 \cdot 10^{-4} \cdot \dfrac{mole}{liter}$	$2.2 \cdot 10^{-6} \cdot \dfrac{mole}{liter \cdot min}$
$5.0 \cdot 10^{-4} \cdot \dfrac{mole}{liter}$	$3.8 \cdot 10^{-6} \cdot \dfrac{mole}{liter \cdot min}$
$1.0 \cdot 10^{-3} \cdot \dfrac{mole}{liter}$	$5.9 \cdot 10^{-6} \cdot \dfrac{mole}{liter \cdot min}$
$1.5 \cdot 10^{-3} \cdot \dfrac{mole}{liter}$	$7.1 \cdot 10^{-6} \cdot \dfrac{mole}{liter \cdot min}$

$$\frac{1}{v} = \frac{1}{v_{max}} + \frac{K_m}{v_{max}} \cdot \frac{1}{S}$$

$$y_i = \frac{1}{v_i} \qquad x_i = \frac{1}{S_i}$$

$$v_{max} = \frac{1}{intercept(x,y)} \qquad v_{max} = 1.32 \cdot 10^{-5} \cdot \frac{mole}{liter \cdot min}$$

$$K_m = v_{max} \cdot slope(x,y) \qquad K_m = 1.24 \cdot 10^{-3} \cdot \frac{1}{liter}$$

10.77

A material balance on the enzyme gives:

Eo = [E] + [ES] + [EI] + [ESI]

At equilibrium, from eq. (10.61), we have:

$$K_m = \frac{[E][S]}{[ES]} \quad \Rightarrow \quad [ES] = \frac{[E][S]}{K_m} \qquad and \qquad K_I = \frac{[E][I]}{[EI]} \quad \Rightarrow \quad [EI] = \frac{[E][I]}{K_I}$$

$$K_I = \frac{[ES][I]}{[ESI]} \quad \Rightarrow \quad [ESI] = \frac{[ES][I]}{K_I} = \frac{[E][S][I]}{K_I K_m}$$

We can plug these into the enzyme balance:

$$Eo = [E] + [ES] + [EI] + [ESI] = [E] + \frac{[E][S]}{K_m} \div \frac{[E][I]}{K_I} + \frac{[E][S][I]}{K_I K_m} = [E]\left(1 + \frac{[S]}{K_m} + \frac{[I]}{K_I} + \frac{[S][I]}{K_I K_m}\right)$$

$$[E] = \frac{Eo}{\left(1 + \dfrac{[S]}{K_m} + \dfrac{[I]}{K_I} + \dfrac{[S][I]}{K_I' K_m}\right)}$$

The rate is given by:

$$v = k_2[ES] = \frac{k_2[E][S]}{K_m} = \frac{k_2[S]\,Eo}{K_m\left(1 + \dfrac{[S]}{K_m} + \dfrac{[I]}{K_I} + \dfrac{[S][I]}{K_I K_m}\right)}$$

$$v = \frac{v_{max}[S]}{K_m + [S] + \dfrac{[I]K_m}{K_I} + \dfrac{[S][I]}{K_I}}$$

SECTION 10.11

10.79

For two molecules of equal size, eq. (10.67) reduceds to:

$$k = \left(\frac{2 \cdot R \cdot T}{3 \cdot \eta}\right) \cdot \left[\frac{\left(r_A - r_B\right)^2}{r_A \cdot r_B}\right] = \frac{8 \cdot R \cdot T}{3 \cdot \eta}$$

We can get the viscosity from Table 9.2:

$$T = 313.15 \cdot K \qquad \eta = 3.41 \cdot 10^{-3} \cdot \frac{gm}{cm \cdot sec} \qquad R = 8.31451 \cdot \frac{joule}{mole \cdot K}$$

$$k = \frac{8 \cdot R \cdot T}{3 \cdot \eta} \qquad k = 2 \cdot 10^{10} \cdot \frac{liter}{mole \cdot sec}$$

10.81

Equation (10.67) is:

$$k = \left(\frac{2 \cdot R \cdot T}{3 \cdot \eta}\right) \cdot \left[\frac{\left(r_A + r_B\right)^2}{r_A \cdot r_B}\right]$$

All other things being equal, at two different temperatures, the relative rates would be:

$$k_1 = \left(\frac{2 \cdot R \cdot T_1}{3 \cdot \eta_1}\right) \cdot \left[\frac{\left(r_A - r_B\right)^2}{r_A \cdot r_B}\right] \qquad k_2 = \frac{2 \cdot R \cdot T_2}{3 \cdot \eta_2} \cdot \left[\frac{\left(r_A - r_B\right)^2}{r_A \cdot r_B}\right] \qquad \frac{k_{50}}{k_{20}} = \frac{T_{50} \cdot \eta_{20}}{T_{20} \cdot \eta_{50}}$$

Using data from Table 9.2,

$$T_{50} = 323.15 \cdot K \qquad T_{20} = 293.15 \cdot K \qquad mp = 10^{-3} \cdot poise \qquad \eta_{50} = 5.494 \cdot mp \qquad \eta_{20} = 10.05 \cdot mp$$

$$k\_ratio = \frac{T_{50} \cdot \eta_{20}}{T_{20} \cdot \eta_{50}} \qquad k\_ratio = 2.016$$

10.83

We can rearrange eq. (10.71) to solve for k1:

$$\tau = \frac{1}{k_1 - k_2 \cdot \left(A_{eq} + B_{eq}\right)} \qquad k_1 = \frac{1}{\tau \cdot \left[1 + \dfrac{k_2}{k_1} \cdot \left(A_{eq} + B_{eq}\right)\right]}$$

We know the ratio of k1 to k2 from the equilibrium:

$$\frac{k_2}{k_1} = K_{eq} = \frac{H2O}{H \cdot OH} \qquad k\_ratio = \frac{55.6 \cdot \dfrac{mole}{liter}}{\left(10^{-7} \cdot \dfrac{mole}{liter}\right)^2} \qquad k\_ratio = 5.56 \cdot 10^{15} \cdot \frac{liter}{mole}$$

$$\mu s = 10^{-6} \cdot sec \qquad A_{eq} = 10^{-7} \cdot \frac{mole}{liter} \qquad B_{eq} = 10^{-7} \cdot \frac{mole}{liter} \qquad \tau = 36 \cdot \mu s$$

$$k_1 = \frac{1}{\tau \cdot \left[1 - k\_ratio \cdot \left(A_{eq} + B_{eq}\right)\right]} \qquad k_1 = 2.5 \cdot 10^{-5} \cdot \frac{1}{sec}$$

$$k_2 = k\_ratio \cdot k_1 \qquad k_2 = 1.4 \cdot 10^{11} \cdot \frac{liter}{mole \cdot sec}$$

10.85

First we write the rate of appearance of A:

$$\frac{d[A]}{dt} = - k_1[A] + k_{-1}[B]$$

We can relate [A] and [B] to the equilibrium concentrations and an extent of reaction, x:

$$[A] = [A]_{eq} - x \qquad \text{and} \qquad [B] = [B]_{eq} + x$$

$$\frac{dx}{dt} = k_1\left([A]_{eq} - x\right) - k_{-1}\left([[B]_{eq} + x\right)$$

$$\frac{dx}{dt} = -\left(k_1 + k_{-1}\right)x + \left(k_1[A]_{eq} - k_{-1}[B]_{eq}\right)$$

Because of the equilibrium condition, the second term is zero.

$$\frac{dx}{dt} = -\left(k_1 + k_{-1}\right)x$$

The definition of the relaxation time is:

$$\tau = - \frac{x}{dx/dt} = \frac{1}{\left(k_1 + k_{-1}\right)}$$

CHAPTER 11   *Quantum Theory*

SECTION 11.1

11.1

$$p = mv = \frac{h\nu}{c} = \frac{h}{\lambda} \implies v = \frac{h}{m\lambda}$$

$h = 6.626 \times 10^{-34}$ J s   and $\lambda = 400$ nm

a) $m = 9.11 \times 10^{-31}$ kg

$$v = \frac{6.626 \times 10^{-34}}{(400 \times 10^{-9})(9.11 \times 10^{-31})} = 1.82 \times 10^{3} \text{ m/s}$$

b) $m = \dfrac{4.003 \times 10^{-3} \text{ kg/mol}}{6.022 \times 10^{23} \text{ mol}^{-1}} = 6.647 \times 10^{-27}$ kg

$v = 0.2492$ m/s

c) $m = 1$ g

$v = 1.657 \times 10^{-24}$ m/s

11.3

$$E = h\nu = \frac{hc}{\lambda} = \frac{1}{2}mv^2 = \frac{M}{2L}\left(\sqrt{\frac{3RT}{M}}\right)^2 = \frac{3RT}{2L} \implies \lambda = \frac{2hcL}{3RT}$$

$T = 300 \text{ K}$

$\lambda = 3.197 \times 10^{-5} \text{ m} = 31.97 \text{ } \mu\text{m}$

11.5

$1 \text{ g } H_2O \times 1°C \times 4.184 \text{ J / g } °C = 4.184 \text{ J}$

$\lambda = 750 \text{ nm}$

$E_{750} = \frac{hc}{\lambda} = 2.649 \times 10^{-19} \text{ J}$

$\# \text{ photons} = \frac{4.184 \text{ J}}{E_{750}} = 1.580 \times 10^{19} \text{ photons}$

11.7

$\theta = 90°$

$\lambda_i = 300 \text{ pm} = 300 \times 10^{-12} \text{ m}$

$\lambda_f = \lambda_i + \Delta\lambda = 302.4 \text{ pm}$, where $\Delta\lambda$ is defined in the problem

$$-\Delta E = \frac{hc}{\lambda_i} - \frac{hc}{\lambda_f} = \frac{1}{2} m_e v^2$$

$$\Rightarrow v^2 = \frac{2hc}{m_e} \left(\frac{1}{\lambda_i} - \frac{1}{\lambda_f} \right)$$

$$\Rightarrow v = \sqrt{\frac{2hc}{m_e} \left(\frac{1}{\lambda_i} - \frac{1}{\lambda_f} \right)} = 3.40 \times 10^6 \text{ m/s}$$

SECTION 11.2

11.9

$$1 \text{ cm} \times \frac{1 \text{ bohr}}{5.2918 \times 10^{-9} \text{ cm}} \times \frac{1 \text{ GB}}{10^9 \text{ bohr}} = 0.189 \text{ GB}$$

$$1 \text{ inch} \times \frac{2.54 \text{ cm}}{\text{inch}} \times \frac{0.189 \text{ GB}}{\text{cm}} = 0.480 \text{ GB}$$

$$1 \text{ g} \times \frac{1 m_e}{9.1094 \times 10^{-28} \text{ g}} \times \frac{1 \text{ MU}}{10^{27} \text{ m}} = 1.098 \text{ MU}$$

$$1 \text{ J} = 1 \text{ kg m}^2/s^2 \times \left(\frac{100 \text{ cm}}{\text{m}} \right)^2 \times \left(\frac{0.189 \text{ GB}}{\text{cm}} \right)^2 \times \left(\frac{1000 \text{ g}}{\text{kg}} \right) \times \left(\frac{1.098 \text{ MU}}{\text{g}} \right) = 3.92 \times 10^5 \text{ MU GB}^2/s^2 = 0.3$$

$$(4.184 \text{ J}/\text{g K}) \times (1 \text{ g}) \times (1 \text{ K}) \times (0.392 \text{ MEU}/\text{J}) = 1.64 \text{ MEU}$$

11.11

E_h = 1 hartree

For Helium, Z = 2 and N = 1

$$IP = -E = \frac{1}{2}\frac{Z^2}{N^2}E_h = \frac{1}{2}\frac{2^2}{1^2}\text{ hartree x }\frac{27.2114\text{ eV}}{\text{hartree}} = 54.4228\text{ eV}$$

11.13

$\Re = 109,737\text{ cm}^{-1}$

$\Re_{He} = Z^2\Re = 4\Re = 438,948\text{ cm}^{-1}$

$$\tilde{\nu} = \Re_{He}\left(\frac{1}{N_1^2} - \frac{1}{N_2^2}\right)$$

$\underline{N_1}$	$\underline{N_2}$	$\tilde{\nu}$ / cm^{-1}
3	4	21,338
3	∞	48,772
4	5	9876
4	∞	27,434
5	6	5365
5	∞	17,558

11.15
_____

$N\hbar = m_e vr$ (de Broglie Hypothesis) and $r = \dfrac{N^2 a_o}{Z}$

$\Rightarrow v = \dfrac{N\hbar}{m_e r} = \dfrac{Z\hbar}{m_e N a_o}$

$\Rightarrow \dfrac{v}{c} = \dfrac{Z\hbar}{m_e c N a_o}$

let $N = 1$

H ($Z = 1$): $v/c = 0.0073$

Hg^{79+} ($Z = 80$) $v/c = 0.584$

U^{91+} ($Z = 92$) $v/c = 0.671$

$\dfrac{v}{c} = \dfrac{Z\hbar}{m_e c N a_o} = 1 \Rightarrow Z = \dfrac{m_e c a_o}{\hbar} = 137.05$

This implies v/c > 1 when Z ≥ 138

11.17
_____

$r = \dfrac{N^2 a_o}{Z} \Rightarrow N = \sqrt{\dfrac{Zr}{a_o}}$

let $Z = 1$, $r = 1$ cm

then $N = 13746.7 \cong 13747$

11.19

let $\lambda = 100$ pm, $m = m_e$

$$\lambda = \frac{h}{mv} \quad \Rightarrow \quad v = \frac{h}{m\lambda}$$

$$T \text{ (kinetic energy)} = \frac{1}{2}mv^2 = \frac{h^2}{2m_e\lambda^2} = 150.4 \text{ eV}$$

SECTION 11.3

11.21

$$\hat{O} \exp[i\theta] = \frac{d}{d\theta} \sin\theta \frac{d}{d\theta} \exp[i\theta] = \frac{d}{d\theta}\left(\frac{\exp[i\theta] - \exp[-i\theta]}{2i}\right) i \exp[i\theta] = \frac{d}{d\theta}\left(\frac{\exp[2i\theta] - 1}{2}\right) = i \exp[2i\theta]$$

Therefore, $e^{i\theta}$ is not an eigenfunction of this operator

11.23

$$\left[\frac{d}{dr}, \frac{1}{r}\right]f(r) = \frac{d}{dr}\left(\frac{f}{r}\right) - \frac{1}{r}\frac{df}{dr} = \frac{1}{r}\frac{df}{dr} - \frac{1}{r^2}f - \frac{1}{r}\frac{df}{dr} = -\frac{1}{r^2}f$$

$$\Rightarrow \left[\frac{d}{dr}, \frac{1}{r}\right] = -\frac{1}{r^2}$$

11.25

$$\hat{O} \, 3\exp[i\phi] = i\frac{d}{d\phi} \, 3\exp[i\phi] = -3\exp[i\phi]$$

Therefore, $3e^{i\phi}$ is an eigenfunction of this operator with eigenvalue = -1

11.27

$$\left(\frac{1}{r^2}\frac{d}{dr}r^2\frac{d}{dr}+\frac{2}{r}\right)A\exp[-br] = \frac{1}{r^2}\frac{d}{dr}\left(-Abr^2\exp[-br]\right) + \frac{2A\exp[-br]}{r}$$

$$= \frac{1}{r^2}\left(-Abr^2(-b\exp[-br]) - 2Abr\exp[-br]\right) + \frac{2A\exp[-br]}{r}$$

$$= Ab^2\exp[-br] + (2A - 2Ab)\frac{\exp[-br]}{r}$$

$$\Rightarrow \quad 2A - 2Ab = 0 \quad \Rightarrow \quad b = 1 \text{ and } A \neq 0$$

eigenvalue $= b^2 = 1$

11.29

$$\nabla^2 = \frac{\partial^2}{\partial x^2} + \frac{\partial^2}{\partial y^2} + \frac{\partial^2}{\partial z^2}$$

$$\nabla^2 \left(x^2 + y^2 + z^2\right) = \frac{\partial^2}{\partial x^2} x^2 + \frac{\partial^2}{\partial y^2} y^2 + \frac{\partial^2}{\partial z^2} z^2 = 2 + 2 + 2 = 6 \quad \text{not an eigenfunction}$$

$$\nabla^2 = \left(\frac{1}{r^2} \frac{\partial}{\partial r} r^2 \frac{\partial}{\partial r}\right) + \left(\frac{1}{r^2 \sin \theta} \frac{\partial}{\partial \theta} \sin \theta \frac{\partial}{\partial \theta}\right) + \left(\frac{1}{r^2 \sin \theta} \frac{\partial^2}{\partial \phi^2}\right)$$

$$\left(x^2 + y^2 + z^2\right) = r^2$$

$$\nabla^2 r^2 = \frac{1}{r^2} \frac{\partial}{\partial r} r^2 \frac{\partial}{\partial r} r^2 = \frac{1}{r^2} \frac{\partial}{\partial r} 2r^3 = \frac{1}{r^2} \left(6r^2\right) = 6$$

SECTION 11.4

11.31

$$p_x \psi = \frac{\hbar}{i} \frac{\partial}{\partial x} A \exp[ikx] = \frac{\hbar}{i} A \, ik \, \exp[ikx] = \hbar k \, A \exp[ikx] = \hbar k \, \psi$$

eigenvalue is $\hbar k$

11.33
___

$$1 = A^2 \int_0^1 \cos^2(n\pi x)dx = A^2 \left(\frac{2n\pi + \sin 2n\pi}{4n\pi} \right) = \frac{A^2}{2} \quad \text{since } \sin 2n\pi = 0 \text{ for integer n}$$

$$\Rightarrow \quad A^2 = 2 \quad \Rightarrow \quad A = \sqrt{2}$$

$$2 \int_0^1 \cos(n\pi x) \cos(m\pi x)dx = 2 \left(\frac{(m+n)\sin(m-n)\pi + (m-n)\sin(m+n)\pi}{(m^2 - n^2)\pi} \right) = 0 \text{ if } m \neq n$$

$$\Rightarrow \quad \sqrt{2} \cos(n\pi x) \text{ are orthonormal}$$

SECTION 11.5

11.35
___

$$\psi = \sqrt{\frac{2}{a}} \sin\frac{n\pi x}{a}$$

$$\text{probability} = \int_0^{a/4} \psi^2 dx = \int_0^a \frac{2}{a} \sin^2\frac{n\pi x}{a} dx \qquad \text{let } u = \frac{n\pi x}{a}$$

$$\Rightarrow \langle x \rangle = \frac{2}{(n\pi)} \int_0^{n\pi/4} \sin^2 u \, du = \frac{2}{(n\pi)} \left(\frac{(n\pi)}{8} - \frac{1}{4}\sin\frac{n\pi}{2} \right) = \frac{1}{4} - \frac{1}{2n\pi}\sin\frac{n\pi}{2}$$

$$\lim_{n \to \infty} \left(\frac{1}{4} - \frac{1}{2n\pi}\sin\frac{n\pi}{2} \right) = \frac{1}{4}$$

11.37

$$\psi = A\sin\frac{n\pi x}{a}$$

$$1 = \int_0^a \psi^2 dx = \int_0^a A^2 \sin^2\frac{n\pi x}{a}\,dx \qquad \text{let } u = \frac{n\pi x}{a}$$

$$\Rightarrow 1 = A^2 \frac{a}{(n\pi)} \int_0^{n\pi} \sin^2 u\,du = A^2 \frac{a}{(n\pi)}\left(\frac{u}{2} - \frac{1}{4}\sin 2u\right)_0^{n\pi} =$$

$$\Rightarrow 1 = A^2 \frac{a}{(n\pi)}\left(\frac{n\pi}{2}\right) = A^2 \frac{a}{2} \Rightarrow A = \sqrt{\frac{2}{a}}$$

11.39

$$E = (n_x^2 + n_y^2)\frac{h^2}{8ma^2}$$

(n_x, n_y)	E (multiples of $h^2/8ma^2$)	degeneracy
1,1	2	1
2,1	5	2
2,2	8	1
3,1	10	2
2,3	13	2
4,1	17	2
3,3	18	1
4,2	20	2
4,3	25	2
5,1	26	2

17 total states, highest energy is 26

11.41

$$E_0 = \frac{3h^2}{8ma^2}$$

a) $m = (2.016 \times 10^{-3} / L) \text{ kg}$
    $a = 1 \times 10^{-3} \text{ m}$

   $E_0 = 4.92 \times 10^{-35} \text{ J}$

b) $m = m_e$
    $a = 1 \times 10^{-10} \text{ m}$

   $E_0 = 1.81 \times 10^{-17} \text{ J}$

11.43

$$\psi = A_x \sin\frac{n_x \pi x}{a} A_y \sin\frac{n_y \pi y}{b} A_z \sin\frac{n_z \pi z}{c}$$

$$H\psi = -\frac{\hbar^2}{2m}\nabla^2\psi = -\frac{\hbar^2}{2m}\left(-\frac{n_x^2\pi^2}{a^2} - \frac{n_y^2\pi^2}{b^2} - \frac{n_z^2\pi^2}{c^2}\right)A_x \sin\frac{n_x \pi x}{a} A_y \sin\frac{n_y \pi y}{b} A_z \sin\frac{n_z \pi z}{c} = E\psi$$

$$\Rightarrow E = -\frac{\hbar^2}{2m}\left(-\frac{n_x^2\pi^2}{a^2} - \frac{n_y^2\pi^2}{b^2} - \frac{n_z^2\pi^2}{c^2}\right) = \frac{h^2}{8m}\left(\frac{n_x^2}{a^2} + \frac{n_y^2}{b^2} + \frac{n_z^2}{c^2}\right)$$

SECTION 11.6

11.45
___

$$1 = \int_{-\infty}^{\infty} |\psi|^2 \, dx = A_1^2 \int_{-\infty}^{\infty} (2y)^2 \, \exp[-y^2] \, dx \qquad \text{let } x = \alpha y$$

$$= 4\alpha \, A_1^2 \int_{-\infty}^{\infty} y^2 \, \exp[-y^2] \, dy = 4\alpha \, A_1^2 \left(\frac{\sqrt{\pi}}{2} \right)$$

$$\Rightarrow \quad A_1^2 = \frac{1}{2\alpha \sqrt{\pi}} \quad \Rightarrow \quad A_1 = \frac{1}{(2\alpha)^{1/2} (\pi)^{1/4}}$$

11.47
___

$$\int_{-\infty}^{\infty} \psi_0^* \psi_1 \, dx = k_0 k_1 \int_{-\infty}^{\infty} \exp[-y^2/2] \, y \, \exp[-y^2/2] \, dy$$

$$= C \int_{-\infty}^{\infty} y \, \exp[-y^2] \, dy = 0 \cdot \text{ since the integrand is odd}$$

$$\int_{-\infty}^{\infty} \psi_0^* \psi_2 \, dx = k_0 k_2 \int_{-\infty}^{\infty} \exp[-y^2/2] \, (4y^2 - 2) \, \exp[-y^2/2] \, dy$$

$$= C \int_{-\infty}^{\infty} (4y^2 - 2) \, \exp[-y^2] \, dy$$

$$= C \left(4 \frac{\sqrt{\pi}}{2} - 2\sqrt{\pi} \right) = 0$$

11.51

$$\langle p \rangle = \int_{-\infty}^{\infty} \psi_1^* p \psi_1 dx = \frac{A_1^2 \int_{-\infty}^{\infty} 2y \, \exp[-y^2/2] \left(\frac{\hbar}{i} \frac{\partial}{\partial x} \right) 2y \, \exp[-y^2/2] \, dx}{A_1^2 \int_{-\infty}^{\infty} 4y^2 \, \exp[-y^2] \, dx} \qquad \text{let } x = \alpha y$$

$$= \frac{\frac{\hbar}{i} \frac{1}{\alpha} \int_{-\infty}^{\infty} y \, \exp[-y^2/2] \left[x \frac{-x}{\alpha^2} + 1 \right] \exp[-y^2/2] \, dx}{\int_{-\infty}^{\infty} \alpha \, y^2 \, \exp[-y^2] \, dy}$$

$$= \frac{\frac{\hbar}{i} \int_{-\infty}^{\infty} y(1 - y^2) \, \exp[-y^2] \, dy}{\int_{-\infty}^{\infty} \alpha \, y^2 \, \exp[-y^2] \, dy} = 0 \quad \text{since the integrand in the numerator is odd}$$

$$\langle p^2 \rangle = \frac{A_1^2 \int_{-\infty}^{\infty} 2y \, \exp[-y^2/2] \left(-\hbar^2 \frac{\partial^2}{\partial x^2} \right) 2y \, \exp[-y^2/2] \, dx}{A_1^2 \int_{-\infty}^{\infty} 4y^2 \, \exp[-y^2] \, dx} \qquad \text{let } x = \alpha y$$

$$= \frac{-\hbar^2 \frac{1}{\alpha} \int_{-\infty}^{\infty} y \, \exp[-y^2/2] \left[\frac{x^3}{\alpha^4} - \frac{3x}{\alpha^2} \right] \exp[-y^2/2] \, dx}{\int_{-\infty}^{\infty} \alpha \, y^2 \, \exp[-y^2] \, dy}$$

$$= \frac{\frac{-\hbar^2}{\alpha^2} \int_{-\infty}^{\infty} (y^4 - 3y^2) \, \exp[-y^2] \, dy}{\int_{-\infty}^{\infty} y^2 \, \exp[-y^2] \, dv} = \frac{-\hbar^2}{\alpha^2} \frac{\left(\frac{3}{4} \sqrt{\pi} - \frac{3}{2} \sqrt{\pi} \right)}{\frac{1}{2} \sqrt{\pi}} = \frac{3}{2} \frac{\hbar^2}{\alpha^2}$$

$$\Delta x \Delta p = \sqrt{\langle x^2 \rangle \langle p^2 \rangle} = \sqrt{\frac{3}{2} \alpha^2 \frac{3}{2} \frac{\hbar^2}{\alpha^2}} = \frac{3}{2} \hbar \geq \frac{\hbar}{2}$$

11.53

a)

$$\hat{A} = y - \frac{d}{dy}$$

$$\hat{B} = y + \frac{d}{dy}$$

$$\left[\hat{A}\hat{B}\right]f = \left(y - \frac{d}{dy}\right)\left(yf + \frac{df}{dy}\right) = y^2f + y\frac{df}{dy} - y\frac{df}{dy} - 1f - \frac{d^2f}{dy^2} = \left(y^2 - \frac{d^2}{dy^2} - 1\right)f = \left(\hat{h} - 1\right)f$$

$$\left[\hat{B}\hat{A}\right]f = \left(y + \frac{d}{dy}\right)\left(yf - \frac{df}{dy}\right) = y^2f - y\frac{df}{dy} + y\frac{df}{dy} + 1f - \frac{d^2f}{dy^2} = \left(y^2 - \frac{d^2}{dy^2} + 1\right)f = \left(\hat{h} + 1\right)f$$

b)

$$\hat{h}\psi = \left(\hat{A}\hat{B} + 1\right)\psi = \hat{A}\hat{B}\psi + 1\psi = \hat{A}0 + 1\psi = 1\psi$$

11.55

This solution uses the answers to problems 11.53 and 11.54

a)

$$\left[\hat{A}, \hat{B}\right] f = \left(\hat{A}\hat{B} - \hat{B}\hat{A}\right) f = \left(\hat{h} - 1\right) f - \left(\hat{h} + 1\right) f = -2f \quad \Rightarrow \quad \left[\hat{A}, \hat{B}\right] = -2$$

b)

$$\hat{A}\hat{B}\psi_n = 2n\psi_n \quad \text{from 11.54}$$

$$\hat{A}\hat{B}\psi_n = \hat{A}(\hat{B}\psi_n) = \hat{A}(\text{const } \psi_{n-1}) = \text{const } (\hat{A}\psi_{n-1}) = \text{const } \psi_n \quad \Rightarrow \quad \text{const} = 2n$$

11.57

$\nu = 8.963 \times 10^{13}$ Hz
$M_H = 1.008 \times 10^{-3}$ kg/mol
$M_{Cl} = 34.969 \times 10^{-3}$ kg/mol

$$\nu = \frac{1}{2\pi} \sqrt{\frac{k}{\mu}} \quad \Rightarrow \quad k = 4\pi^2 \mu \nu^2 = 4\pi^2 \frac{m_1 m_2}{m_1 + m_2} \nu^2 = \frac{4\pi^2}{L} \frac{M_H M_{Cl}}{M_H + M_{Cl}} \nu^2 = 5.16 \times 10^2 \text{ kg}/\text{s}^2$$

$x = 1$ cm
$g = 9.81$ m/s^2

$F_{spring} = F_{gravity}$

SECTION 11.7

11.59
_____

m_H = mass of hydrogen atom
r_{CH} = 0.11 nm
$\phi = 180° - \theta_T = 70.53°$

$R = r_{CH} \sin \phi = 0.104$ nm

$I = 3 \, m_H \, R^2 = 5.4 \times 10^{-47}$ kg m^2

$$kT = \frac{m^2 h^2}{8\pi^2 I} \Rightarrow m^2 = \frac{kT 8\pi^2 I}{h^2} = 40 \cong 36$$
$$\Rightarrow m = 6$$

11.61

$T = 300$ K
m_H = mass of hydrogen atom
$r_{OH} = 95.8$ pm
$\phi = 180° - 120° = 60°$

$R = r_{OH} \sin \phi = 83.0$ pm

$I = m_H R^2 = 1.15 \times 10^{-47}$ kg m^2

$$\frac{E_m}{kT} = \frac{m^2 h^2}{8\pi^2 I k T}$$

m	E
0	0
1	0.117
2	0.467
3	1.051
4	1.869
5	2.920
6	4.205

11.63

$$L_- = L_x - iL_y = i\left(\sin\phi\frac{\partial}{\partial\theta} + \cot\theta\cos\phi\frac{\partial}{\partial\phi}\right) - i\left(-i\left(\cos\phi\frac{\partial}{\partial\theta} - \cot\theta\sin\phi\frac{\partial}{\partial\phi}\right)\right)$$

$$= -(\cos\phi - i\sin\phi)\frac{\partial}{\partial\theta} + i\,\cot\theta(\cos\phi - i\sin\phi)\frac{\partial}{\partial\phi}$$

$$= -\exp[-i\phi]\frac{\partial}{\partial\theta} + i\,\cot\theta\,\exp[-i\phi]\frac{\partial}{\partial\phi}$$

$$L_-\cos\theta = -\exp[-i\phi]\frac{\partial}{\partial\theta}\cos\theta + i\,\cot\theta\,\exp[-i\phi]\frac{\partial}{\partial\phi}\cos\theta = \exp[-i\phi]\sin\theta$$

$$L_+L_-\cos\theta = L_+(\exp[-i\phi]\sin\theta) = \exp[i\phi]\frac{\partial}{\partial\theta}(\exp[-i\phi]\sin\theta) + i\,\cot\theta\,\exp[i\phi]\frac{\partial}{\partial\phi}(\exp[-i\phi]\sin\theta)$$

$$= \cos\theta + i\,\cot\theta(\sin\theta)\exp[i\phi](-i\,\exp[-i\phi]) = 2\cos\theta$$

11.65

$$L_z\sin\theta\exp[i\phi] = -i\frac{\partial}{\partial\phi}\sin\theta\exp[i\phi] = -i\sin\theta(i\exp[i\phi]) = \sin\theta\exp[i\phi]$$

eigenvalue = m = 1

$$L_z\sin\theta\exp[-i\phi] = -i\frac{\partial}{\partial\phi}\sin\theta\exp[-i\phi] = -i\sin\theta(-i\exp[-i\phi]) = -\sin\theta\exp[-i\phi]$$

eigenvalue = m = −1

$$L_z\cos\theta = -i\frac{\partial}{\partial\phi}\cos\theta = 0$$

eigenvalue = m = 0

11.67

$$L_-L_+\Psi_{l,m} = C^2\Psi_{l,m} = (L^2 - L_z^2 - L_z)\Psi_{l,m} = \left[l(l+1) - m^2 - m\right]\Psi_{l,m}$$
$$= \left[(l(l+1) - m(m+1)\right]\Psi_{l,m}$$

$$\Rightarrow C^2 = \left[(l(l+1) - m(m+1)\right]^{1/2}$$
$$C = 0 \text{ if } m = l$$

11.69

$$I = \frac{9}{2}$$
$$m_I = \frac{9}{2}, \frac{7}{2}, \frac{5}{2}, \frac{3}{2}, \frac{1}{2}, \frac{-1}{2}, \frac{-3}{2}, \frac{-5}{2}, \frac{-7}{2}, \frac{-9}{2}$$

10 Orientations

$$\cos\theta = \frac{m_I}{|I|} = \frac{m_I}{\sqrt{I(I+1)}} = \frac{9/2}{\sqrt{\frac{9}{2}\left(\frac{11}{2}\right)}} = \frac{9}{\sqrt{99}} \Rightarrow \theta = \arccos\left(\frac{9}{\sqrt{99}}\right) = 25.24°$$

11.71

$$I = \frac{9}{2}$$

$$m_I = \frac{9}{2}, \frac{7}{2}, \frac{5}{2}, \frac{3}{2}, \frac{1}{2}, \frac{-1}{2}, \frac{-3}{2}, \frac{-5}{2}, \frac{-7}{2}, \frac{-9}{2}$$

10 Orientations

$$\cos\theta = \frac{m_I}{|I|} = \frac{m_I}{\sqrt{I(I+1)}} = \frac{9/2}{\sqrt{\frac{9}{2}\left(\frac{11}{2}\right)}} = \frac{9}{\sqrt{99}} \Rightarrow \theta = \arccos\left(\frac{9}{\sqrt{99}}\right) = 25.24°$$

11.73

$$I = 1.448 \times 10^{-46} \text{ kg·m}^2$$
$$T = 300 \text{ K}$$

$$E = kT = \frac{J(J+1)h^2}{8\pi^2 I} \quad \Rightarrow \quad J(J+1) = \frac{8\pi^2 I kT}{h^2} = 108 \cong 110 \quad \Rightarrow \quad J = 10$$

11.75

$I = 2.644 \times 10^{-46}$ kg·m^2
$T = 300$ K

$$\frac{E}{kT} = \frac{J(J+1)h^2}{8\pi^2 IkT} = J(J+1)(0.05077)$$

J	E/kT
0	0
5	1.523
10	5.585
15	12.18
20	21.32

SECTION 11.8

11.77

$I = 2.644 \times 10^{-46}$ kg·m^2
$J_i = 1$
$J_f = 2$

$$\Delta E = h\nu = \frac{hc}{\lambda} = \frac{h^2}{8\pi^2 I}\left[J_f(J_f+1) - J_i(J_i+1)\right]$$

$$\Rightarrow \quad \lambda = \frac{8\pi^2 Ic}{h}\left[J_f(J_f+1) - J_i(J_i+1)\right]^{-1} = 0.2362 \text{ nm}$$

11.79

$m = m_e$
$a = 10$ bohr

$$E = (n_x^2 + n_y^2 + n_z^2)\frac{h^2}{8ma^2}$$

$$\Delta E = E_{211} - E_{111}$$

$$\Rightarrow \bar{\nu} = \frac{\Delta E}{hc} = \frac{(2^2 - 1^2)h}{8ma^2c} = 3.249 \times 10^4 \text{ cm}^{-1}$$

11.81

$m = m_e$
$a = 0.424$ nm

$$E = n^2\frac{h^2}{8ma^2}$$

$$\Delta E = E_3 - E_2$$

$$\Rightarrow \bar{\nu} = \frac{\Delta E}{hc} = \frac{(3^2 - 2^2)h}{8ma^2c} = 84{,}352 \text{ cm}^{-1} = 10.46 \text{ eV}$$

11.83

$$I_{12}^2 \equiv \left[\int_0^a x \sin\frac{\pi x}{a} \sin\frac{2\pi x}{a} \, dx \right]^2 = \left(\frac{8a}{9\pi}\right)^2 = \frac{64a^2}{81\pi^2}$$ This should then be scaled to 1

$$I_{13}^2 \equiv \left[\int_0^a x \sin\frac{\pi x}{a} \sin\frac{3\pi x}{a} \, dx \right]^2 = 0$$

$$I_{14}^2 \equiv \frac{\left[\int_0^a x \sin\frac{\pi x}{a} \sin\frac{2\pi x}{a} \, dx \right]^2}{I_{12}^2} = 0.0064$$

11.85

$$I_x \equiv \iint (\sin\theta \cos\phi) \cos\theta(3\cos^2\theta - 1) \sin\theta \, d\theta d\phi = \int_0^{2\pi} \cos\phi d\phi \int_0^\pi \cos\theta(3\cos^2\theta - 1) \sin^2\theta \, d\theta$$

$$= 0 \left(\int \ldots d\theta \right) = 0$$

$$I_y \equiv \iint (\sin\theta \sin\phi) \cos\theta(3\cos^2\theta - 1) \sin\theta \, d\theta d\phi = \int_0^{2\pi} \sin\phi d\phi \int_0^\pi \cos\theta(3\cos^2\theta - 1) \sin^2\theta \, d\theta$$

$$= 0 \left(\int \ldots d\theta \right) = 0$$

$$I_z \equiv \int_0^{2\pi} d\phi \int_0^\pi R \cos\theta \cos\theta(3\cos^2\theta - 1) \sin\theta \, d\theta = 2\pi R \int_0^\pi (3\cos^4\theta - \cos^2\theta) \sin\theta \, d\theta$$

$$= 2\pi R \int_{-1}^1 3u^4 - u^2 \, du \quad \text{if } u = \cos\theta$$

$$= 2\pi R \left(\frac{3}{5} - \frac{1}{3} + \frac{3}{5} - \frac{1}{3}\right) = \frac{16\pi R}{15} \Rightarrow I_z^2 = \frac{256\pi^2 R^2}{225}$$

SECTION 11.8

11.87
___

$$v_{tunnel} = \frac{\Delta E}{h} = \frac{E_2 - E_1}{h}$$

V_0	v_{tunnel}
50	3.14×10^{16} Hz
100	1.24×10^{16} Hz
500	2.63×10^{14} Hz

11.89
___

$$V(x) = 4 \, V_0 \, x \, (1-x)$$

$$E_1^1 = E_1^0 + \int_0^1 \psi_1^* V(x) \psi_1 dx = \frac{\pi^2}{2} + 4 \, V_0 \, (2) \int_0^1 \sin^2 (\pi x) \, x \, (1-x) dx$$

$$= 4.93 + 0.869 \, V_0$$

$$E_2^1 = E_2^0 + \int_0^1 \psi_2^* V(x) \psi_2 dx = 2\pi^2 + 8 \, V_0 \int_0^1 \sin^2 (2\pi x) \, x \, (1-x) dx$$

$$= 19.74 + 0.717 \, V_0$$

11.91

$V(x) = V_0 \sin^2 (\pi x)$

$$E_1^1 = E_1^0 + \int_0^1 \psi_1^* V(x) \psi_1 dx = \frac{\pi^2}{2} + 2 V_0 \int_0^1 \sin^4 (\pi x) dx$$

$$= \frac{\pi^2}{2} + \frac{3}{4} V_0$$

$$E_2^1 = E_2^0 + \int_0^1 \psi_2^* V(x) \psi_2 dx = 2\pi^2 + 2 V_0 \int_0^1 \sin^2 (2\pi x) \sin^2 (\pi x) dx$$

$$= 2\pi^2 + \frac{1}{2} V_0$$

CHAPTER 12   *Atoms*

SECTION 12.1

12.1

$$r = \sigma a_o$$

$$1 = \int |\psi|^2 d\tau = \iiint_{\text{all space}} A^2\sigma^2 \exp[-\sigma]\cos^2\theta(r^2)\sin\theta d\phi d\theta dr$$

$$= 2\pi A^2 \int_0^\pi \cos^2\theta\sin\theta d\theta \int_0^\infty \sigma^2 \exp[-\sigma]r^2 dr = 2\pi A^2\left(\frac{2}{3}\right)\left(24a_o^3\right) \text{ (using u - substitution to evaluate each integral}$$

$$\Rightarrow A^2 = \frac{1}{32\pi a_o^3} \Rightarrow A = \frac{1}{\sqrt{32\pi a_o^3}}$$

12.3
___

$$R = (2 - \sigma) \exp[-\frac{\sigma}{2}]$$

$$\frac{-1}{2} \frac{1}{\sigma^2} \frac{\partial}{\partial \sigma} \sigma^2 \frac{\partial}{\partial \sigma} (2 - \sigma) \exp[-\frac{\sigma}{2}] - \frac{1}{\sigma} (2 - \sigma) \exp[-\frac{\sigma}{2}] + \frac{0(0+1)}{2\sigma^2} (2 - \sigma) \exp[-\frac{\sigma}{2}]$$

$$= \frac{-1}{2} \frac{1}{\sigma^2} \frac{\partial}{\partial \sigma} \sigma^2 \left[(2 - \sigma) \left(\frac{-1}{2} \right) - 1 \right] \exp[-\frac{\sigma}{2}] - \frac{1}{\sigma} (2 - \sigma) \exp[-\frac{\sigma}{2}]$$

$$= \frac{-1}{2} \frac{1}{\sigma^2} \left[\left(\frac{\sigma^3}{2} - 2\sigma^2 \right) \left(\frac{-1}{2} \right) + \left(\frac{3\sigma^3}{2} - 4\sigma \right) \right] \exp[-\frac{\sigma}{2}] - \frac{1}{\sigma} (2 - \sigma) \exp[-\frac{\sigma}{2}]$$

$$= \left(\frac{-1}{4} + \frac{1}{8} \sigma \right) \exp[-\frac{\sigma}{2}] = \frac{-1}{8} (2 - \sigma) \exp[-\frac{\sigma}{2}] = \frac{-1}{8} R$$

$$\varepsilon = \frac{-1}{8}$$

12.5
___

$$R = A \exp[-3\sigma]$$

$r = \sigma a_o$ (use in u - substitution to evaluate integrals)

$$\text{probability} = \frac{A^2 \int_0^{a_o} \exp[-6\sigma] r^2 dr}{A^2 \int_0^{\infty} \exp[-6\sigma] r^2 dr} = \frac{a_o^3}{a_o^3} \frac{\int_0^1 \exp[-6\sigma] \sigma^2 d\sigma}{\int_0^{\infty} \exp[-6\sigma] \sigma^2 d\sigma} = 0.938031$$

evaluate integrals by tables or numerically

12.7

$R = 2A\sigma \exp[-\sigma]$

$r = \sigma a_o$ (use in u - substitution to evaluate integrals)

$$\langle r \rangle = \frac{\int rR^2 r^2 dr}{\int R^2 r^2 dr} = \frac{4A^2 \int_0^\infty \sigma^2 \exp[-2\sigma]r^3 dr}{4A^2 \int_0^\infty \sigma^2 \exp[-2\sigma]r^2 dr} = \frac{a_o^4}{a_o^3} \frac{5!/2^6}{4!/2^5} = \frac{5}{2} a_o$$

12.9

$$R = A(2 - Z\sigma)\exp[-\frac{Z\sigma}{2}] = 0$$

$$\Rightarrow 2 - Z\sigma = 0 \Rightarrow \sigma = \frac{2}{Z} \Rightarrow r_{node} = a_o \sigma = \frac{2a_o}{Z}$$

$r = \sigma a_o$ (use in u - substitution to evaluate integrals)

$w = Z\sigma$ (use in u - substitution to evaluate integrals)

$$\text{probability(inside)} = \frac{\int_0^{r_{node}} R^2 r^2 dr \int r}{\int R^2 r^2 dr} = \frac{A^2 \int_0^{\frac{2a_o}{Z}} (2 - Z\sigma)^2 \exp[-Z\sigma]r^2 dr}{A^2 \int_0^\infty (2 - Z\sigma)^2 \exp[-Z\sigma]r^2 dr} = \frac{a_o^3}{a_o^3} \frac{\int_0^{\frac{2}{Z}} (2 - Z\sigma)^2 \exp[-Z\sigma]\sigma^2 d\sigma}{\int_0^\infty (2 - Z\sigma)^2 \exp[-Z\sigma]\sigma^2 d\sigma}$$

$$= \frac{\frac{1}{Z^3} \int_0^2 (2 - w)^2 \exp[-w]w^2 dw}{\frac{1}{Z^3} \int_0^\infty (2 - w)^2 \exp[-w]w^2 dw} = 0.052653$$

Probability (outside) = 1 - probability (inside) = 0.947347

12.11

For 2s and 2p, n = 2

Energy is a function of n only. Therefore $E_{2s} = E_{2p}$. They are degenerate

12.13

$R = A \exp[-3\sigma]$

$r = \sigma a_o$ (use in u - substitution to evaluate integrals)

$$\langle r^2 \rangle = \frac{\int r^2 R^2 r^2 dr}{\int R^2 r^2 dr} = \frac{A^2 \int\limits_0^\infty \exp[-6\sigma] r^4 dr}{A^2 \int\limits_0^\infty \exp[-6\sigma] r^2 dr} = \frac{a_o^5 \frac{4!}{6^5}}{a_o^3 \frac{2!}{6^3}} = \frac{1}{3} a_o^2$$

12.15

$d_1 + d_{-1} = R(r) \sin\theta \cos\theta \exp[i\phi] + R(r) \sin\theta \cos\theta \exp[-i\phi]$

$= R(r) \sin\theta \cos\theta (2\cos\phi) = 2R(r) \cos\theta (\sin\theta \cos\phi) = 2R(r)\left(\frac{z}{r}\right)\left(\frac{x}{r}\right)$

$= xzF(r) = d_{xz}$

$d_2 - d_{-2} = R(r) \sin^2\theta \exp[2i\phi] - R(r) \sin^2\theta \exp[-2i\phi]$

$= R(r) \sin^2\theta (2i\sin 2\phi) = 4iR(r) \sin\theta \sin\theta (\sin\phi \cos\phi) = 4iR(r) \sin\theta \cos\phi (\sin\theta \sin\phi)$

$= 4iR(r)\left(\frac{x}{r}\right)\left(\frac{y}{r}\right) = xyF(r) = d_{xy}$

Note: R(r) and F(r) are not the same in different cases

330

12.17

$$f_2 - f_{-2} = R(r)\sin^2\theta\cos\theta\exp[2i\phi] - R(r)\sin^2\theta\cos\theta\exp[-2i\phi]$$
$$= R(r)\sin^2\theta\cos\theta(2i\sin 2\phi) = 4iR(r)\sin\theta\sin\theta\cos\theta(\sin\phi\cos\phi) = 4iR(r)\sin\theta\cos\phi(\sin\theta\sin\phi)(\cos\theta)$$
$$= 4iR(r)\left(\frac{x}{r}\right)\left(\frac{y}{r}\right)\left(\frac{z}{r}\right) = xyzF(r) = F_{xyz}$$

12.19

$$\psi^*\psi \propto p_1^*p_1 + p_0^*p_0 + p_{-1}^*p_{-1} \propto \left|\frac{1}{\sqrt{2}}\sin\theta\exp[i\phi]\right|^2 + \left|\cos\theta\right|^2 + \left|\frac{1}{\sqrt{2}}\sin\theta\exp[-i\phi]\right|^2$$
$$= \frac{1}{2}\sin^2\theta + \cos^2\theta + \frac{1}{2}\sin^2\theta = 1$$

SECTION 12.2

12.21

$$S_y = \frac{1}{2i}\left(S_+ - S_-\right)$$

$$S_y\alpha = \frac{1}{2i}\left(0 - \beta\right) = \frac{i}{2}\beta$$

$$S_y\beta = \frac{1}{2i}\left(\alpha - 0\right) = -\frac{i}{2}\alpha$$

12.23

$$S_x S_y \beta = S_x\left(-\frac{i}{2}\alpha\right) = -\frac{i}{2}\frac{1}{2}\beta = -\frac{i}{4}\beta$$

12.25

let $\chi = \alpha + i\beta$

$$S_y\chi = S_y(\alpha + i\beta) = \frac{i}{2}\beta + i\frac{-i}{2}\alpha = \frac{1}{2}(\alpha + i\beta) = \frac{1}{2}\chi$$

let $\xi = \alpha - i\beta$

$$S_y\xi = S_y(\alpha - i\beta) = \frac{i}{2}\beta - i\frac{i}{2}\alpha = -\frac{1}{2}(\alpha - i\beta) = -\frac{1}{2}\xi$$

SECTION 12.3

12.27
___

$Z = 3$
$E_h = 27.2$ eV

$Li^{1+} \rightarrow Li^{2+} + e^-$

$E_{var} = -\left(Z - \dfrac{5}{16}\right)^2 E_h = -7.223\ E_h$

$IP = -\dfrac{Z^2}{2}E_h - E_{var} = 2.723\ E_h = 74.1$ eV

$Li^{2+} \rightarrow Li^{3+} + e^-$

$IP = \dfrac{Z^2}{2}E_h = 122.4$ eV   (Hydrogen-like ion)

SECTION 12.4

12.29

$$\int |\chi_s|^2 d\tau_1 d\tau_2 = \frac{1}{2} \int (\alpha_1\beta_2 + \alpha_2\beta_1)^2 d\tau_1 d\tau_2 = \frac{1}{2} \int \alpha_1^2\beta_2^2 + 2\alpha_1\beta_2\alpha_2\beta_1 + \alpha_2^2\beta_1^2 d\tau_1 d\tau_2$$

$$= \frac{1}{2}\left[1(1) + 2(0)(0) + 1(1)\right] = \frac{1}{2}(2) = 1$$

$$\int |\chi_a|^2 d\tau_1 d\tau_2 = \frac{1}{2} \int (\alpha_1\beta_2 - \alpha_2\beta_1)^2 d\tau_1 d\tau_2 = \frac{1}{2} \int \alpha_1^2\beta_2^2 - 2\alpha_1\beta_2\alpha_2\beta_1 + \alpha_2^2\beta_1^2 d\tau_1 d\tau_2$$

$$= \frac{1}{2}\left[1(1) - 2(0)(0) + 1(1)\right] = \frac{1}{2}(2) = 1$$

$$\int \chi_s\chi_a d\tau_1 d\tau_2 = \frac{1}{2} \int (\alpha_1\beta_2 + \alpha_2\beta_1)(\alpha_1\beta_2 - \alpha_2\beta_1) d\tau_1 d\tau_2 = \frac{1}{2} \int \alpha_1^2\beta_2^2 + \alpha_1\beta_2\alpha_2\beta_1 - \alpha_1\beta_2\alpha_2\beta_1 - \alpha_2^2\beta_1^2 d\tau_1 d\tau_2$$

$$= \frac{1}{2}\left[1(1) - 0 - 1(1)\right] = \frac{1}{2}(0) = 0$$

12.31

$$S_{x1}S_{x2}\alpha_1\alpha_2 = S_{x1}\alpha_1\left(\frac{1}{2}\beta_2\right) = \left(\frac{1}{2}\beta_1\right)\left(\frac{1}{2}\beta_2\right) = \frac{1}{4}\beta_1\beta_2$$

not an eigenvalue equation

12.33

$$X_s = \frac{\alpha_1\beta_2 + \alpha_2\beta_1}{\sqrt{2}}$$

$$X_a = \frac{\alpha_1\beta_2 - \alpha_2\beta_1}{\sqrt{2}}$$

$$S_T^2 \frac{\alpha_1\beta_2}{\sqrt{2}} = \left\{S_1^2 + S_2^2 + 2\left(S_{1x}S_{2x} + S_{1y}S_{2y} + S_{1z}S_{2z}\right)\right\}\frac{\alpha_1\beta_2}{\sqrt{2}}$$

$$= \frac{1}{\sqrt{2}}\left(\frac{3}{4}\alpha_1\beta_2 + \frac{3}{4}\alpha_1\beta_2 + 2\left(\frac{1}{4}\beta_1\alpha_2 - \frac{1}{4i^2}\beta_1\alpha_2 - \frac{1}{4}\alpha_1\beta_2\right)\right) = \frac{1}{\sqrt{2}}\left(\alpha_1\beta_2 + \beta_1\alpha_2\right) = X_s$$

$$S_T^2 \frac{\alpha_2\beta_1}{\sqrt{2}} = \frac{1}{\sqrt{2}}\left(\alpha_1\beta_2 + \beta_1\alpha_2\right) = X_s$$

$$S_T^2 X_s = 2X_s = s_T(s_T + 1)X_s$$
$$\Rightarrow s_T = 1$$

$$S_T^2 X_a = X_s - X_s = 0$$
$$\Rightarrow s_T = 0$$

12.35

$\lambda_1 = 58.44$ nm
$\lambda_2 = 2058.2$ nm

$$\Delta E = \frac{1}{\lambda_1} - \frac{1}{\lambda_2} = 166.3 \times 10^3 \text{ cm}^{-1}$$

SECTION 12.6

12.37
_____

$j_1 = 3/2$
$j_2 = 5/2$

$J = (j_1 + j_2), (j_1 + j_2 - 1), \ldots, |j_1 - j_2| = 4, 3, 2, 1$

degeneracy $= 2J + 1$

4	$2(4)+1 = 9$
3	7
2	5
1	3

Total $= 9 + 7 + 5 + 3 = 24$

12.39
_____

$j_1 = 1$
$j_2 = 1/2$
$j_3 = 1/2$

$J_{12} = (j_1 + j_2), (j_1 + j_2 - 1), \ldots, |j_1 - j_2| = 3/2, 1/2$

$J_{123} = (J_{12} + j_3), (J_{12} + j_3 - 1), \ldots, |J_{12} - j_3| = 2, 1 \text{ and } 1, 0$

degeneracy $= \Sigma\, 2J + 1 = 5 + 2(3) + 1(1) = 12$

12.41

a)

^4S	$L = 0$	$g_L = 2L + 1 = 1$
	$S = 3/2$	$g_S = 2S + 1 = 4$

$$g = g_L (g_S) = 1(4) = 4$$

b)

^4G	$L = 4$	$g_L = 2L + 1 = 9$
	$S = 3/2$	$g_S = 2S + 1 = 4$

$$g = g_L (g_S) = 9(4) = 36$$

c)

^3P	$L = 1$	$g_L = 2L + 1 = 3$
	$S = 1$	$g_S = 2S + 1 = 3$

$$g = g_L (g_S) = 3(3) = 9$$

d)

^2D	$L = 2$	$g_L = 2L + 1 = 5$
	$S = 1/2$	$g_S = 2S + 1 = 2$

$$g = g_L (g_S) = 5(2) = 10$$

SECTION 12.7

12.43

d orbital - The maximum number of electrons in the d orbitals is 10.

The number of configurations for 5 electrons in a d orbital is:

$$\frac{10!}{(10-5)!5!} = \frac{10(9)(8)(7)(6)}{5(4)(3)(2)(1)} = 252$$

ground state: ↑   ↑   ↑   ↑   ↑
       l: 2   1   0   -1   -2

For this configuration, L = 0 and S = 5/2

^{6}S   L = 0 $g_L = 2L + 1 = 1$
         S = 5/2 $g_S = 2S + 1 = 6$

                    $g = g_L (g_S) = 1(6) = 6$

12.45

p orbital - The maximum number of electrons in the p orbitals is 6

The number of configurations for 3 electrons in a p orbital is:

$$\frac{6!}{(6-3)!\,3!} = \frac{6(5)(4)}{3(2)(1)} = 20$$

$\underline{p_1}$	$\underline{p_0}$	$\underline{p_{-1}}$	$\underline{M_L}$	$\underline{M_S}$	$\underline{^4S}$	$\underline{^2D}$	$\underline{^2P}$
↑↓	↑		2	1/2		X	
↑↓	↓		2	-1/2		X	
↑↓		↑	1	1/2		X	
↑↓		↓	1	-1/2		X	
↑	↑↓		1	1/2			X
↓	↑↓		1	-1/2			X
	↑↓	↑	-1	1/2		X	
	↑↓	↓	-1	-1/2		X	
↑		↑↓	-1	1/2			X
↓		↑↓	-1	-1/2			X
	↑	↑↓	-2	1/2		X	
	↓	↑↓	-2	-1/2		X	
↑	↑	↑	0	3/2	X		
↑	↑	↓	0	1/2	X		
↑	↓	↑	0	1/2			X
↓	↑	↑	0	1/2		X	
↓	↓	↑	0	-1/2	X		
↓	↑	↓	0	-1/2		X	
↑	↓	↓	0	-1/2			X
↓	↓	↓	0	-3/2	X		

12.47

d^2	$S = 1, L = 3$	^3F
$f^9 = f^5$	$S = 5/2, L = 5$	^6H
$f^{14} = f^0$	$S = 0, L = 0$	^1S
s^1d^5	$S = 1/2 + 5/2 = 3, L = 0$	^7S
f^3	$S = 3/2, L = 6$	^4I
g^2	$S = 1, L = 7$	^3K

12.49

^6H $S = 5/2, L = 5$ $g = 6(11) = 66$

$J = L + S, L + S - 1, \ldots, |L - S| = 15/2, 13/2, 11/2, 9/2, 7/2, 5/2$

$g_J = \Sigma \, 2J + 1 = 16 + 14 + 12 + 10 + 8 + 6 = 66$

12.51

$E_2 = 0 \text{ cm}^{-1}$
$E_1 = 158.265 \text{ cm}^{-1}$
$E_0 = 226.997 \text{ cm}^{-1}$

$$\frac{E_J - E_{J-1}}{hc} = \frac{1}{2}A\big(J(J+1) - L(L+1) - S(S+1)\big) - \frac{1}{2}A\big((J-1)J - L(L+1) - S(S+1)\big) = \frac{A}{2}\big[J^2 + J - (J^2 - J)\big] =$$

$$\Rightarrow A = \frac{E_J - E_{J-1}}{hcJ}$$

$$J = 2: \quad A = \frac{E_2 - E_1}{2hc} = -79.13 \text{ cm}^{-1}$$

$$J = 1: \quad A = \frac{E_1 - E_0}{hc} = -68.73 \text{ cm}^{-1}$$

12.53

$E_4 = 0$ cm^{-1}
$E_3 = 415.932$ cm^{-1}
$E_2 = 704.003$ cm^{-1}
$E_1 = 888.132$ cm^{-1}
$E_0 = 978.076$ cm^{-1}

$$\frac{E_J - E_{J-1}}{hc} = \frac{1}{2}A\big(J(J+1) - L(L+1) - S(S+1)\big) - \frac{1}{2}A\big((J-1)J - L(L+1) - S(S+1)\big) = \frac{A}{2}\Big[J^2 + J - (J^2 - J)\Big] =$$

$$\Rightarrow A = \frac{E_J - E_{J-1}}{hcJ}$$

$J = 4$: $A = \dfrac{E_4 - E_3}{4hc} = -103.98$ cm^{-1}

$J = 3$: $A = \dfrac{E_3 - E_2}{3hc} = -96.02$ cm^{-1}

$J = 2$: $A = \dfrac{E_2 - E_1}{2hc} = -92.06$ cm^{-1}

$J = 1$: $A = \dfrac{E_1 - E_0}{hc} = -89.94$ cm^{-1}

SECTION 12.8

12.55

The ground state electronic configuration for K is $[Ar]4s^1$. This gives a term symbol of 4^2S. The selection rules for allowed transitions are $\Delta S = 0$ and $\Delta L = \pm 1$. This implies the excited state is 4^2P, which is separated by spin-orbit coupling into $4^2P_{3/2}$ and $4^2P_{1/2}$.

$\nu_{3/2} = 393.366$ nm
$\nu_{1/2} = 396.847$ nm

$$\frac{E_J - E_{J-1}}{hc} = \frac{1}{2}A\big(J(J+1) - L(L+1) - S(S+1)\big) - \frac{1}{2}A\big((J-1)J - L(L+1) - S(S+1)\big) = \frac{A}{2}\Big[J^2 + J - (J^2 - J)\Big] = AJ$$

$$\Rightarrow \quad A = \frac{E_J - E_{J-1}}{hcJ} = \frac{\frac{1}{\nu_J} - \frac{1}{\nu_{J-1}}}{J}$$

$$J = 3/2: \quad A = \frac{\frac{1}{\nu_{3/2}} - \frac{1}{\nu_{1/2}}}{\frac{3}{2}} = 148.7 \text{ cm}^{-1}$$

12.57

The ground state has energy levels $3^2P_{1/2}$ (1) and $3^2P_{3/2}$ (2)

The first excited state has energy level $4^2S_{1/2}$ (3)

The second excited state has energy levels $3^2D_{3/2}$ (4) and $3^2D_{5/2}$ (5)

line	transition
ν_1	(1) to (3)
ν_2	(2) to (3)
ν_3	(1) to (4)
ν_4	(2) to (4)
ν_5	(2) to (5)

$$\Delta E(^2P) = \nu_2 - \nu_1 = 112.1 \text{ cm}^{-1} = A\left(\frac{3}{2}\right) \Rightarrow A = 74.7 \text{ cm}^{-1}$$

SECTION 12.9

12.59

$$g_L = 1 + \frac{J(J+1) + S(S+1) - L(L+1)}{2J(J+1)}$$

$^4S_{3/2}$ $J = 3/2$, $S = 3/2$, $L = 0$ $g_L = 1 + \dfrac{\frac{3}{2}(\frac{3}{2}+1) + \frac{3}{2}(\frac{3}{2}+1) - 0(0+1)}{2\frac{3}{2}(\frac{3}{2}+1)} = 1 + \dfrac{2\left(\frac{15}{4}\right)}{\frac{15}{2}} = 1 + 1 = 2$

$^4P_{5/2}$ $J = 5/2$, $S = 3/2$, $L = 1$ $g_L = 1 + 6/10 = 1.6$

$^4P_{3/2}$ $J = 3/2$, $S = 3/2$, $L = 1$ $g_L = 1 + 22/30 = 1.733$

$^4P_{1/2}$ $J = 1/2$, $S = 3/2$, $L = 1$ $g_L = 1 + 5/3 = 2.667$

12.61

$^3P_2$ splits into 5 lines (M_j = -2 to 2); $^3S_1$ splits into 3 lines (M_j = -1 to 1)

Parallel - $\Delta M_j = \pm 1$ 6 lines

Perpendicular - $\Delta M_j = \pm 1, 0$ 9 lines

12.63

$$g_L = 1 + \frac{J(J+1) + S(S+1) - L(L+1)}{2J(J+1)}$$

$^3H_4$ $J = 4, S = 1, L = 5$ $g_L = 1 - 1/5 = 0.8$

$$|\mu_J| = g_L \sqrt{J(J+1)} \mu_B = 3.58 \mu_B$$

The lower bound for the magnetic moment would be the spin-only moment

$$|\mu_S| = 2\sqrt{S(S+1)} \mu_B = 2.83 \mu_B$$

Therefore, the range of values would be 2.83 μ_B to 3.58 μ_B

12.65

$g = 2$
$B_o = 1.4$ Tesla

$$\nu = \frac{g\mu_B B_o}{h} = 39.2 \times 10^9 \text{ Hz}$$

12.67

In the absence of an electric field, there will be a single line for the transition.

In the presence of the field, the $^1P_1$ line will be split into two levels, one for $M_j = 0$ and one for $M_j = \pm 1$, since $(\pm 1)^2 = 1$. This will lead to two lines in the transition

12.69

$g_N = 5.585486$
$\mu_N = 5.0508 \times 10^{-27}$ J / Tesla
$\nu = 500 \times 10^6$ Hz

$$\nu = \frac{g_N \mu_N B_o}{h} \quad \Rightarrow \quad B_o = \frac{h\nu}{g_N \mu_N} = 11.74 \text{ Tesla}$$

SECTION 12.10

12.71

$$\frac{I(P_{3/2})}{I(P_{1/2})} = \frac{g_{3/2}}{g_{1/2}} \frac{\exp\left[\dfrac{-E_{3/2}}{kT}\right]}{\exp\left[\dfrac{-E_{1/2}}{kT}\right]} = \frac{g_{3/2}}{g_{1/2}} \exp\left[\frac{\Delta E}{kT}\right]$$

$g_J = 2J + 1$

The relative intensity will be less for the 4d anf 4f lines because the energy difference is less, meaning the exponential is closer to 1, and the ratio of degeneracies is also closer to 1 for larger J.

The high energy peak is due to $4d_{3/2}$ and the low energy peak is due to $4d_{5/2}$.

SECTION 12.11

12.73

^{23}Na = [Ne] $3s^1$ L = 0,   S = 1/2

I = 3/2

T = I + S, ..., |I-S| = 2, 1

12.75

$$H\psi = \left\{-g\mu_B B_o S_z + AhI_z S_z\right\}\psi = \left(-g\mu_B B_o m_s + Ahm_I m_s\right)\psi = E\psi$$
$$\Rightarrow E = -g\mu_B B_o m_s + Ahm_I m_s$$

There will be two transitions, $m_s = \pm 1/2$ when $m_I = 1/2$ and $m_s = \pm 1/2$ when $m_I = -1/2$.

$$\nu_1 = \frac{\frac{1}{2}g\mu_B B_o + \frac{1}{4}Ah - \left(-\frac{1}{2}g\mu_B B_o - \frac{1}{4}Ah\right)}{h} = \frac{g\mu_B B_o}{h} + \frac{1}{2}A$$

$$\nu_2 = \frac{\frac{1}{2}g\mu_B B_o - \frac{1}{4}Ah - \left(-\frac{1}{2}g\mu_B B_o + \frac{1}{4}Ah\right)}{h} = \frac{g\mu_B B_o}{h} - \frac{1}{2}A$$

$$\Delta\nu = \nu_1 - \nu_2 = A$$

CHAPTER 13 *Diatomic Molecules*

SECTION 13.2

13.1

$$\mu = \frac{m_1 m_2}{m_1 + m_2}$$

$\mu^{35\text{-}35} = 2.903359 \times 10^{-23}$ g
$\mu^{35\text{-}37} = 2.983962 \times 10^{-23}$ g
$\mu^{35\text{-}35} = 3.069168 \times 10^{-23}$ g

$\omega_e^{35\text{-}35} = 564.9$ cm^{-1}

$$\omega_e^{35\text{-}37} = \omega_e^{35\text{-}35} \sqrt{\frac{\mu^{35\text{-}35}}{\mu^{35\text{-}37}}} = 557.2 \text{ cm}^{-1}$$

$$\omega_e^{37\text{-}37} = \omega_e^{35\text{-}35} \sqrt{\frac{\mu^{35\text{-}35}}{\mu^{37\text{-}37}}} = 549.4 \text{ cm}^{-1}$$

13.3

$$\mu = \frac{m_1 m_2}{m_1 + m_2}$$

$$k_e = \left(2\pi c \omega_e\right)^2 \mu = \left(2\pi c \omega_e\right)^2 \frac{m_C m_O}{m_C + m_O} = 1902.5 \text{ kg/s}^2$$

$$m = 1 \text{ lb} = 0.4536 \text{ kg}$$

$$mg = k_e x \implies x = \frac{mg}{k_e} = 0.234 \text{ cm}$$

13.5

$$\Delta_f H_0 = \frac{1}{2} D_o(Cl_2) + \frac{1}{2} D_o(H_2) - D_o(HCl) = -0.9599 \text{ eV} = -92.61 \text{ kJ} / \text{mol}$$

13.7

Note: All unprimed varibles refer to H_2; All primed variables refer to D_2; All double-primed variables refer to HD.

$$\omega_e{}' = \omega_e \sqrt{\frac{\mu}{\mu'}} = 3109.1 \text{ cm}^{-1}$$

$$\omega_e{}'' = \omega_e \sqrt{\frac{\mu}{\mu''}} = 3806.8 \text{ cm}^{-1}$$

$$D_e = D_o + \frac{1}{2}\omega_e = 38309.4 \text{ cm}^{-1}$$

$$D_o{}' = D_e - \frac{1}{2}\omega_e{}' = 36754.9 \text{ cm}^{-1} = 4.5570 \text{ eV}$$

$$D_o{}'' = D_e - \frac{1}{2}\omega_e{}'' = 36406.0 \text{ cm}^{-1} = 4.5138 \text{ eV}$$

$$H_2 + D_2 \rightarrow 2HD$$

$$\Delta H_0 = D_o + D_o{}' - 2D_o{}'' = 6.7 \times 10^{-3} \text{ eV} = 647 \text{ J} / \text{mol}$$

13.9
───

$$D_e = D_o + \frac{1}{2}\omega_e - \frac{1}{4}\omega_e\chi_e = 90674 \ cm^{-1} = 11.242 \ eV$$

$$\mu = \frac{m_1 m_2}{m_1 + m_2}$$

$$k = (2\pi c\omega_e)^2\mu \implies \sqrt{k} = 2\pi c\omega_e\sqrt{\mu} = 2\pi c\omega_e\sqrt{\frac{m_C m_O}{m_C + m_O}} = 1379.34 \ \sqrt{g}/s$$

$$\beta = \sqrt{\frac{k}{2D_e}} = 22.9816 \ nm^{-1}$$
$$R_e = 0.1128 \ nm$$

$$E = D_e\left(1 - \exp\left[-\beta(R - R_e)\right]\right)^2$$

R (nm)	E (eV)
0.06	62.88 eV
0.11	0.0497
0.2	8.42
0.3	10.94
5	11.24

13.11
───

$$n_{max} = \frac{\omega_e}{2\omega_e\chi_e} = 81.61 \approx 81$$

$$D_e = \frac{\omega_e^2}{4\omega_e\chi_e} = 96288.04 \ cm^{-1} = 11.94 \ eV$$

13.13

$$\tilde{\nu}_0 = \frac{E_1 - E_0}{hc} = \omega_e - 2\omega_e\chi_e = 3958.38 \text{ cm}^{-1} \text{ (fundamental)}$$

$$\tilde{\nu}_1 = \frac{E_2 - E_0}{hc} = 2\omega_e - 6\omega_e\chi_e = 7736.63 \text{ cm}^{-1} \text{ (first overtone)}$$

$$\tilde{\nu}_2 = \frac{E_2 - E_1}{hc} = \omega_e - 4\omega_e\chi_e = 3778.24 \text{ cm}^{-1} \text{ (hot band)}$$

13.15

$$\tilde{\nu}_0 = \frac{E_1 - E_0}{hc} = \omega_e - 2\omega_e\chi_e = 381.25 \text{ cm}^{-1} \text{ (fundamental)}$$

$$\tilde{\nu}_2 = \frac{E_2 - E_1}{hc} = \omega_e - 4\omega_e\chi_e = 378.32 \text{ cm}^{-1} \text{ (hot band)}$$

$$T = 300 \text{ K}$$

$$\text{Re lative Intensity} \equiv \frac{N_1}{N_0} = \exp\left[-\frac{hc\omega_e}{kT}\right] = 0.158$$

13.17

$\omega_e = 159.12 \text{ cm}^{-1}$

$\omega_e\chi_e = 0.725 \text{ cm}^{-1}$

$\omega_e\gamma_e = 0.0011 \text{ cm}^{-1}$

$$E_n = \omega_e(n+\frac{1}{2}) - \omega_e\chi_e(n+\frac{1}{2})^2 - \omega_e\gamma_e(n+\frac{1}{2})^3$$

$$E_n = E_{n-1} \Rightarrow \omega_e(n+\frac{1}{2}) - \omega_e\chi_e(n+\frac{1}{2})^2 - \omega_e\gamma_e(n+\frac{1}{2})^3 = \omega_e(n-\frac{1}{2}) - \omega_e\chi_e(n-\frac{1}{2})^2 - \omega_e\gamma_e(n-\frac{1}{2})^3$$

$$\Rightarrow (\frac{1}{4}\omega_e\gamma_e - \omega_e) + 2\omega_e\chi_e n + 3\omega_e\gamma_e n^2 = 0$$

Use the quadratic equation to solve for n.

$n = 90.9 \cong 90$

$\Rightarrow D_o = E_{90} - E_0 = 7567.7 \text{ cm}^{-1} = .9383 \text{ eV}$

SECTION 13.3

13.19

$$\Delta E_J^n = E_{J+1}^n - E_J^n = (J+1)(J+2)B_n - (J+1)^2(J+2)^2 D_c - \left\{ (J)(J+1)B_n - (J)^2(J+1)^2 D_c \right\}$$

$$\Delta E_J^n = 2(J+1)B_n - 4(J+1)^3 D_c$$

$\omega_e = 1405.65 \text{ cm}^{-1}$

$\widetilde{B}_e = 7.5131 \text{ cm}^{-1}$

$\alpha_e = 0.2132 \text{ cm}^{-1}$

$D_c = \dfrac{4\widetilde{B}_e^3}{\omega_e^2} = 8.58545 \times 10^{-4} \text{ cm}^{-1}$

$B_0 = \widetilde{B}_e - \dfrac{1}{2}\alpha_e = 7.4065 \text{ cm}^{-1}$

$$\Delta E_J^0 = 2(J+1)B_0 - 4(J+1)^3 D_c$$

J	ΔE_J^0 (GHz)
0	443.98
1	887.33
2	1329.5
3	1769.7

13.21

$$\Delta E_J^n = E_{J+1}^n - E_J^n = (J+1)(J+2)B_n - (J+1)^2(J+2)^2 D_c - \left\{ (J)(J+1)B_n - (J)^2(J+1)^2 D_c \right\}$$

$$\Delta E_J^n = 2(J+1)B_n - 4(J+1)^3 D_c$$

$$\Delta E_1^0 = 4B_0 - 32D_c = 230.53797 \text{ GHz}$$

$$\Delta E_3^0 = 8B_0 - 256D_c = 461.0468 \text{ GHz}$$

$$8\Delta E_1^0 - \Delta E_3^0 = 24B_0 = 1383.2567 \text{ GHz} \implies B_0 = 57.6357 \text{ GHz}$$

$$2\Delta E_1^0 - \Delta E_3^0 = 192D_c = 0.02914 \text{ GHz} \implies D_c = 151.8 \text{ MHz}$$

13.23

$T = 300 \text{ K}$

$\mu = 1.656 \times 10^{-24} \text{ g}$

$R_e = 1.414 \times 10^{-8} \text{ cm}$

$$I = \mu R_e^2$$

$$\frac{N'}{N_0} = (2J'+1) \exp\left[-\frac{J'(J'+1)}{kT} \frac{h^2}{8\pi^2 \mu R_e^2} \right]$$

J'	N'/N_0
0	1
4	4.0
10	0.24
20	1.6×10^{-6}

13.25

Note: The unprimed variables refer to the 35 isotope of chlorine; the primed variables refer to the 37 isotope of chlorine

$B_e = 15.58369$ GHz

$B'_e = 15.18922$ GHz

$\mu = 2.044172 \times 10^{-23}$ g

$\mu' = 2.083803 \times 10^{-23}$ g

$I = \mu R_e^2$

$$B_e = \frac{h}{8\pi^2 I} = \frac{h}{8\pi^2 \mu R_e^2} \implies R_e = \left(\frac{h}{8\pi^2 \mu B_e}\right)^{1/2} = 0.1623078 \text{ nm}$$

$$B'_e = \frac{h}{8\pi^2 I'} = \frac{h}{8\pi^2 \mu' R_e'^2} \implies R'_e = \left(\frac{h}{8\pi^2 \mu' B'_e}\right)^{1/2} = 0.1628311 \text{ nm}$$

13.27

$\omega_e = 384.18 \text{ cm}^{-1}$
$\omega_e\chi_e = 1.465 \text{ cm}^{-1}$
$B_e = 0.1142 \text{ cm}^{-1}$
$\alpha_e = 0.00053 \text{ cm}^{-1}$
$D_c \cong 0 \text{ cm}^{-1}$

$$E_J^n = \omega_e(n+\tfrac{1}{2}) - \omega_e\chi_e(n+\tfrac{1}{2})^2 + (J)(J+1)\left(B_e - (n+\tfrac{1}{2})\alpha_e\right) - (J)^2(J+1)^2 D_c$$

$$R0 = E_1^1 - E_0^0 = \frac{3}{2}\omega_e - \frac{9}{4}\omega_e\chi_e + 2B_e - 3\alpha_e - 4D_c - \left(\frac{1}{2}\omega_e - \frac{1}{4}\omega_e\chi_e\right) = \omega_e - 2\omega_e\chi_e + 2B_e - 3\alpha_e = 381.48 \text{ cm}$$

$$P1 = E_0^1 - E_1^0 = \frac{3}{2}\omega_e - \frac{9}{4}\omega_e\chi_e - \left(\frac{1}{2}\omega_e - \frac{1}{4}\omega_e\chi_e + 2B_e - \alpha_e - 4D_c\right) = \omega_e - 2\omega_e\chi_e - 2B_e + \alpha_e = 381.02 \text{ cm}^{-1}$$

13.31

$J_{max} = 8$
$B_e \cong B_o = 1.9227 \text{ cm}^{-1}$

J_{max} is found by maximizing the expression for the relative population in the Jth rotational energy level, as given by Boltzmann's Law.

$$J_{max} = \frac{1}{2}\left(\left(\frac{2kT}{B_e ch}\right)^{1/2} - 1\right)$$

$$\Rightarrow \frac{2kT}{B_e ch} = (2J_{max} + 1)^2 \Rightarrow T = \frac{B_e ch}{2k}(2J_{max} + 1)^2 \cong 400 \text{ K}$$

13.33

$\omega_e = 2170.21$ cm^{-1}

$\omega_e \chi_e = 13.461$ cm^{-1}

$B_e = 1.9314$ cm^{-1}

$\alpha_e = 0.01748$ cm^{-1}

$$E_J^n = \omega_e(n+\tfrac{1}{2}) - \omega_e\chi_e(n+\tfrac{1}{2})^2 + (J)(J+1)\left(B_e - (n+\tfrac{1}{2})\alpha_e\right)$$

$$E_0^1 = E_J^0 \Rightarrow \tfrac{3}{2}\omega_e - \tfrac{9}{4}\omega_e\chi_e = \left(\tfrac{1}{2}\omega_e - \tfrac{1}{4}\omega_e\chi_e + J(J+1)B_e - \tfrac{1}{2}J(J+1)\alpha_e\right) \Rightarrow J(J+1) = \frac{\omega_e - 2\omega_e\chi_e}{B_e - \alpha_e} = 1114.75$$

$$\Rightarrow J = 33$$

13.35

$\omega_e = 498.8$ cm^{-1}

$$\frac{I}{I_o} = C_1 \exp\left[-\frac{n\omega_e hc}{kT}\right] \Rightarrow \ln\left(\frac{I}{I_o}\right) = \ln(C_1) - \frac{n\omega_e hc}{kT}$$

Plot $\ln\left(\dfrac{I}{I_o}\right)$ vs. n and do linear regression; slope $= -\dfrac{\omega_e hc}{kT}$

n	I/I$_o$ (mm) - from figure	ln (I/I$_o$)
0	27.6	3.32
1	15.2	2.72
2	9.4	2.24
3	6.1	1.81
4	4.9	1.59
5	3.0	1.10

slope = -0.4263

T = 1686 K $\cong$ 1700 $\pm$ 100 K (within standard error)

SECTION 13.4

13.37

$(2s\sigma_g)^1(2s\sigma_u^*)^1$ $\Lambda = 0$ $S = 0$ or 1 $^1\Sigma_u^+$ or $^3\Sigma_u^+$

$(2p\sigma_g)^2(2p\pi_u)^2$ first term gives $\Lambda = 0$ $S = 0$
second term gives $\Lambda = 0$ $S = 0$ or 1 **or**
$\Lambda = \pm 2$ $S = 0$

Total symbols are $^1\Sigma_g^+$ or $^3\Sigma_g^-$ or $^1\Delta_g$

$(2p\sigma_g)^1(2p\pi_u)^3$ $\Lambda = \pm 2$ $S = 0$ or 1 $^1\Pi_u$ or $^3\Pi_u$

$(2p\sigma_g)^1(2p\pi_u)^1(2p\pi_g^*)^1$ $\Lambda = \pm 2$ $S = 0$ or 1 **or**
$\Lambda = 0$ $S = 0$ or 1

Total symbols are $^1\Sigma_u^{+/-}$ or $^3\Sigma_u^{+/-}$ or $^1\Delta_u$ or $^3\Delta_u$

Student's Solutions Manual

SECTION 13.5

13.39

$$(1s\sigma_g)^2 \rightarrow (1s\sigma_g)^1(2p\pi_u)^1 \qquad \Lambda = \pm 1 \quad S = 0 \text{ or } 1$$

term symbols are $^1\Pi_u$ or $^3\Pi_u$

13.41

$^1\Sigma^+$:	$1 \times 1 = 1$	$^1\Sigma^-$:	$1 \times 1 = 1$
$^2\Pi$:	$2 \times 2 = 4$	$^2\Delta$:	$2 \times 2 = 4$
$^3\Sigma_g$:	$3 \times 1 = 3$	$^3\Sigma_u$:	$3 \times 1 = 3$

13.43

O_2^+ $\quad (1s\sigma_g)^2(1s\sigma_u{}^*)^2(2s\sigma_g)^2(2s\sigma_u{}^*)^2(2p\pi_u)^4(2p\sigma_g)^2(2p\pi_g{}^*)^1$ $\quad$ bond order = 5/2

O_2 $\quad (1s\sigma_g)^2(1s\sigma_u{}^*)^2(2s\sigma_g)^2(2s\sigma_u{}^*)^2(2p\pi_u)^4(2p\sigma_g)^2(2p\pi_g{}^*)^2$ $\quad$ bond order = 2

O_2^- $\quad (1s\sigma_g)^2(1s\sigma_u{}^*)^2(2s\sigma_g)^2(2s\sigma_u{}^*)^2(2p\pi_u)^4(2p\sigma_g)^2(2p\pi_g{}^*)^3$ $\quad$ bond order = 3/2

As the bond order decreases, the strength of the bond also decreases. This is consistent with the fact that the bond length increases as the bond strength decreases (as shown by the data in the problem).

13.45

$X^1\Sigma_g$ $\quad (1s\sigma_g)^2(1s\sigma_u{}^*)^2(2s\sigma_g)^2(2s\sigma_u{}^*)^2(2p\pi_u)^4$

$^3\Pi_u$ $\quad (1s\sigma_g)^2(1s\sigma_u{}^*)^2(2s\sigma_g)^2(2s\sigma_u{}^*)^2(2p\pi_u)^3(2p\sigma_g)^1$

$^3\Sigma_g$ $\quad (1s\sigma_g)^2(1s\sigma_u{}^*)^2(2s\sigma_g)^2(2s\sigma_u{}^*)^2(2p\pi_u)^2(2p\sigma_g)^2$

13.47

BO is a near-homonuclear molecule with 13 electrons

$(1s\sigma)^2(1s\sigma^*)^2(2s\sigma)^2(2s\sigma^*)^2(2p\pi)^4(2p\sigma)^1$ $^2\Sigma^+$ Bond order = 5/2

13.49

CH	$^2\Pi$	7 electrons	$(1\sigma)^2\,(2\sigma)^2(3\sigma)^2(1\pi)^1$
NH	$^3\Sigma$	8 electrons	$(1\sigma)^2\,(2\sigma)^2(3\sigma)^2(1\pi)^2$
OH	$^2\Pi$	9 electrons	$(1\sigma)^2\,(2\sigma)^2(3\sigma)^2(1\pi)^3$
HF	$^1\Sigma$	10 electrons	$(1\sigma)^2\,(2\sigma)^2(3\sigma)^2(1\pi)^4$

13.51

O_2^- $(1s\sigma_g)^2(1s\sigma_u^*)^2(2s\sigma_g)^2(2s\sigma_u^*)^2(2p\pi_u)^4(2p\sigma_g)^2(2p\pi_g^*)^3$ $^2\Pi_g$

13.53

The selection rules are: $\Delta\Lambda = 0, \pm1$
                                  $\Delta S = 0$
                                  + does not go to -

The allowed transitions are: $X^3\Sigma^- \leftrightarrow A^3\Pi$
                                  $b^1\Sigma^+ \leftrightarrow c^1\Sigma^+$

13.55

The selection rules are:

$\Delta \Lambda = 0, \pm 1$

$\Delta S = 0$

$g \leftrightarrow u$

State	Λ	S
$^3\Sigma_g$	0	1
$^1\Delta_g$	2	0

13.57

The selection rules are:

$\Delta \Lambda = 0, \pm 1$

$\Delta S = 0$

$g \leftrightarrow u$

vertical mirror symmetry (+/-) does not change

N_2 ground state is $^1\Sigma_g^+$ Allowed transiotions are to $^1\Sigma_u^+$ and $^1\Pi_u$

O_2 ground state is $^3\Sigma_g^-$ Allowed transiotions are to $^3\Sigma_u^-$ and $^3\Pi_u$

13.59

$$v_{v'-0} = v_{00} + v'(\omega'_e - \omega'_e\chi'_e) - (v')^2\omega'_e\chi'_e$$

$$\Rightarrow \frac{v_{v'-0} - v_{00}}{v'} = (\omega'_e - \omega'_e\chi'_e) - (v')\omega'_e\chi'_e$$

$$v_{00} = 39,699.1 \text{ cm}^{-1}$$

Plotting $\frac{v_{v'-0} - v_{00}}{v'}$ vs. v' and using linear regression gives

(slope) $-7.280 = -\omega'_e\chi'_e \Rightarrow \omega'_e\chi'_e = 7.3 \text{ cm}^{-1}$

(int ercept) $1094.84 = (\omega'_e - \omega'_e\chi'_e) \Rightarrow \omega'_e = 1102.1 \text{ cm}^{-1}$

13.61

a)   Plotting the data and doing linear regression gives an intercept of 57,214 cm^{-1}. This value is the dissociation energy of O_2.

b) D_0 equals dissociation energy minus atomic energy

$$D_0 = 57,214 \text{ cm}^{-1} - 15,868 \text{ cm}^{-1} = 41,346 \text{ cm}^{-1} = 5.13 \text{ eV}$$

13.63

N_2^+ $\quad (1s\sigma_g)^2(1s\sigma_u*)^2(2s\sigma_g)^2(2s\sigma_u*)^2(2p\pi_u)^3(2p\sigma_g)^2$ $\qquad ^2\Pi_u$

13.65

$Rb \quad \rightarrow \quad Rb^+ + e^- \quad \Delta E = IP = 4.176 \text{ eV}$

$e^- + Br \quad \rightarrow \quad Br^- \qquad \Delta E = -EA = -3.364 \text{ eV}$

$Rb + Br \quad \rightarrow \quad Rb^+ + Br^- \quad \Delta E = 0.812 \text{ eV} = 78.3 \text{ kJ} / \text{mol}$

13.67

$\omega_e = 364.6 \text{ cm}^{-1}$
$R_e = 2.3609 \times 10^{-8} \text{ cm}$
$IP-EA = 1.523 \text{ eV}$
$\mu = 2.30328 \times 10^{-23} \text{ g}$

For equations, see example 13.12

$$B = 2 + \frac{k_e R_e^3 (4\pi\varepsilon_o)}{e^2} = 8.1966$$

$$k_e = \mu(2\pi c\omega_e)^2 = 108636.1 \text{ g} / s^2$$

$$A = \frac{e^2 \exp[B]}{(4\pi\varepsilon_o)BR_e} = 4.32556 \times 10^{-16} \text{ J} = 2700.04 \text{ eV}$$

$$D_e = -E_e(R_e) = 3.83 \text{ eV}$$

13.69

$\mu_0 = 6.32$ debye

$R_e = 1.5639 \times 10^{-10}$ m

$\%$ ionic $= \dfrac{100\mu_0}{eR_e} = 84.14\%$

13.71

$P = \dfrac{\varepsilon-1}{\varepsilon+2}\dfrac{M}{\rho} = \dfrac{4\pi L}{3}\left(\alpha + \dfrac{\mu_o^2}{3kT}\right)$ (equation 13.55)

Plotting P vs. $\dfrac{1}{T}$ and using linear regression gives

(slope) $\dfrac{4\pi L}{3}\dfrac{\mu_o^2}{3k} = 7060.6$ cm^3 K $\Rightarrow \mu_o = 1.08$ D   and

(y - intercept) $\dfrac{4\pi L}{3}\alpha = 25.547$ cm$^3 \Rightarrow \alpha = 1.013 \times 10^{-23}$ cm^3

13.73
___

$\mu = 2.3033 \times 10^{-26}$ kg

$R_e = 2.3609 \times 10^{-10}$ m

$I = \mu R_e^2 = 1.2838 \times 10^{-45}$ kg·m^2

$\varepsilon = 250{,}000 \dfrac{V}{m}$

$\mu_0 = 9.00$ Debye

$$\Delta E(M = 0 \text{ to } 1) = \frac{3I\mu_0^2\varepsilon^2}{10\hbar^2} = 1.9506 \times 10^{-24} \text{ J}$$

$$\nu = \frac{\Delta E}{h} = 2.94 \text{ GHz}$$

13.75
___

a)   non-zero   b)   0   c)   non-zero   d)   0

b) and d) are odd functions

CHAPTER 14 *Polyatomic Molecules*

SECTION 14.2

14.1
___

C_2H_2: $D_{\infty h}$ C_2HCl: $C_{\infty v}$

HCN: $C_{\infty v}$ CCl$_4$: T_d

14.3
___

a) C_{2v} C_{2h} C_{2v}

b) C_{2v} C_{2v} D_{2h}

c) C_{2v} C_{2v} C_s

14.5
___

The S_8 molecule has the following symmetry elements:

> C_4 principal axis
>   4 C_2 axes perpendicular to C_4
>   no σ_h
>   4 σ_d and S_8 (the symmetry element, not the actual molecule) parallel to C_4

Therefore, the point group is D_{4d}.

14.7
_____

a) D_{3h} b) D_{3d} c) C_2 d) C_{2h}

14.9
_____

Hexamethylbenzene (see the first figure in problem 14.11) has the following symmetry elements:

C_3 principal axis
3 C_2 axes perpendicular to C_3
no σ_h
3 σ_d and S_6 parallel to C_3

Therefore, the point group is D_{3d}.

14.11

a)

Figure 1 has the following symmetry elements:

C_3 principal axis
3 C_2 axes perpendicular to C_3
no σ_h
3 σ_d and S_6 parallel to C_3

Therefore, the point group is D_{3d}.

Figure 2 has the following symmetry elements:

C_2 principal axis
2 σ_v parallel to C_2

Therefore, the point group is C_{2v}.

b)

The cyclohexane "chair" is the same as figure 1 (D_{3d}), and the cyclohexane "boat" is the same as figure 2 (C_{2v}).

SECTION 14.3 and 14.4

14.13

From the C_{3v} character table, $A_2 \otimes E$ is

A_2	1	1	-1
E	2	-1	0
	2	-1	0

This direct product is E.

From the D_{3h} character table, $A_1'' \otimes A_2''$ is

A_1''	1	1	1	-1	-1	-1
A_2''	1	1	-1	-1	-1	1
	1	1	-1	1	1	-1

This direct product is A_2'.

From the T_d character table, $A_2 \otimes T_1$ is

A_2	1	1	1	-1	-1
T_1	3	0	-1	1	-1
	3	0	-1	-1	1

This direct product is T_2.

14.15

For the integrals to be non-zero, the integrand must by symmetric, i.e.

$\Gamma_\psi \otimes (x, y, z) \otimes \Gamma_{\psi^*} = \Gamma_s$ (Γ_s means all 1's in the representation)

The point groups where x, y, or z fall in Γ_s are C_s and C_{nv}.

SECTION 14.5

14.17

a) $C(gr) + 2 H_2 + 1/2 O_2 \rightarrow CH_3OH$

$\Delta H = \Sigma(\text{bonds broken}) - \Sigma(\text{bonds made})$

$C(gr) \rightarrow C(g)$	+ 717 kJ/mol
2 H-H	+ 2 (435 kJ/mol)
1/2 O=O	+.5 (492 kJ/mol)

3 C-H	- 3 (415 kJ/mol)
1 C-O	- 350 kJ/mol
1 O-H	- 464 kJ/mol

$\Delta H = 1833 - 2059 = -226$ kJ/mol

b) $C(gr) + Cl_2 + 1/2 O_2 \rightarrow COCl_2$

$\Delta H = 1202 - 1376 = -174$ kJ/mol

c) $2 C(gr) + 2 H_2 \rightarrow C_2H_4$

$\Delta H = 2304 - 2275 = 29$ kJ/mol

SECTION 14.6, 14.7, and 14.8

14.19

The point group for this hypothetical molecule is C_{2h}.

function	E	C_2	i	σ_h	symmetry
X_1	1	1	1	1	a_g
X_2	1	-1	-1	1	b_u
X_3	1	1	-1	-1	a_u
X_4	1	-1	1	-1	b_g

14.21

The point group for butadiene is C_{2h}.

$$X_1 = (C_A p_z) + (C_B p_z) + (C_C p_z) + (C_D p_z)$$
$$X_2 = (C_A p_z) + (C_B p_z) - (C_C p_z) - (C_D p_z)$$
$$X_3 = (C_A p_z) - (C_B p_z) - (C_C p_z) + (C_D p_z)$$
$$X_4 = (C_A p_z) - (C_B p_z) + (C_C p_z) - (C_D p_z)$$

function	E	C_2	i	σ_h	symmetry	nodes
X_1	1	1	-1	-1	$1a_u$	0
X_2	1	-1	1	-1	$1b_g$	1
X_3	1	1	-1	-1	$2a_u$	2
X_4	1	-1	1	-1	$2b_g$	3

		$1a_u$	$1b_g$	$2a_u$	$2b_g$	State
	4	↑↓	↑		↓	^1A_g
	3	↑↓	↑		↑	^3A_g
E ↑	2	↑↓	↑	↓		^1B_u
	1	↑↓	↑	↑		^3B_u
	gr	↑↓	↑↓			^1A_g

The total state comes from the direct product of the configuration for each electron. For example, the state of the ground level is $a_u \otimes a_u \otimes b_g \otimes b_g = a_g$. $S = 0$ implies it is a singlet state. Thus, the total state is ^1A_g.

The selection rules for electric dipole transitions require $\Delta S = 0$ and $g \longleftrightarrow u$, so the only allowed transition is ^1A_g to ^1B_u.

14.23

BH$_3$ has eight electrons

BH$_3$	$(1a_1')^2(2a_1')^2(1e')^4$	term symbol $^1A_1'$
BH$_3^+$	$(1a_1')^2(2a_1')^2(1e')^3$	term symbol $^2E'$
BH$_3^-$	$(1a_1')^2(2a_1')^2(1e')^4(1a_2'')^1$	term symbol $^2A_2''$

14.25

NH$_2^+$	$(1a_1)^2(2a_1)^2(1b_2)^2(3a_1)^2$	term symbol $^1A_1$
NH$_2^-$	$(1a_1)^2(2a_1)^2(1b_2)^2(3a_1)^2(1b_1)^2$	term symbol $^1A_1$
BH$_2^-$	$(1a_1)^2(2a_1)^2(1b_2)^2(3a_1)^2$	term symbol $^1A_1$
CH$_2^-$	$(1a_1)^2(2a_1)^2(1b_2)^2(3a_1)^2(1b_1)^1$	term symbol $^2B_1$

14.27

In the first excited state, 1 electron is promoted from e_{1g} to b_{2g}

e_{1g}	2	1	-1	-2	0	0	2	1	-1	-2	0
b_{2g}	1	-1	1	-1	-1	1	1	-1	1	-1	-1
e_{2g}	2	-1	-1	2	0	0	2	-1	-1	2	0

The final state is $^1E_{2g}$ or $^3E_{2g}$

SECTION 14.9 and 14.10

14.29

| linear molecules: | 3N-5 normal modes |
| non-linear molecules | 3N-6 normal modes |

H_2O_2: $3(4) - 6 = 6$ vibrational modes

C_2H_2: $3(4) - 5 = 7$ vibrational modes

C_2H_4: $3(6) - 6 = 12$ vibrational modes

C_2H_3Cl: $3(6) - 6 = 12$ vibrational modes

C_6H_5Cl: $3(12) - 6 = 30$ vibrational modes

14.31

$$\omega_{e1} = 830 \text{ cm}^{-1}$$
$$\omega_{e2} = 490 \text{ cm}^{-1}$$
$$\omega_{e3} = 1110 \text{ cm}^{-1}$$

$$\nu = \frac{E_{vib}}{hc} = \sum_{1}^{3} (v_i + 1/2)\omega_{ei}$$

(0, 0, 0)	$\nu = 1215 \text{ cm}^{-1}$
(0, 1, 0)	$\nu = 1705 \text{ cm}^{-1}$
(1, 0, 0)	$\nu = 2045 \text{ cm}^{-1}$
(0, 2, 0)	$\nu = 2195 \text{ cm}^{-1}$
(0, 0, 1)	$\nu = 2325 \text{ cm}^{-1}$
(1, 1, 0)	$\nu = 2535 \text{ cm}^{-1}$
(0, 3, 0)	$\nu = 2685 \text{ cm}^{-1}$
(0, 1, 1)	$\nu = 2815 \text{ cm}^{-1}$
(2, 0, 0)	$\nu = 2875 \text{ cm}^{-1}$

14.33

Point group C_{2v}
ground state is given by representation a_1

(0, 0, 0, 1, 0, 0) is b_1 $a_1 \otimes b_1 = b_1$ B_1 is an allowed transition

(1, 0, 0, 0, 0, 1) is $a_1 \otimes b_2 = b_2$ $a_1 \otimes b_2 = b_2$ B_2 is an allowed transition

(0, 0, 0, 0, 1, 1) is $b_1 \otimes b_2 = a_2$ $a_1 \otimes a_2 = a_2$ A_2 is NOT an allowed transition

14.35

A mode will be IR active if the dipole moment changes. From Figure 14.12, modes Q_3 and Q_5 show a change in dipole moment, so they are IR active.

14.37

Point group D_{2h}

IR active - b_{1y}, $2b_{2u}$, and $2b_{3u}$

Raman active - $3a_g$, $2b_{1g}$, and $1b_{2g}$

a_u is inactive

$b_{1u} \otimes b_{1u} = a_g$, so this overtone is not IR active

$a_u \otimes b_{1g} = B_{1u}$, so this overtone is IR active

14.39

$$\tilde{v}_i = \frac{1}{\lambda} = \frac{1}{600 \text{ nm}} = 16667 \text{ cm}^{-1}$$

Scattered frequency will equal the incident frequency minus the fundamental

$$\lambda_1 = \frac{1}{\tilde{v}_i - \omega_1} = \frac{1}{(16667 - 1151)\text{cm}^{-1}} = 645 \text{ nm}$$

$$\lambda_2 = \frac{1}{\tilde{v}_i - \omega_2} = \frac{1}{(16667 - 524)\text{cm}^{-1}} = 620 \text{ nm}$$

$$\lambda_3 = 654 \text{ nm}$$

SECTION 14.11

14.41

Axis 1 is along the C=C double bond.

Axis 2 is perpendicular to the plane of the molecule through the center of the C=C bond.

Axis 3 is perpendicular to the C=C bond in the plane of the molecule and through the center of the C=C bond.

14.43
___

Case 1:   Axis through 1 corner and center of mass

$$I = m(0)^2 + 2m\left(\frac{a}{2}\right)^2 = \frac{1}{2}ma^2$$

Case 2:   Axis through center of mass and parallel to one side

let x = distance from corner to axis

let $.5x$ = distance from parallel side to axis

$$x^2 = \left(\frac{x}{2}\right)^2 + \left(\frac{a}{2}\right)^2 \Rightarrow \frac{3}{4}x^2 = \frac{1}{4}a^2 \Rightarrow x^2 = \frac{1}{3}a^2$$

$$I = mx^2 + 2m\left(\frac{x}{2}\right)^2 = m\left(\frac{1}{3}a^2\right) + 2m\left(\frac{1}{12}a^2\right) = \left(\frac{1}{3}+\frac{1}{6}\right)ma^2 = \frac{1}{2}ma^2$$

14.45
___

Axis a is perpendicular to the plane of the molecule and through the center of the molecule.

Axes b,c are in the plane of the molecule. These moments are the same - see problem 14.43

$I_a = 3\ m_y\ (R)^2 = 3m_yR^2$

$I_b = I_c = 2\ m_y\ (x)^2 = {}^3/_2\ m_y\ R^2$ - To calculate the value of x, consider a right triangle with legs x and .5R and hypotenuse R.

None of these moments can be measured with microwave spectroscopy because the molecule lacks a permanent dipole moment.

14.47

$$R_{12} = 162.9 \times 10^{-12} \text{ m}$$
$$R_{23} = 116.3 \times 10^{-12} \text{ m}$$

m_1 = mass of chlorine atom
m_2 = mass of carbon atom
m_3 = mass of nitrogen atom

$$I = \frac{(m_1 m_2 + m_1 m_3)R_{12}^2 + 2m_1 m_3 R_{12} R_{23} + (m_1 m_3 + m_2 m_3)R_{23}^2}{m_1 + m_2 + m_3} = 1.405 \times 10^{-45} \text{ kg} \cdot \text{m}^2$$

14.49
___

$R_{12} = R_{HC}$

$R_{23} = R_{CN}$

molecule 1: $^1H^{12}C^{14}N$

$B = 44.3160$ GHz

$I = \dfrac{h}{8\pi^2 B} = 0.11404$ amu·nm^2

$m_1 = $ mass of hydrogen atom

$m_2 = $ mass of carbon-12 atom

$m_3 = $ mass of nitrogen atom

$$I = \frac{\left(m_1 m_2 + m_1 m_3\right) R_{12}^2 + 2 m_1 m_3 R_{12} R_{23} + \left(m_1 m_3 + m_2 m_3\right) R_{23}^2}{m_1 + m_2 + m_3} = 0.11404 \text{ amu·nm}^2$$

molecule 2: $^1H^{13}C^{14}N$

$B' = 43.1698$ GHz

$I' = \dfrac{h}{8\pi^2 B'} = 0.11707$ amu·nm^2

$m_1' = $ mass of hydrogen atom

$m_2' = $ mass of carbon-13 atom

$m_3' = $ mass of nitrogen atom

$$I' = \frac{\left(m_1 m_2 + m_1 m_3\right)' R_{12}^2 + 2 m_1' m_3' R_{12} R_{23} + \left(m_1 m_3 + m_2 m_3\right)' R_{23}^2}{\left(m_1 + m_2 + m_3\right)'} = 0.11707 \text{ amu·nm}^2$$

Solve the equations I and I' simultaneously using a computer to get R_{12} and R_{23}

$R_{12} = 0.1067$ nm $R_{23} = 0.1156$ nm

14.51

$R_{12} = R_{OC}$

$R_{23} = R_{CS}$

molecule 1: $^{16}O^{12}C^{32}S$

$B = 6.08149$ GHz

$I = \dfrac{h}{8\pi^2 B} = 0.83101 \ amu \cdot nm^2$

m_1 = mass of oxygen - 16 atom

m_2 = mass of carbon atom

m_3 = mass of sulfur atom

$$I = \frac{\left(m_1 m_2 + m_1 m_3\right)R_{12}^2 + 2m_1 m_3 R_{12} R_{23} + \left(m_1 m_3 + m_2 m_3\right)R_{23}^2}{m_1 + m_2 + m_3} = 0.83101 \ amu \cdot nm^2$$

molecule 2: $^{18}O^{12}C^{32}S$

$B' = 5.70483$ GHz

$I' = \dfrac{h}{8\pi^2 B'} = 0.88588 \ amu \cdot nm^2$

m_1' = mass of oxygen - 18 atom

m_2' = mass of carbon atom

m_3' = mass of sulfur atom

$$I' = \frac{\left(m_1 m_2 + m_1 m_3\right)' R_{12}^2 + 2m_1' m_3' R_{12} R_{23} + \left(m_1 m_3 + m_2 m_3\right)' R_{23}^2}{\left(m_1 + m_2 + m_3\right)'} = 0.88588 \ amu \cdot nm^2$$

Solve the equations I and I' simultaneously using a computer to get R_{12} and R_{23}

$R_{12} = 0.1155$ nm $\qquad R_{23} = 0.1566$ nm

14.53
***

CO_2: $O = C = O$

R_{CO} = bond length
m_O = mass of oxygen atom
$B = 0.3937 \text{ cm}^{-1} = 1.180 \times 10^{10} \text{ Hz}$

$$I = 2m_O R_{CO}^2$$

$$B = \frac{h}{8\pi^2 I}$$

$$\Rightarrow I = \frac{h}{8\pi^2 B} = 2m_O R_{CO}^2$$

$$\Rightarrow R_{CO} = \left(\frac{h}{16 m_O \pi^2 B}\right)^{1/2} = 115.7 \text{ pm}$$

14.55
***

$m_H = 1.67 \times 10^{-27} \text{ kg}$

From Figure 14.16, the distance between successive peaks is approximately 2B

distance between peaks = 4.0 mm
scale = 42.5 mm / 100 cm^{-1}

$$2B = \frac{4.0 \text{mm}}{42.5 \text{mm} \Big/ 100 \text{cm}^{-1}} \Rightarrow B = 9.4 \text{ cm}^{-1}$$

$$B = \frac{h}{8\pi^2 Ic} \Rightarrow I = \frac{h}{8\pi^2 Bc} = \frac{8}{3} m_H R^2 \quad \text{(from problem 14.54)}$$

$$\Rightarrow R = \left(\frac{3h}{64\pi^2 m_H Bc}\right)^{1/2} = 0.82 \text{ angstroms}$$

14.57
___

The A axis is along the C=O bond in formaldehyde

$A = 282.106 \times 10^9$ Hz
$m_H = 1.6735 \times 10^{-27}$ kg
$R_{CH} = 0.107 \times 10^{-9}$ m
α = HCH angle

$$I_A = \frac{h}{8\pi^2 A} = 2m_H \left(R_{CH} \sin\frac{\alpha}{2} \right)^2 \Rightarrow \sin\frac{\alpha}{2} = \left(\frac{h}{16 m_H \pi^2 A} \right)^{1/2} \frac{1}{R_{CH}}$$

$$\Rightarrow \alpha = 2 \arcsin\left[\left(\frac{h}{16 m_H \pi^2 A} \right)^{1/2} \frac{1}{R_{CH}} \right] = 2 \arcsin(0.8811) = 123.54°$$

SECTION 14.12

14.59
___

$$I_{CCl3} = \frac{8}{3} m_{Cl} R_{C-Cl}^2 = \frac{8}{3} \left(\frac{34.96885 \times 10^{-3} \text{ kg}}{L} \right) (0.178 \times 10 \text{ m})^2 = 4.906 \times 10^{-45} \text{ kg} \cdot \text{m}^2$$

$$I_{CH3} = 5.302 \times 10^{-47} \text{ kg} \cdot \text{m}^2$$
(see problem 14.56)

$$I_{rel} = \frac{I_1 I_2}{I_1 + I_2} = \frac{I_{CH3} I_{CCl3}}{I_{CH3} + I_{CCl3}} = 5.245 \times 10^{-47} \text{ kg} \cdot \text{m}^2$$

14.61

$$H_{m,n} = \int \psi_m^*(\phi)[H]\psi_n(\phi)d\phi$$

$$= \frac{1}{2\pi}\int_0^{2\pi} \exp[-im\phi]\left[\frac{h^2}{8\pi^2 I}\frac{\partial^2}{\partial\phi^2} + \frac{1}{2}V_3(1-\cos 3\phi)\right]\exp[in\phi]d\phi$$

$$= \frac{1}{2\pi}\left(\frac{h^2 n^2}{8\pi^2 I} + \frac{1}{2}V_3\right)\int_0^{2\pi}\exp[i(n-m)\phi]d\phi - \frac{1}{4\pi}\frac{V_3}{2}\int_0^{2\pi}\exp[i(n-m)\phi](\exp[3i\phi]+\exp[-3i\phi])d\phi$$

$$= \frac{1}{2\pi}\left(\frac{h^2 n^2}{8\pi^2 I} + \frac{1}{2}V_3\right)2\pi\delta_{m,n} - \frac{1}{4\pi}\frac{V_3}{2}2\pi\left[\delta_{m,n+3} + \delta_{m,n-3}\right]$$

$$\Rightarrow H_{m,m} = \left(\frac{h^2 m^2}{8\pi^2 I} + \frac{1}{2}V_3\right) \text{ and } H_{m,m\pm 3} = -\frac{V_3}{4}$$

CHAPTER 15   *Statistical Machanics*

SECTION 15.1-15.3

15.1

$g(\varepsilon) \propto \varepsilon^{1/2}$ (equation 15.26)

$$\int_0^\infty x^n \exp[-ax]dx = \frac{\Gamma(n+1)}{a^{n+1}}$$ (use a standard set of tables to evaluate the gamma function)

$$\langle \varepsilon \rangle = \frac{\displaystyle\int_0^\infty \varepsilon^{3/2} \exp\left[-\varepsilon/kT\right]d\varepsilon}{\displaystyle\int_0^\infty \varepsilon^{1/2} \exp\left[-\varepsilon/kT\right]d\varepsilon} = \frac{\dfrac{\Gamma\left(\dfrac{5}{2}\right)}{\left(\dfrac{1}{kT}\right)^{5/2}}}{\dfrac{\Gamma\left(\dfrac{3}{2}\right)}{\left(\dfrac{1}{kT}\right)^{3/2}}} = kT\frac{\dfrac{3\sqrt{\pi}}{4}}{\dfrac{\sqrt{\pi}}{2}} = \frac{3}{2}kT$$

SECTION 15.4-15.6

15.3

$S_m = R \ln (g_N)$
$g_N = 2I + 1$

$I_B = 3/2$
$I_F = 1/2$

$g_B = 4$
$g_F = 2$

$g(BF_3) = g_B g_F^3 = 32$

$S_m (BF_3) = R \ln (32) = 28.815 \text{ J/K}$

15.5

$$z_{nsr} = \sum_{even\ J} (2J+1) \exp\left[\frac{-J(J+1)\theta_r}{T}\right] + 3\sum_{odd\ J} (2J+1) \exp\left[\frac{-J(J+1)\theta_r}{T}\right]$$

or

$$z_{nsr} = \frac{g_N T}{\sigma \theta_r} \quad \text{(approximate)}$$

$\theta_r = 85.35\ K$

$\sigma = 2$

$g_N = 4$

T (K)	z (exact)	z (approximate)
100	2.663	2.343
500	12.4067	11.7165

15.7

$\theta_r = 85.35\ K$

$T = 200\ K$

$$\langle E_{rot} \rangle = RT^2 \frac{\partial \ln z_{nsr}}{\partial T} = \frac{RT^2}{z_{nsr}} \frac{\partial z_{nsr}}{\partial T}$$

$$\frac{\partial z_{nsr}}{\partial T} = \sum_{even\ J} \frac{J(J+1)(2J+1)\theta_r}{T} \exp\left[\frac{-J(J+1)\theta_r}{T}\right] = 5.021 \times 10^{-3}\ /K$$

$z_{nsr} = 1.3881$ (from problem 15.6)

$$\Rightarrow \langle E_{rot} \rangle = 1203\ J$$

15.9

J" = initial state
J' = final state

Stokes (J' = J" + 2)

$$\tilde{v}_s(S) = \tilde{v}_i - \tilde{B}\left[(J''+2)(J''+3) - J''(J''+1)\right] = \tilde{v}_i - \tilde{B}(4J''-2)$$

Anti-Stokes (J' = J" - 2)

$$\tilde{v}_s(aS) = \tilde{v}_i - \tilde{B}\left[(J''-2)(J''-1) - J''(J''+1)\right] = \tilde{v}_i + \tilde{B}(4J''-2)$$

15.11

$I_1 = 3/2$ and $I_2 = 3/2$

Let T_I = total molecular spin = $I_1 + I_2, I_1 + I_2 - 1, ..., |I_1 - I_2|$ = 3, 2, 1, 0

$g_T = 2T_I + 1$

$g_{odd} = \Sigma g_T$ (T_I odd) = 3 + 7 = 10
$g_{even} = \Sigma g_T$ (T_I even) = 1 + 5 = 6

$$z_{nsr} = 6 \sum_{even\ J} (2J+1) \exp\left[\frac{-J(J+1)\theta_r}{T}\right] + 10 \sum_{odd\ J} (2J+1) \exp\left[\frac{-J(J+1)\theta_r}{T}\right]$$

$$\frac{odd}{even} = \frac{10}{6} = \frac{5}{3}$$

15.13

$$\frac{g_N}{z_{nsr}} = \left(\frac{N_{J''}}{N}\right)\frac{1}{2J+1}\exp\left[\frac{J(J+1)\theta_r}{T}\right]$$

$\theta_r = 3.0\ \text{K}$

$T = 300\ \text{K}$

$\left(\dfrac{N_{J''}}{N}\right)$ = relative int ensity

$\dfrac{g_N}{z_{nsr}} = 1.000$ for even J'' and $\dfrac{g_N}{z_{nsr}} \cong 1.67$ for odd J''

$$\Rightarrow \frac{\text{odd}}{\text{even}} = \frac{1.67}{1} = \frac{5}{3} \Rightarrow I = \frac{3}{2} \quad \text{(as in problem 15.11)}$$

SECTION 15.7

15.15

$T = 5\ \text{K}$

$C_{pm} = 0.0012\ \text{J/K}$

$$\theta_D = \left(\frac{36\pi^4 R}{15 C_{vm}}\right)^{1/3} T$$

$C_{pm} \cong C_{vm} \Rightarrow \theta_D = 587\text{K}$

rearranging the equation above and using $T = 10\ \text{K}$ gives

$$C_{vm} = \frac{36\pi^4 R}{15\theta_D^3}T^3 = 0.0096\ \text{J / K}$$

15.17

$$\theta_E = \frac{h\nu_E}{k}$$

$$\theta_D = \frac{h\nu_m}{k}$$

$$\nu_E = \langle \nu_m \rangle = \frac{\displaystyle\int_0^{\nu_m} \nu g(\nu)d\nu}{\displaystyle\int_0^{\nu_m} g(\nu)d\nu} = \frac{\displaystyle\int_0^{\nu_m} \nu^3 d\nu}{\displaystyle\int_0^{\nu_m} \nu^2 d\nu} = \frac{3}{4}\nu_m$$

$$\Rightarrow \theta_E = \frac{3}{4}\theta_D$$

$$u = \frac{3}{4}\theta_D\frac{1}{T} = \frac{9}{4}$$

$$\frac{C_{vm}}{3R} = \frac{u^2 \exp[u]}{(\exp[u]-1)^2} = 0.667 \quad \text{(Einstein Theory)}$$

$$\frac{C_{vm}}{3R} = D\left(\frac{\theta_D}{T}\right) = 0.67 \quad \text{(Debye Theory and Figure 15.10)}$$

SECTION 15.8

15.19

$$u = \frac{h\nu}{kT}$$

$$\rho(\nu)d\nu = \frac{8\pi k^4 T^4 u^3 du}{h^3 c^3 (\exp[u] - 1)}$$

$$\Rightarrow \rho(u) = \frac{8\pi k^3 T^3 u^3}{h^2 c^2 (\exp[u] - 1)}$$

$$\frac{\partial \rho}{\partial u} = 0 \Rightarrow u = 3(1 - \exp[-u])$$

Solving this equation numerically gives

$$u = 2.8214$$

$$\Rightarrow \nu_{max} = 2.8214 \frac{kT}{h}$$

Anti-Stokes (J' = J" - 2)

$$\tilde{\nu}_s(aS) = \tilde{\nu}_i - \tilde{B}\left[(J" - 2)(J" - 1) - J"(J" + 1)\right] = \tilde{\nu}_i + \tilde{B}(4J" - 2)$$

Student's Solutions Manual

CHAPTER 16   *Structure of Condensed Phases*

SECTION 16.1

16.1
___

$a = b \neq c$
$\alpha = \beta = 90$ degrees
$\gamma = 120$ degrees

$$V = \vec{a} \cdot \vec{b} \times \vec{c} = abc\left[1 - \cos^2(\alpha) - \cos^2(\beta) - \cos^2(\gamma) + 2\cos(\alpha)\cos(\beta)\cos(\gamma)\right]^{1/2} = abc\sqrt{0.75} = 0.866a^2c$$

16.3
___

$V \cong abc = 7.079 \times 10^{-22}$ cm^3

$m = \rho V = 1.097 \times 10^{-21}$ g

M.W. $= 314.3$ g / mol

$$n = \frac{(m)L}{M.W.+18(x)} = 2.0 \text{ if } x = 1$$

This implies that the salt has one attached water molecule and two molecules per unit cell

16.5
___

C_{5v} (Schoenflies) = 5m (herman-Mauguin)

5-fold rotation axis implies 5
mirror plane implies m

SECTION 16.2

16.9

n = 1
θ = 10°
λ = 1.54 A

$$d = \frac{\lambda}{2\sin\theta} = 4.43 \text{ A}$$

As θ, and thus sin θ, decreases, d increase further.

16.11

For a body-centered cell, $h + k + l$ is even:

       110, 200, 211, 220, 301, 222, 321, 400

For a face-centered cell, h, k, and l are all even or all odd:

       111, 200, 220, 311, 222, 400

16.13

$a = 3.4$ angstroms
$\lambda = 1.8$ angstroms

$$d_{hkl} = \frac{a}{\left(h^2 + k^2 + l^2\right)^{1/2}}$$

$$\theta = \sin^{-1}\left(\frac{\lambda}{2d_{hkl}}\right)$$

Σhkl must be even

hkl	d_{hkl} (angstroms)	θ
110	2.40	22
200	1.70	32
211	1.39	40
220	1.20	49
301	1.08	57
222	0.98	67
321	0.91	82
400	0.85	---

16.15

a = 5.64 angstroms

Na-Cl separation = $\frac{1}{2}$ a = 2.82 angstroms

Na-Na separation = $\frac{\sqrt{2}}{2}$ a = 3.99 angstroms

m_{Na} = 22.9 g / mol

m_{Cl} = 35.45 g / mol

$$\rho = \frac{\frac{1}{2}\left(m_{Na} + m_{Cl}\right)}{\left(\frac{1}{2}a\right)^3 L} = 2.164 \text{ g / cm}^3$$

SECTION 16.3

16.17

a = 5.462 angstroms

m_{Ca} = 40.08 g / mol

m_{Cl} = 18.99 g / mol

$$\rho = \frac{\left(4m_{Na} + 8m_{Cl}\right)}{\left(a\right)^3 L} = 3.182 \text{ g / cm}^3$$

16.19

$$\Delta_f H = -435.9 \text{ kJ / mol}$$
$$\Delta_{sub} H = 90.0 \text{ kJ / mol}$$
$$D_o(Cl_2) = 238.8 \text{ kJ / mol}$$
$$IP(K) = 418.8 \text{ kJ / mol}$$
$$EA(Cl) = 348.8 \text{ kJ / mol}$$

$$\Delta_c U = -\Delta_f H + \Delta_{sub} H + \frac{D_o(Cl_2)}{2} + IP(K) - EA(Cl) = 715.3 \text{ kJ / mol}$$

16.21

$$\frac{\partial E(R)}{\partial R}\bigg|_{R=R_e} = 0 = \frac{q^2 ML}{R_e^2} - \frac{A}{\rho}\exp\left[-\frac{R_e}{\rho}\right] \Rightarrow A = \frac{\rho q^2 ML}{R_e^2}\exp\left[\frac{R_e}{\rho}\right]$$

$$\Delta U_c = -E(R_e) = \frac{q^2 ML}{R_e} - \frac{\rho q^2 ML}{R_e^2}\exp\left[\frac{R_e}{\rho}\right]\exp\left[-\frac{R_e}{\rho}\right] = \frac{q^2 ML}{R_e}\left(1 - \frac{\rho}{R_e}\right)$$

16.23

$$m = 137.22 \text{ g/mol}$$

$$\rho = \frac{2m}{La^3} = 3.595 \text{ g / cm}^3$$

16.25

$$E = \frac{n^2 h^2}{8 m V^{2/3}}$$

let $n = n_f$ at $E = E_f$ with N electrons in a volume V

$$\left(\frac{4}{3} \pi n_f^3\right) \frac{1}{8} = \frac{NV}{2}$$

(1/8 for the positive octant, N/2 due to electron + or -)

$$\Rightarrow \frac{n_f^3}{V} = \frac{3N}{\pi} \Rightarrow E_f = \frac{h^2}{8m} \left(\frac{3N}{\pi}\right)^{2/3}$$

sodium has one valence electron per atom

$$\frac{N}{L} = \frac{\rho}{M.W} \Rightarrow N = 2.54 \times 10^{22} \; e^{-1}/cm^3$$

$$\Rightarrow E_f = 3.15 \; eV$$

SECTION 16.4

16.29
_____

a)   methyl group on polyethylene

$\delta = 8.85 + 2(9.51) + 2(-2.34) + 2(0.28) + 2(0.03) + (-0.96) - 2.35 = 20.5$ ppm

b)   ethyl group on polyethylene

$\delta = 8.85 + 9.51 + 2(-2.34) + 2(0.28) + 2(0.03) - 2.35 = 11.8$ ppm

c)   propyl group on polyethylene

$\delta = 8.85 + 9.51 - 2.34 + 2(0.28) + 2(0.03) - 2.35 = 14.3$ ppm

d)   butyl group on polyethylene

$\delta = 8.85 + 9.51 - 2.34 + 0.28 + 2(0.03) - 2.35 = 14.0$ ppm

e)   For longer groups, the shift will stay near 14.0 ppm

16.31

$I_{38} = 15$
$I_{30} = 747$

$$\%\text{branching} = 100\left(\frac{I_{38}}{I_{30} + 7I_{38}}\right) = 1.76\%$$

$$\text{average run length} = \frac{I_{30} + 6I_{38}}{I_{38}} = 56$$

16.33

Most bonds are m (meso), except where there is a change in stereochemistry, which are r (racemic)

Here there would be 98 m followed by 2 r (on average).

Per hundred units this would mean 2 mrr, 2 mmr, and 96 mmm groups

16.35

a)   BBB

$\delta = 3(8.85) + 3(9.51) + 4(-2.34) + 4(0.28) + 3(-3.04) - 2.35 = 35.4$ ppm

EBB

$\delta = 3(8.85) + 3(9.51) + 3(-2.34) + 3(0.28) + 3(-3.04) - 2.35 = 37.4$ ppm

EBE

$\delta = 3(8.85) + 3(9.51) + 2(-2.34) + 2(0.28) + 3(-3.04) - 2.35 = 39.5$ ppm

b)   frequency equals $\Sigma n_i P_i$, where n_i is the degeneracy of permutations. $P = 0.5$ in both cases

$f(BBB) = (0.5)^3 = 0.125$
$f(EBB) = 2(0.5)^2(0.5) = 0.250$
$f(EBE) = (0.5)^3 = 0.125$

Relative intensities are 1:2:1

c) $f(BBB) = 8$
$f(EBB) = f(BEE) = 2$
$f(EBE) = 0$

Relative intensities are 4:1:0

16.37

$$\text{compliance} = \frac{\text{area}}{\text{force}} = \frac{m^2}{N} = \frac{m^2}{kg \cdot m \cdot s^{-2}} = \frac{m \cdot s^2}{kg}$$

NOTES

NOTES

NOTES

NOTES

NOTES

NOTES